D0521766

Invasive Species in the Pacific Northwest

Invasive Species in the PACIFIC NORTHWEST

Edited by

P. D. BOERSMA, S. H. REICHARD,
and A. N. VAN BUREN

Rebecca L. Gamboa, Photo Editor

UNIVERSITY OF WASHINGTON PRESS *Seattle and London*

Invasive Species in the Pacific Northwest is published with the assistance of a generous grant from the PENDLETON AND ELISABETH CAREY MILLER CHARITABLE FOUNDATION.

This book is published in memory of MARSHA L. LANDOLT (1948–2004), Dean of the Graduate School and Vice Provost, University of Washington, with the support of the University of Washington Press Endowment.

Printed in China
Design by Veronica Seyd
12 11 10 09 08 07 06 05 5 4 3 2 1

University of Washington Press
P.O. Box 50096, Seattle, WA 98145
www.washington.edu/uwpress

Library of Congress Cataloging-in-Publication Data

Invasive species in the Pacific Northwest /
edited by P. D. Boersma, S. H. Reichard, and
A. N. Van Buren.
p. cm.
Includes bibliographical references and index.
ISBN 0-295-98596-8 (pbk. : alk. paper)
1. Biological invasions—Northwest, Pacific.
2. Introduced organisms—Northwest, Pacific.
I. Boersma, P. D. (P. Dee). II. Reichard, S. H.
(Sarah H.) III. Van Buren, A. N. (Amy N.)
QH353.I598 2006
577'.1809795—dc22 2006008628

The paper used in this publication is acid-free and meets the minimum requirements of American National Standard for Information Sciences—Permanence of Paper for Printed Library Materials, ANSI Z39.48-1984.

Contents

FRESHWATER INVERTEBRATES

MARINE INVERTEBRATES

TERRESTRIAL INVERTEBRATES

FRESHWATER VERTEBRATES

MARINE VERTEBRATES

TERRESTRIAL VERTEBRATES

DISEASES

THREATS

Appendixes

Acknowledgments

As living organisms, our survival, reproduction, and longevity are governed by biology. If we don't understand biology we don't understand life. Humans are driving the biology of earth. We change the composition of species, the interactions among species, and the course of evolution. The invasive species that come with human presence, either intentionally or unintentionally, alter the course of evolution and change the world forever. Humans—by our numbers, consumption, habitat alteration, pollution, and the invasive species that we bring—alter the planet faster than biological processes can repair it or can evolve to withstand it. This book explores some of the wonders of the natural world, celebrates the marvelous natural history that species have evolved, and details what species are or may become problems in the Pacific Northwest and what can be done to slow the success of invasive species.

We are indebted to Carla Geyer for her attention to detail and organization. She constructed the appendixes and tables, and she and Linda Garnett were remarkable in keeping all our pieces together. Rebecca Gamboa was the czar of the pictures. She found the images, convinced people to allow their images to be used, and formatted the pictures. Her eye and choices show the invasive species of the Pacific Northwest. Without her, we would not have the faces to go with the names.

In the best tradition of a university, this book started with a class and became a learning experience for everyone who helped. The authors and the reviewers of these species accounts provided the natural history, distribution, and incredible facts that make each species account accurate and fun to read. We thank the hundreds of people who helped the authors and reviewed accounts. Four women brought this project to life: Vim Wright, Vivian Boersma, Sue Moore, and Estella Leopold. Their support and encouragement provided the passion that fueled this book.

Sarah Reichard oversaw the plants. Amy Van Buren provided the leadership for designing the questions that assess a species's impact on the environment, making the maps, and organizing the book. In the end, Sarah, Amy, and I are responsible for the content. You are responsible for what happens to invasive species in the Pacific Northwest.

P. D. BOERSMA

Invasive Species Around the World

P. D. BOERSMA

An understanding of ecology is fundamental to managing the Earth's resources. Humans' impact on the Earth's resources by overpopulation and consumption is overwhelming the natural rates of biological processes. Right now more species are going extinct than during the great die-off of dinosaurs. Introductions of mammals, and particularly humans as predators, are the driving force of 60% of extinctions in three groups of vertebrates. We have virtually stopped the evolution of large carnivores and most large herbivores because our parks and wilderness areas are too small for many species to continue to survive, let alone evolve. Big fierce animals are rare because humans have removed them. The scale and time frame of these changes are often difficult for us to detect with our own senses, because we live in environments that are constantly changing and are often already transformed.

Only 100 to 150 years ago, the Pacific Northwest (PNW) was largely virgin forest. Now the virgin forests are over 90% gone. We have transformed the landscape from wilderness to cities and suburbs. In 1969 the tallest building in Seattle was the Smith Tower, now dwarfed in a landscape of skyscrapers. Not only have we cut, scraped, dammed, paved, and leveled the land, we have knocked down the barriers that prevented most species from colonizing the rest of the world. As a consequence, we are causing a massive homogenization of the world's species.

One of the biggest threats to biodiversity, ecosystem function, and community interactions is invasive species. Today's extensive global trade, waterway engineering, conversion of natural habitats, and travel are homogenizing the world and taking species to places they never could have arrived at on their own. History documents the disastrous consequences of moving species. Some, usually only a few, become invasive and a problem. We know little about which ones will be problems and which ones will not. Sometimes a new species dominates native species because it comes without its natural predators and diseases. Invasive species transform the world just as yeast changes dough. We add species for a variety of purposes: for food, shelter, beauty, to control predators, for pets, and to make a landscape feel like home. Sometimes they are added by accident. When new species are added to ecosystems, changes occur, and often the environmental costs are huge. Once the genie is out of the bottle, it may never again be contained. Systems and species will be fundamentally changed and evolution and extinctions set on a slightly different trajectory.

The World Conservation Union (IUCN) published a list of 100 of the world's most destructive species (Appendix 1). Of the 10 worst species, the PNW has 4: Feral Pig, Dutch Elm Disease, Gray Squirrel, and Japanese Knotweed. The first on the IUCN list, the Nile Perch, was introduced in 1954 into Lake Victoria, Africa, to improve fishing. The perch contributed to the extinction of over 200 endemic fish. Water Hyacinth from South America clogs waterways and is now found on five continents. *Caulerpa*, an alga called the "killer seaweed," introduced into

Gypsy Moth *(Lymantria dispar)*

the Mediterranean, smothers its competitors and is unpalatable to grazers. Fourth on the list, the Crazy Ant, named for its frenetic movements, has changed communities on remote islands, killing millions of crabs. The Small Indian Mongoose is a voracious predator. Introduced to control rats, it ate birds, reptiles, and amphibians, causing several species to go extinct. Numbers 6 to 9 are the four that are already in the PNW. The tenth, the Giant African Snail, introduced to islands in the Pacific and Indian Oceans as food for humans, ate 500 different kinds of plants, spread to parts of South America, and became an agricultural pest. Appendix 1 shows the IUCN list of 100 of the world's major invasive species and the ones that are here in the PNW.

Adverse effects of introductions extend from the tropics to regions as cold as South Georgia Island in the South Atlantic, where whalers introduced Norwegian Reindeer and rats. Polynesians, when they colonized Hawaii 1,500 years ago, brought as many as 30 species of plants as well as dogs, pigs, rats, and chickens. Captain Cook and other Europeans brought others, and the result is that many species found in no other place in the world became extinct. On Laysan I., HI, rabbits eliminated 26 species of plants between 1903 and 1923. It is not just on land where invasive species are problems. Ballast water carried in oceangoing ships also brings new species. A plankton sample of water from Japan released in OR contained 367 taxa. San Francisco Bay has had multiple waves of invasions. Many invasive species get "naturalized," meaning they have lived in their new habitat so long that we think they are native. Dandelions in the non-native lawns we spend millions to feed and weed are but one example. Many of these "naturalized" invaders mostly thrive in altered habitats. Our worries are the ones that invade what few remnants of native habitats remain or the ones that have a profound impact on native species.

Introduced species change communities in a variety of ways—by altering nutrient cycles, rates of decomposition, composition of the soil, and population dynamics and by changing interactions among species. Invasive species are changing the course and rate of

Cane Toad *(Bufo marinus)*

evolution. The Cane Toad, native to South America, was introduced to Australia in the 1930s to control an insect crop pest. Pest control was not achieved, but the toads multiplied and became a nuisance. They have a poison gland, and many would-be predators die trying to eat them. Interestingly, two species of native snakes that are killed by toad toxins (the Red-bellied Black Snake and the Green Tree Snake) developed a defense. Evolution in 70 years (20 to 25 generations) favored snakes with small heads. The snakes show a steady increase in body size coupled with a decline in head size. Small-headed snakes cannot eat adult toads, which are too big to swallow, so they are not poisoned.

This is one of the first examples of native organisms adapting to the invader and reducing the severity of the impact. Although evolution will fix some problems, invasive species are likely to raise havoc in ecosystems, cause the extinction of species, change the cycling of nutrients, and alter ecosystem services on which humans depend because evolution is not rapid enough for the system, community, or species to adapt. One example in the PNW is the introduction and control of Spotted Knapweed, which may end up causing an increase in hantavirus. To control Spotted Knapweed, an insect gall-forming species was introduced—which mice eat, resulting in a dramatic increase in virus-spreading mouse populations. The likely outcome in the short term is more hantavirus in people. We have opened a Pandora's box: the experiments we have started will continue to run for generations, and the outcomes will be known only long after we are gone.

Here is another example of how one species from one part of the world can dramatically—and irreversibly—change the ecological balance where it invaded. The Brown Tree Snake, indigenous to the Solomon Is., New Guinea, and northern Australia, was introduced to Guam shortly after World War II, probably as a stowaway on a cargo ship. By the early 1980s, 9 of 12 native birds were extinct. Now Guam is largely silent, as the snakes have eaten most of the birds and depleted the lizards. There are as many as 3 million snakes on the island, making more than 5,000 snakes/km^2, or 22 snakes for every person on the island. Eventually the Brown Tree Snake will reach HI and take hold, which will likely be the final blow for the islands' imperiled endemic avifauna. One survived in the wheel-well of an airplane that flew at 33,000 feet from Guam to Oahu, HI, but it was spotted and killed as it slithered off the runway.

Our book concentrates on how you can recognize and control these unwanted species. Our first line of defense is knowledge, followed by political action to slow their spread, and finally, control. Keep your eyes open and your shovel handy. Knowing what species are problems and taking action against them are the best defenses against the spread of invasive species. When these defenses are overwhelmed, we can hope that evolution is rapid enough to reduce the invaders' impacts on native species and natural habitats. People have already changed the world and the course of evolution. Individually and collectively, we can make a meaningful difference in catching, stopping, and controlling invasive species so as to enrich and foster the diversity of the native species and habitats of the PNW.

References: 9a, 51, 59a, 146a, 198a, 223a, 230a, 246a, 329a, 329b

Invasive Species in the Pacific Northwest

P. D. BOERSMA, S. H. REICHARD, and A. N. VAN BUREN

The world is a constantly changing place. For most of us, what we knew as children about the environment and where we lived set a reference point for the rest of our lives. We measure environmental change against our known world. Some environments and some species may change rapidly because of human modifications, and others may be set on new courses because humans alter community compositions or relationships among species. The introduction of non-native species by humans is at the very heart of the invasive species problem.

Teasel (*Dipsacus fullonum*)

What Is an Invasive Species?

Non-native invasive species are defined in many ways. The US government defines them as "an alien species whose introduction does or is likely to cause economic or environmental harm or harm to human health." In other words, they must be non-native to the ecosystem they are invading, though they may be native to nearby geographic areas, and they must be those species that cause harm. The many non-native species that are intentionally raised but that do not escape cultivation are not included. Although it is not explicit in this definition, the species of concern are those that are capable of reproducing and forming self-sustaining populations. If they cannot survive, reproduce, and increase in numbers, they are unlikely to spread far from the points of introduction.

In this book we refer to species moved to new places by humans as "non-native" because they wouldn't be where they are without the help of people. If the species got to a place on its own, we don't consider it non-native. So this book does not include Brown-headed Cowbirds, American Crows, Virginia Opossums, or other species for which humans have facilitated their movement but were not the reason for their arrival. Not all of the species that humans have introduced either intentionally or accidentally survive transport or become established in their new homes, and many intentionally transported species are beneficial. But many others reproduce in their new locations with unfortunate consequences for the local flora and fauna. For this book, we are defining an invasive species as a non-native organism that causes harm to native habitats or species.

The Nature of the Problem. Species have always migrated over time, spreading naturally to new locales by a variety of vectors from animals to wind. We know that around 10,000 years ago much of the Pacific Northwest, especially the northern section, was covered with a huge ice sheet. Therefore, whatever native plants and animals exist here now colonized the area following the recession of the ice. However, natural migrations occur at a very slow rate and usually over relatively short distances. As in many other circumstances, humans have changed the scale of natural processes. Humans are intentionally and inadvertently moving species of plants, animals, fungi, and microbes at increasing rates. Thus, the migration of an insect across the ocean, a process that might have happened historically only during rare severe windstorms, might occur several times a year through the movement of shipped trade goods. This rapid widespread movement of species allows them a greater chance of survival, with lower mortality during migration and higher probability of adapted genetic types being the ones introduced.

Non-native invasive species have been recognized as a biogeographical issue since Charles Darwin's voyages on the *Beagle*. Darwin realized that some introduced species threatened native species. In his chapter on geographical distributions, he reports, "many European productions cover the ground in La Plata, and in a lesser degree in Australia, and have to a certain extent beaten the natives." This may be the first scientific comment on invasive species, although it received little attention or concern.

Darwin also provides the mechanism for us to understand some of the concern about invasive species. One of the observations that stimulated Darwin to conceive of evolution by natural selection was that each new continent or island he visited, despite similar environments, had different species. He wrote: "In considering the distribution of organic beings over the face of the globe, the first great fact which strikes us is that neither the similarity nor the dissimilarity of the inhabitants of various regions can be wholly accounted for by climatic and other physical conditions. . . . There is hardly a climate or condition in the Old World which cannot be paralleled in the New . . . [yet] how widely different their organic productions!" (that is, their species). In their natural settings, species are found with friends and enemies. Darwin realized that this ecological context provided the ultimate check on unfettered population growth. Thus, there is another

Kudzu *(Pueraria montana* var. *lobata)*

important perspective to the invasion problem: we often transport species without their coevolutionary context. This may result in poor performance (e.g., if a plant lacks a pollinator), but it may also result in the potential for rapid population growth (e.g., if the species is freed from natural enemies or other growth constraints).

Not All Non-Native Species Are Created Equal. Many introduced non-native species are beneficial to humans. Most of the important crops and livestock in the US are non-native. If we were to eat only native plants in the US, breakfast, lunch, and dinner might be sunflowers, berries (e.g., cranberries, blueberries, and huckleberries), walnuts, and Jerusalem Artichokes—not a very interesting diet. The European Honey Bee forms the basis of much of the agricultural industry through its pollination of crops; if absent, there would be fewer apples, plums, and peaches. We would not have bagels, muffins, or cake, because wheat is native to the Mediterranean. Most meats and many fish, clams, and the Pacific Oyster from Japan would not be in our diet. Some of the mainstay plants in our yards and gardens would also be missing, such as Forsythia, the bright yellow-flowered bush that announces spring, the Weeping Willow, or the Ginkgo, the odd-shaped tree with the fan-shaped leaf from China.

One criticism about the interest in preventing non-native species invasions is that it represents racial purity arguments that some organisms are more "right" than others. This idea has been widely circulated among those who do not want restrictions on the introduction of non-native species. We emphasize again that not all non-native species are harmful and that non-native species provide great benefit to us. Unfortunately, a relatively small number cause harm that is greater than the benefits they bring. It is important to recognize which ones are or could be problems. In this book you'll find many of the species that worry the biologists of the PNW.

Although humans have undoubtedly benefited from introductions of non-native species, this book focuses on the negative impacts of non-native species in native habitats, or to native species wherever they are found. We are only beginning to understand the impacts of these harmful introduced species, but we know that they may alter relationships among species and change community characteristics. They do this by eating native species, introducing diseases, altering sedimentation or erosion, changing soil chemistry, altering the frequency or intensity of fires, modifying hydrology, and changing the survival, reproductive success, and growth of other species. In the accounts included in this book, we discuss what is known about the impact of these highlighted species. The impacts, however, may be much greater than we currently realize.

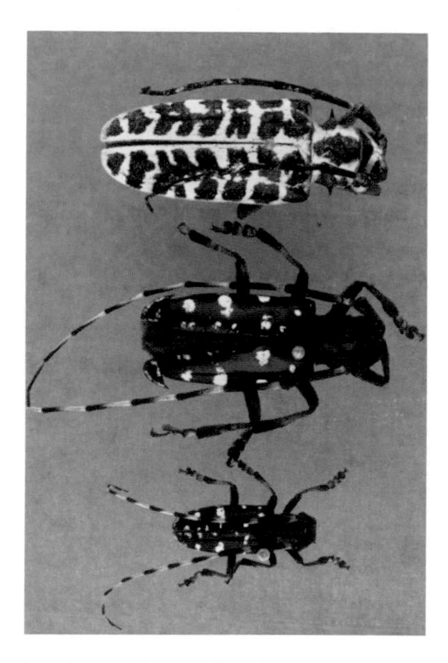

Longhorned beetles *(Anoplophora glabripennis)*

The How, Where, and What of Invasive Species

In the pages of this book you will find accounts of many species that invaded the PNW, including descriptions of where they came from, how they arrived in the PNW, and the harm they do. Species arrive in a new place either intentionally or accidentally, and the range of impacts is as wide as the range of species. Invasive plants compete with native plants for resources, increase siltation in mudflats, change nutrient regimes, and increase fires. Invasive fish eat native fish and change the relative abundance or composition of fish communities, and can eliminate some species of amphibians. In recent years, the concept of "ecosystem services" has become popular. Ecosystem services are processes by which the environment provides things that are beneficial to humans, including timber, clean water, habitat for animals, or pollination of native or agricultural plants. We often think of these services as free but they are not: they depend on healthy functioning systems. Invasive species may alter the delivery of ecosystem services and in some cases completely disrupt them through changes in nutrients or disturbances such as fire. The changes may last forever.

Accidental and Intentional Introductions. Accidental introductions generally come as stowaways and contaminants and they continue to arrive. Ballast transport (wet or dry) introduced a variety of species, from Common Dandelions in early years to Zebra Mussels within the last 30. The European Ruffe arrived in the Great Lakes in the 1980s in ballast water, and like the Alewife that preceded it, it too will likely cause the decline of commercially valuable fish. Only laws that require ballast water to be sterilized or exchanged far offshore are likely to solve the problem. About 9 million liters/hr of foreign ballast water are discharged daily in US ports; introductions of aquatic species are inevitable without some sort of treatment. Other accidental introductions might include insects such as the Asian Longhorned Beetle found in shipping material in crates, or Serrated Tussock Grass seeds found in imported lawn grass seeds.

Intentional introductions, including those for agriculture, aquaculture, horticulture, or hunting and fishing, account for many new non-native species. As recently as the mid-1970s, a new game bird, the Tinamou, was introduced by the state of WA into the Hoh Rainforest because the Department of Game wanted another species for hunting. Fortunately, it didn't survive, or it might have become yet another chapter in this book. New plant species, however, continue to be introduced annually, and many of them will be the invaders of the future. Lastly, species escape or are released. For example, people dump home aquaria containing the algae *Caulerpa taxifolia*, or Northern Snakehead fish into lakes and streams. These releases change the community because *Caulerpa* is toxic and inedible by wildlife and Snakeheads are voracious predators that eat anything slightly smaller than their mouths.

Pets and Domestic Animals Gone Wild. Pets, such as cats, rabbits, and turtles, can become problems. When pets are under human control, they are generally all right. Unfortunately pets and other domesticated animals may not stay with humans. The cold, wet winters in the PNW, with the occasional hard freeze, prevent a number of released animals from surviving, but not all. Cats escape or are abandoned and form wild colonies, depleting the native bird, rodent, lizard, and invertebrate species they hunt to survive. Feral populations of domesticated mammals such as goats and pigs are opportunistic survivors and can trample, uproot, and eat plants, increase erosion, and cause great ecological damage. Amazon Parrots breed at Seattle's Seward Park, and Mynah birds from Asia breed in Vancouver, BC, but neither has moved out into native habitats—*yet.* Some non-natives are caught before the populations breed and disperse. A caiman released into Green Lake in Seattle in 1986 was detected and removed when it was about a meter long, but not before it scared swimmers out of the lake and brought comparisons to the Loch Ness Monster.

In more benign climates, such as FL and HI, more former pets have established populations. There are caimans, iguanas, piranhas, monitor lizards, geckos, macaque and capuchin monkeys, parrots, and cockatiels. A few, like Burmese Pythons in FL, are now found in the Everglades and are probably breeding. A tourist filmed an alligator eating a large python, so native species may control this invader. The Cuban Treefrog eats the native frogs in FL. We have a similar story in the PNW with the introduction of the American Bullfrog, a distasteful prey avoided by a number of predators, but which happily devours our native frogs. In HI, non-native birds are more common than native birds. Our message is, *never* release pets into the wild. At best it is cruel, and, on occasion, they might succeed in reproducing and establishing populations that cause serious environmental problems.

The Inevitable Unknown Future? Some species are relatively new to the PNW, such as West Nile Virus. We know that not all birds are affected equally, European birds are less susceptible to the virus, and crows and jays are more vulnerable than most species. How this disease alters bird communities remains a work in progress. We do know that the disease was unknown in the US until 1999, when it arrived from the Middle East. By 2002 it was found in 42 of the lower 48 states. The number of human cases exploded from 66 in 2001 to 1,460 in 2002. It is now found in 70 bird species and more than 40 mosquito species, and we don't know the ultimate result of this introduction on wildlife or human health.

Other species haven't arrived in the PNW, but we know they are likely to arrive and be a problem. We selected a hit list of those species biologists are particularly worried about. For example, Zebra Mussels entered the Great Lakes in the ballast water of a ship in the mid-1980s and are likely to get to the PNW. Adventurers following the Lewis and Clark trail could bring the mussels west. Dead mussels have been found on recreational boats here so they may already be here and be undetected. The impacts in the Great Lakes are environmental, such as excessive filtering of microinvertebrates, and economic, requiring continual cleaning of hydroelectric plants and sewage-outfall and water-intake pipes.

How do we get the genie back into the bottle? Often we can't, though control efforts may try to correct the problems. The principal methods of control are mechanical (harvesting, cutting, or trapping), chemical (pesticides and herbicides), and biological (introduction of something to control the non-native species, which is usually another non-native species). Better than these control methods, however, is to prevent new introductions from occurring in the first place. Non-native species are probably arriving in the PNW every day. In September 2004, a new invasive tunicate for the Northwest was collected from Edmonds, WA. Awareness about the problem of invasive species and a

Agricultural Inspection

rapid response to potential new invaders are crucial for success in the battle against non-native invasive species. This book is intended to inform about current invasions so that some future invasions may be prevented.

Selecting Species for the Book

This book evolved with the assistance of many people, including the numerous authors of our species accounts. We began in the spring of 2003 with a seminar of faculty and graduate students at the University of Washington, addressing the invasive species problem in the PNW. We first had to decide how to define our parameters. The PNW is usually thought of in political and geographical terms as the states of Washington, Oregon, and Idaho, along with parts of British Columbia. However, we thought it more useful to use ecological criteria for our definition, based on World Wide Fund for Nature (WWF) Ecoregions (Olsen et al. 2001). From an ecological perspective, our ecoregions overlap the states.

Ecoregions are relatively large areas of land or water that "contain geographically distinct assemblages of natural communities. These communities (1) share a large majority of their species, dynamics, and environmental conditions, and (2) function together effectively as a conservation unit at global and continental scales" (Ricketts et al. 1999). Jonathan Hoekstra developed and designed a useful map showing the ecoregions of the PNW (see p. xviii). Having established the PNW in these terms, we then compiled a list of all the non-native species known to exist in the region. This list was distributed to scientists and government agencies throughout the region, for review and for additions.

Ranking Species: The Questions. To prevent a xenophobic bias in our selection of the species for the book, all proposed species went through a strenuous evaluation process. While in many cases clear scientific studies of harm are lacking, in every case there is solid evidence that the species included have a negative impact on native species and ecosystems. Nonetheless, reasonable people may disagree with our list. For example, some horticulturists in the nursery trade see Purple Loosestrife or Butterfly Bush as desirable plants not only because they are attractive but also because they are important nectar sources. Through the evaluation process we attempted to eliminate bias and to stick to known facts about each of the species.

The evaluation criteria were originally developed for plants by The Nature Conservancy and NatureServe and were adapted by the California Invasive Plant Council. Because our list was to include more than plants, we took on the difficult task of adapting it to include all organisms, from viruses to mammals. Our modified criteria, "The Questions," are included in the back of the book as Appendix 4. Using these questions, we ranked each species to determine how big of a problem it was or would likely become. Ranking, although based on scientific papers, data, and expert opinion, has a subjective component. Ranking was sometimes difficult due to incomplete information about some species, but those that rose to the top of the list as problem invaders in native ecosystems generally were those that were repeatedly mentioned as we vetted the list. The degree of concern for the species is shown in the left-hand margin of each account as an invasive score. Appendix 5 gives the invasive score for each species and shows how the score was determined from the results of the questions for each species.

What Species Were Not Included. Agricultural and urban pests are often on state lists of non-native or invasive species. We elected to not include species when their main impact was on agricultural, horticultural,

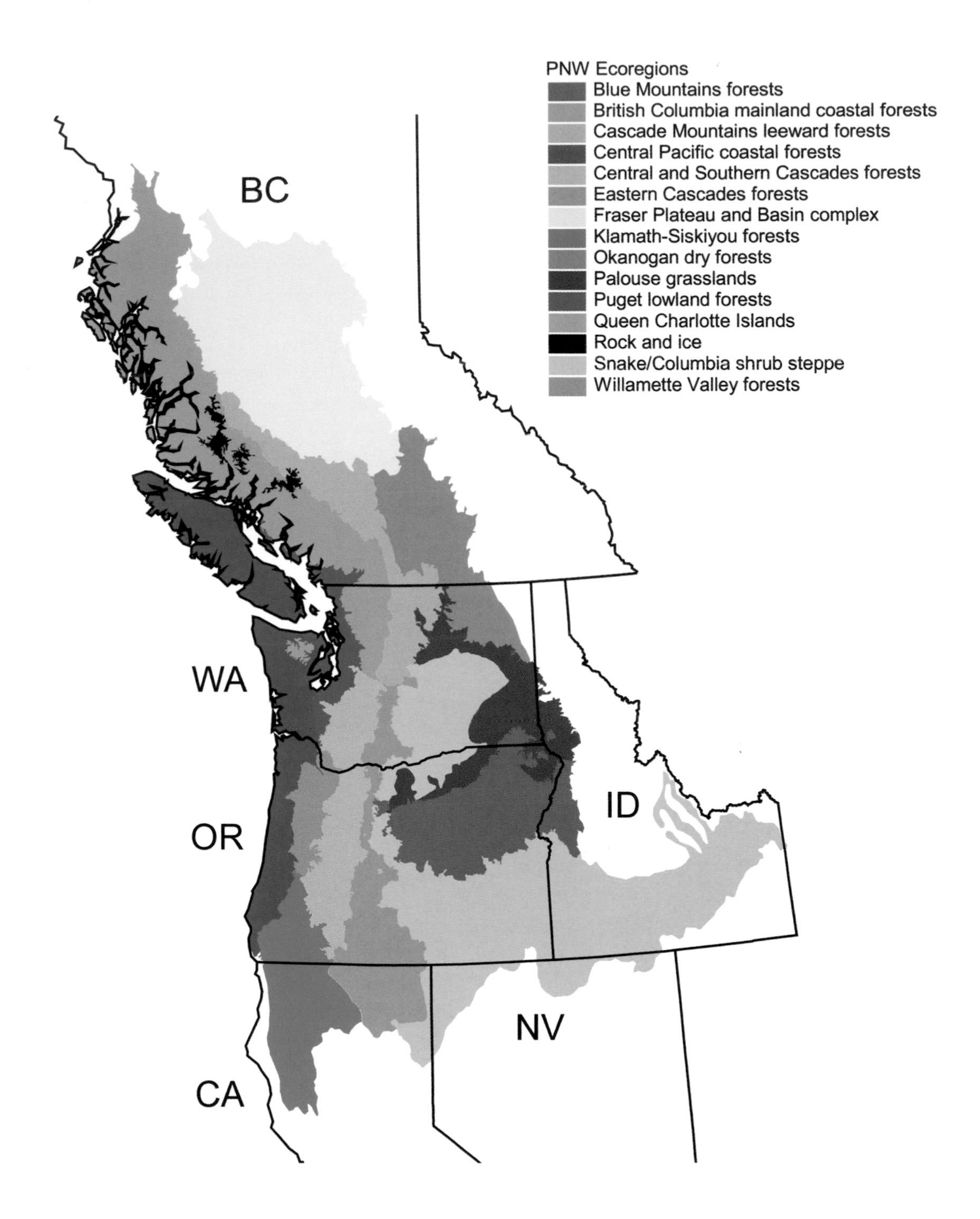

Map of Ecoregions of the Pacific Northwest, by Jonathan Hoekstra.

or timber lands or in urban areas. Not only would the list be a long one but it would detract from our message: that many non-native species are invasive and harming native species and native habitats. We have not included nuisance species like Crabgrass or the German Cockroach that live in close proximity to humans, because they depend partly on human modification of the environment or tend not to invade native habitats. Also excluded are many non-native species that have been here so long that they have become "naturalized" and are largely restricted to where humans modify the habitat or plant them. Common Dandelions and feral pigeons (Rock Doves) fall in this category.

Many experts pointed out that humans are the worst invasive species of all. Certainly, humans irrevocably alter ecosystems and are a serious pest species. We chose not to include humans as a non-native invasive species because, unlike an insect or a plant, we have the ability to recognize harm in our behavior and change our ways. After reading this book, you will be equipped to change your status from a harmful pest species.

The Final List of PNW Invasive Species. As you peruse the list you may come away believing that plants and fish are the most invasive and harmful of all invasive organisms, or that we have a particular dislike of plants and fish. It is true that these taxonomic groups make up a substantial portion of our treatments, reflecting the large number of plants introduced for landscaping and the many fish repeatedly placed in fresh waters to enhance fishing. Almost no lake, even alpine lakes, is free from the dumping of fish by anglers. Our government stocked the remotest of lakes by helicopter, including national parks and wilderness areas like the Alpine Lakes in the Cascades. Few other taxonomic groups have had as many deliberate introductions, inevitably resulting in some of those introductions becoming problematic.

By design, we have included species from a variety of groups, trying to select the worst invaders within these groups. We have included some species that may not cause a great deal of harm, such as the widely hunted game species the Ring-necked Pheasant and the Chukar Partridge, so people will know they are non-native species. We hope that as they flip through this book, even some of the most seasoned biologists will be surprised to learn that some species they thought were native to the PNW are not. Whether the non-native is a slug, a worm, a fish, or a plant, we hope the book provides a better understanding of the non-natives that are here and why they are a problem for our native species and native habitats.

Conclusion

Human values and knowledge determine how we view the world. If you're a beekeeper and use Purple Loosestrife as a honey source or are a fisherman who loves perch, you may not see these species as the problems they are for native species and native ecosystems. By reading this book, you'll learn how interesting biology is, how incredible life is, how non-native species can change native species and native habitats, and how you can mitigate and slow these invaders. The PNW is an interesting and biologically rich place to live. The lives of the species in this book may appear as science fiction but they are not. Natural history is interesting and the ways that each of these species survives—from exploding seeds to toxic chemistry—may amaze or surprise you. You'll more fully enjoy the outdoors if you know more about both the unwanted non-native species and the wanted native species that share the world with us. Armed with knowledge of what should and shouldn't be here, you can help slow immigration of the unwanted and thereby slow the extinction of some of our wanted species.

The Costs of Invasive Species. We think biology is interesting, and reading the accounts provides a sense of why everyone should know problem non-native from native species. But if the biology doesn't sway your viewpoint, economics may. Non-native species cause annual losses of millions of dollars to agriculture, forestry, rangelands, and fisheries. Losses, more broadly defined, reach billions of dollars, with conservative estimates of $137 billion/yr in the US. In the US we spend over $7 billion for pesticides alone each year.

In addition to these costs, the side effects of these applications on native species, habitats, and human health are little publicized but are often harmful and long-lasting. The lack of native butterflies and moths compared to 50 years ago likely derives from pesticides used to control non-native pests and the introduction of viruses and inappropriate host plants. The harm to non-target species from chemical applications is widespread, but more importantly, the chemicals we use promote resistance in the target organism. The use of chemicals needs to be considered carefully because they can do more harm than good to us and to native species and native ecosystems.

More Than Biological and Economic Loss. In the early 1900s, the American Chestnut was among the common trees in the eastern forests and was the most economically important hardwood tree. It was an important food source for deer, squirrels, birds (including the now-extinct Passenger Pigeon), and people and was used extensively in urban plantings. Chestnut Blight, brought in from China on diseased horticultural stock, killed as many as 1 billion trees. The demise of the American Chestnut increased Oak Wilt Disease and is believed to have caused five native insect species to disappear.

Once the chestnuts were gone the oaks increased, providing most host plants for the disease. The biological loss of this species is only part of the story, because its loss is also an artistic one. No longer can someone like Henry Wadsworth Longfellow write a poem such as "The Village Blacksmith" about a man toiling under "the spreading chestnut tree," because there are no large chestnut trees. Our world is poorer for the loss. Do we know what we have missed? What losses will future generations suffer?

Whether it is a remote island or an entire continent, the intentional or accidental introduction of species from somewhere else is altering and homogenizing the biology of the globe. Biological distinctiveness of regions should be protected and embraced, as we celebrate and protect cultural heterogeneity. Based on what we have observed, as Darwin did, over the last two centuries, there is no doubt that our future world will be more homogeneous than it is now. But we can work to slow down the rate of change, if each individual becomes more aware of the environment and what needs to be done to protect it.

References: 15, 238, 251, 269, 334

A Closer Look: Invasive Species on the Queen Charlotte Islands

JOANNA L. SMITH

Given the current scale of invasions and our lack of effective policies to prevent or control them, biotic invasions have joined the ranks of atmospheric and land-use change as major agents of human-driven global change.

—Richard Mack, 2000

The earth is home to more than 100,000 islands. The combined landmass of the 150 largest islands equals the size of Europe. In these isolated, natural laboratories, species evolve over thousands to millions of years, eventually becoming so different from their mainland ancestors that they are endemics; they do not occur anywhere else in the world. The Madagascar Lemurs, the Galápagos Tortoises, and the Hawaiian Silverswords are endemics. Physical isolation and unique geological formations or biological environments give islands more than their share of endemic species. For example, Madagascar has 8,000 endemic plants, twice the number found in the entire US. The 7,100 islands in the Philippines contain 460 endemic vertebrates, more than anywhere else on earth. Islands have a disproportionately higher number of endangered species than continental habitats because of this high rate of endemism. Compared to the other four oceans, the Pacific Ocean has the highest per capita number of rare, threatened, and endangered species. Dubbed the "endangered species capital of the world," the Hawaiian Is. have 72% of all plant and animal extinctions recorded in the US, this in a state that is less than 0.02% of the nation's land area.

In the northeast Pacific Ocean, most offshore islands were pristine areas, free from human disturbance until 10–12,000 years ago. When glaciers retreated, humans settled in small villages along ice-free coastal areas. Today, 1 in 10 people live on an island—a total of more than 600 million people. In the PNW, people occupy many of the largest islands, including most of the San Juan and Gulf Islands. Rabbits, sheep, goats, cows, cats, rats, invertebrates, and an assortment of invasive or agricultural plants and animals came with human settlers. Island species are vulnerable to human activities, especially agriculture and alien species, because of their restricted geographic range. Domestic animals, like pigs and cows, destroy native plant communities. Introduced predators like foxes and cats kill endemic nesting birds. Introduced species and habitat loss have caused oceanic centers of evolution like Hawaii, New Zealand, and the Caribbean Islands to lose more than 50% of their native and endemic species.

The Queen Charlotte Islands—or Haida Gwaii—form the largest island chain in the PNW. Literally, "Islands of the People," the archipelago spans 280 km and occupies 9,596 sq km with 150 large and 1,300 small islands. Only 50 km from the coast of AK, the snow-capped peaks of the Alaskan panhandle are visible from the northeast tip of the islands.

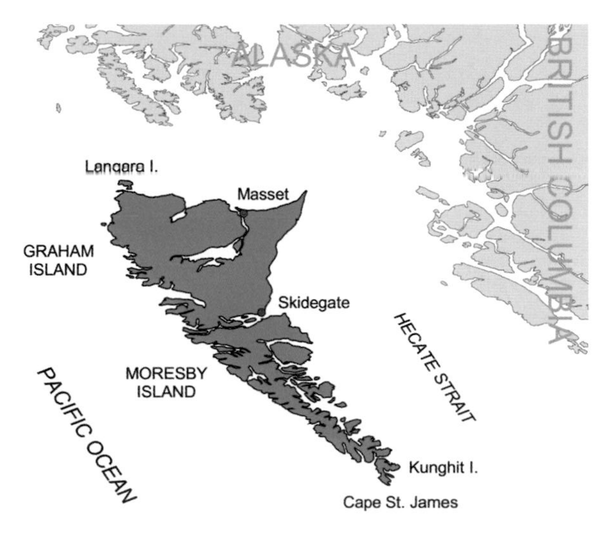

Haida Gwaii (Queen Charlotte Islands)

Well known for their spruce and cedar forests, the islands are a rich marine community and the traditional territory of the Haida people.

Since the retreat of the glaciers 10,000+ years ago, the Haida occupied these islands. Their lifestyle, art, and traditional ceremonies reflect their connection to the land and sea. Until the late 1800s, native plants and animals supported at least 14,000 Haida in 126 coastal villages with few non-native agricultural plants. After the devastating effects of smallpox introduced by European settlers, no more than 590 Haida remained at 3 villages by 1911. Today, 3,000 people live on the archipelago, most arriving from the mainland in the last 100 years.

Large-scale ecosystem changes to Haida Gwaii began in the 1700s but increased dramatically after 1850. Europeans cleared entire forests for permanent settlements and roads, and harvested native species for food, fuel, and trade. A growing local economy created whaling stations, salmon canneries, mines, and logging camps. Beginning in 1878, dwindling food supplies inspired people to introduce the Sitka Black-tailed Deer, rabbits, and domestic cattle. Over the next century, more than 19 species were released either accidentally or intentionally on this island chain. Some were native to the PNW, others originated in Europe. Rocky Mountain Elk, European Red Deer, muskrats, mice, pheasants, cats, and beavers arrived between 1901 and 1936. The commercial extinction of the native sea otters in the 1940s led trappers to release raccoons to enhance a faltering fur trade. In 1947, Red Squirrels were put on five islands in hopes that their stockpiles of valuable Sitka Spruce seeds could be used for forest silviculture. Seeds were not collected but squirrels ate thousands of songbird eggs and nestlings. Further introductions included goats, Pacific Tree Frogs, bullfrogs, slugs, wasps, scale insects, aphids, and weedy plants. Once on the islands, impacts rarely remained local. Cows escaped into sensitive sand dune habitats, deer populations exploded southward, threatening endemic and rare plants from sea level to the subalpine, and raccoons and rats quickly discovered an abundant food source—seabird colonies.

Haida Gwaii is an important breeding area for seabirds because of its proximity to nutrient-rich waters and numerous nesting locations. Before the arrival of raccoons and rats, more than 1.5 million birds, from 12 species, nested on 37 islands in 13 major colonies. Half of the world's population of Ancient Murrelets—500,000 birds—used to breed on these predator-free, offshore islands. Recently, the Canadian Wildlife Service calculated that raccoons can access 80% of all small islands in the archipelago. Raccoons or rats are already on 20 seabird colonies and have destroyed or seriously damaged at least 10 large ones. Capable of swimming more than 1 km, raccoons are now at the southernmost tip of the archipelago, traveling more than 100 km across some of the strongest currents on the West Coast.

In the 1800s, prominent offshore islands became lighthouse stations to protect shipping traffic from treacherous rocks and reefs. Most of these islands are now infested with rats that kill and eat seabirds' eggs, young, and even adults. For at least 100 years, rats have been disturbing Tufted Puffin, Rhinoceros Auklet, Cassin's Auklet, Storm-petrel, and Ancient Murrelet colonies on Haida Gwaii. Adult seabirds weighing as little as 45 to 200 g are easy prey for a 480 g adult Norway Rat. Rats nearly destroyed the largest Ancient Murrelet colony in the world on Langara I., causing it to decline from 400,000 breeding adults before 1920, to fewer than 20,000 by 1988. Remarkably, the Cape Flattery Lighthouse station, on Tatoosh I., WA, remains rat- and mice-free despite having a post office, school, and 50 inhabitants for four decades.

While raccoons or rats affect island species at population levels, deer cause damage at the ecosystem level. On Haida Gwaii, deer eliminate understory vegetation and limit the regeneration of trees, especially Red and Yellow Cedar. They severely restrict the range of endemic coastal fauna, alter nutrient cycling, reduce soil stability, and increase erosion and landslides. The decline in structural complexity alters the abundance and diversity of beetles, wasps, and other insects, eliminates nesting habitat for songbirds, and encourages introduced species previously limited by competition or shade. Deer swim up to 6 km to access offshore islands and now occupy all but 8 of the 150 islands greater than 50 ha. While the economic costs to the forestry industry are high—nearly $2 million/yr to protect seedlings—the ecological costs are staggering. On Haida Gwaii and around the world, fewer than 10% of introduced species cause 90% of the damage to island and continental ecosystems.

The extinction of endemic species or rare island isolates is a serious conservation concern because the loss is permanent. Endemics are irreplaceable. In Madagascar, the extinction of 16 species of endemic lemurs gives them the unenviable distinction of being the most gravely endangered primate group in the world. On the Galápagos Is., five races of giant tortoise are extinct because of human activities in the 1800s. On Pinta I., introduced goats ate the native vegetation so quickly that all but one tortoise died. Today, the sole survivor, "Lonesome George," lives at the Darwin Station on Santa Cruz I., awaiting a mate that will never arrive.

Protecting island-dwelling species is challenging. Populations can be small or hard to find, may occupy unusual or rare habitats, or become more vulnerable to diseases and predators over time. Species that adapted to defend themselves against predators lose physical or chemical characteristics in the safety of island isolation. Western Red Cedar on Haida Gwaii, for example, lost most of its natural, chemical defenses against herbivory; when deer were introduced in the early 1900s, they ate the young trees.

Land designations creating national parks, reserves, or international sites of significance are one option to protect islands and associated native species. The Galápagos National Park covers 95% of the land area, protecting at least 96% of the original biological diversity. The Gwaii Haanas National Park Reserve protects 15% of the Haida Gwaii archipelago, and throughout the PNW most islands that contain marine birds and mammals or endemic plant communities are protected under provincial, state, or federal laws. But land designations do not protect island flora and fauna from introduced species. Control or eradication plans reduce or remove particularly harmful species, but unfortunately these programs are costly or require indefinite commitments. A five-year, $1 million project

Table 1. Vertebrates introduced to Haida Gwaii (Queen Charlotte Islands) since the 1700s

Introduced species	Scientific name	Year	Current distribution
Norway Rat	*Rattus norvegicus*	1700–1900	Localized
Black Rat	*Rattus rattus*	1700–1900	Localized
Sitka Black-tailed Deer	*Odocileus hemionus sitkensis*	1878	Widespread
House Mouse	*Mus musculus domesticus*	1880s	Widespread
European Rabbit	*Oryctolagus cuniculus*	1880s	Widespread?
Feral Cattle	*Bos taurus*	1893	Localized
Ring-necked Pheasant	*Phasianus colchicus*	1913	Extinct?
European Red Deer	*Cervus elaphus elaphus*	1919	Extinct
Muskrat	*Ondatra zibethica osoyoosensis*	1924	Widespread?
Rocky Mountain Elk	*Cervus elaphus nelsoni*	1929	Localized
Pacific Tree Frog	*Hyla regilla*	1933	Widespread
Beaver	*Castor canadensis leucondontus*	1936	Widespread
Raccoon	*Procyon lotor*	1940s	Widespread
Red Squirrel	*Tamiasciurus hudsonicus anuginosus*	1947	Localized
Goat	*Capra hirus*	1976	Localized
European Starling	*Sturnus vulgaris*	1980s	Localized
American Bullfrog	*Rana catesbeiana*	1990s	Localized
Green Frog	*Rana clamitans*	1990s	Localized
European House Sparrow	*Passer domesticus*	2000	Localized

on Haida Gwaii attempted to eradicate deer to quantify long-term effects of deer pressure on plants, insects, and songbirds. Hunters killed more than 400 animals, allowing native plant and bird communities to recover, but only for a few years, at two locations. Eradicating rats from a single seabird colony in BC cost $2 million. Raccoon control on Haida Gwaii, by comparison, is inexpensive but new animals near seabird colonies must be killed every year. Prevention is the first step to protect island ecosystems from introduced species. Quarantine programs and an educated public are sentinels to thwart or identify new introductions.

In the last 400 years, islands worldwide had 50% of all animal extinctions and 90% of all bird species extinctions. This unprecedented rate of species extinctions has been described as "one of the swiftest and most profound biological catastrophes in the history of the earth." Islands are special places. Taking care of them is essential if we are to maintain their remarkable wildness.

Other Sources of Information: 40, 51, 82
References: 106, 141, 143, 196, 202

How to Use This Book

The book is divided into 5 categories: plants, invertebrates, vertebrates, diseases, and threats. The sections on plants, invertebrates, and vertebrates are further divided into freshwater, marine, and terrestrial subsections. They are color-coded:

Plants (freshwater, marine, and terrestrial): shades of green

Invertebrates (freshwater, marine,and terrestrial): shades of brown

Vertebrates (freshwater, marine, and terrestrial): shades of blue

Species are organized alphabetically by common name; Latin names appear in the header of each species' account.

Maps

In each account, we include a map showing a gray outline of all the counties in our defined area of the Pacific Northwest (PNW). The counties in which that species' invasion, suspected presence, or eradication is known are indicated in color. For species not yet detected in the PNW, the map has a question mark. Four categories describe each species or group of species:

Confirmed present (likely established: confirmed identification, herbarium specimen, expert opinion)

Suspected present (one/several individuals observed: unconfirmed identification, anecdotal observation, assumed present)

Eradicated (either intentionally eradicated or has naturally died out)

Likely throughout (assumed but not confirmed throughout PNW)

All map data are reported by county (California, Idaho, Oregon, Nevada, and Washington) or district (British Columbia) (see p. xxvi). For marine species, the county or district is adjacent to the area occupied. For freshwater species, the county or district contains the water body occupied. For accounts with multiple species, a county or district is reported as occupied if at least one of the species is present.

Invasive Species Ranking

Each species' ability and likelihood to cause damage to native ecosystems is ranked according to a series of 47 questions (Appendix 4). As an index to the damage these invaders may cause, we provide an invasive score in the top left-hand margin of the species account:

H = High (currently causing large-scale ecological damage)

HA = High Alert (high potential for causing ecological damage)

M = Medium (currently causing ecological damage)

MA = Medium Alert (medium potential for causing ecological damage)

L = Low (currently causing small-scale ecological damage)

Not Listed (unknown)

Appendix 5 gives the invasive score for each species and shows how the score was determined. The introduction explains more fully how the various factors were weighted.

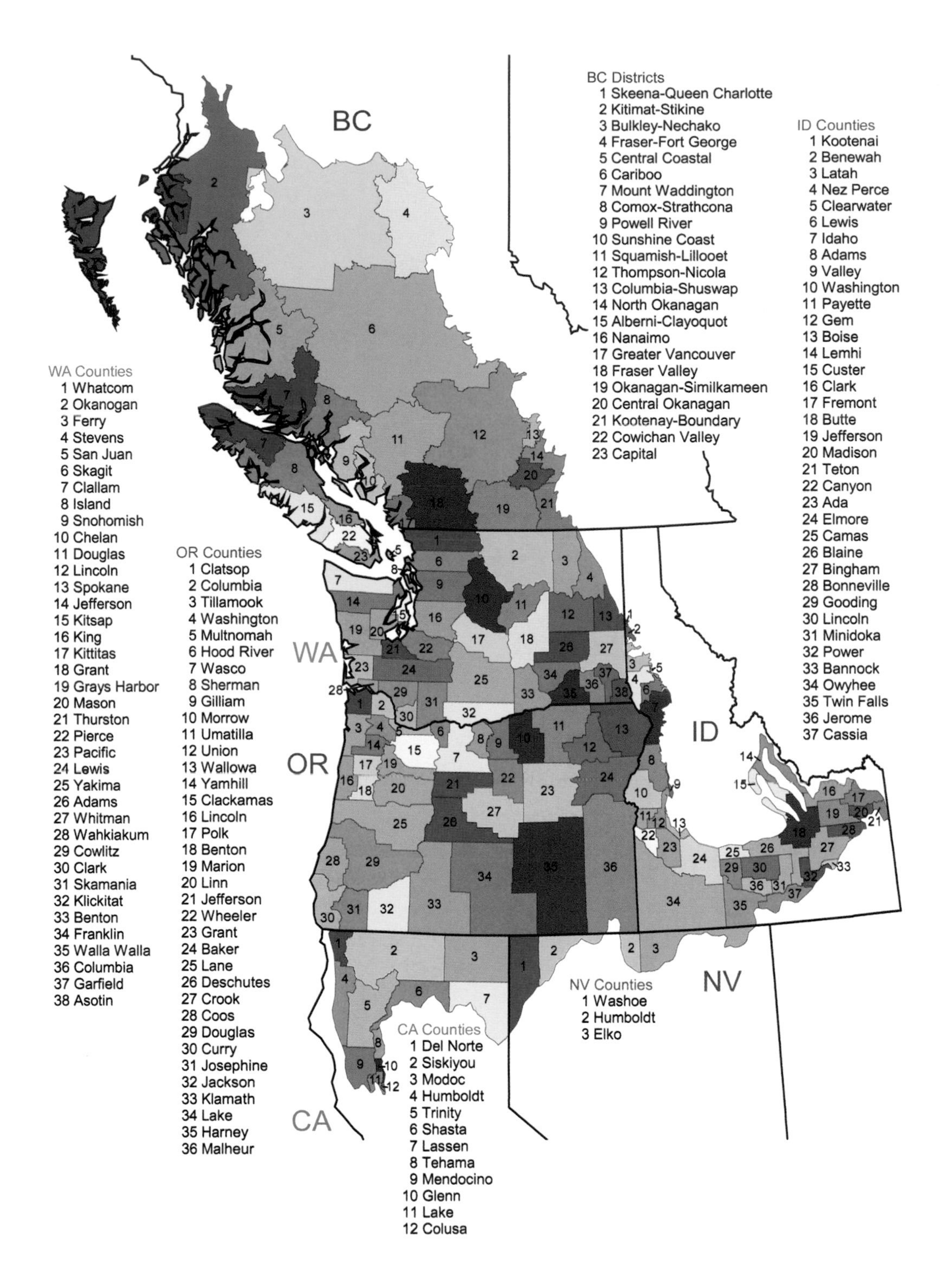

Counties and districts of the Pacific Northwest

Species Accounts

Brazilian Elodea *Egeria densa*

Species Description and Current Range

Brazilian Elodea, a bright green aquatic weed native to South America, roots in the bottom of freshwater bodies. Highly branched stems grow up to 6 m to the water surface. Mature leaves are 1.4–2.5 mm long and 1.6–2.8 mm wide; they typically radiate from the stems in sets of four. Diminutive three-petal white blooms (less than 1 cm in diameter) float on the surface. Brazilian Elodea is often confused with the native species, *Elodea canadensis*.

Brazilian Elodea is found primarily in freshwater lakes, ponds, and reservoirs in western CA, OR, WA, and southwestern BC. It is also found in relatively low-velocity streams, rivers, and canals. Distribution across the landscape is spotty, though it can significantly impact individual waterways, inhibiting all other water plants. Brazilian Elodea's ability to adapt makes it a successful invader. In shade or high turbidity, it grows long stems, forming a canopy near the water's surface.

Impact on Communities and Native Species

Brazilian Elodea impacts communities and native species through competition and habitat modification. It grows in a wide array of conditions and can form dense stands that hinder other aquatic plants. Its impact on native species and plant diversity affects the whole food web. Dense stands also slow water flow and trap sediment and nutrients. In some systems with an overabundance of nutrients, Brazilian Elodea acts as a "nutrient filter" to trap nutrient-rich particles and inhibit algae blooms. It causes sediment to build up, buries seeds and eggs, and increases the rate at which a water body is filled. It can also reduce nighttime dissolved oxygen and cause fluctuations in pH, both of which may be detrimental to aquatic organisms. Brazilian Elodea also significantly impedes recreation. Dense stands inhibit boating, swimming, and waterskiing, posing a serious threat to those who become entangled. Officials in OR estimate that Brazilian Elodea's recreational cost alone is over $3.5 million/yr.

Control Methods and Management

There are three major strategies for aquatic weed control: mechanical, chemical, and biological. Mechanical methods include bottom barriers and diver dredging. Both have been effective at controlling small-scale invasions. However, mechanical methods are not recommended for large-scale invasions because they are highly labor intensive. Also, diver dredging can produce copious plant fragments, which root and establish new plants, spreading the invasion.

There are several effective chemical controls, including diquat and fluridone; however, because the plants are submerged, the herbicide must be applied to the water, potentially affecting other species. Additionally, permits, management plans, and licensed herbicide applicators may be required, depending on state and regional requirements. A toxin more specific to Brazilian Elodea is being developed from a fungus (*Fusarium* sp.) found in Brazilian Elodea's native range.

The most widely explored biological control is the Grass Carp (*Ctenopharyngoden idella*). Limited experiments in the PNW and elsewhere have shown that Grass Carp can remove Brazilian Elodea from a water body, and that, upon carp removal, revegetation by native species is possible. However, Grass Carp should not be introduced to any water body supporting salmon. Also, results varied across lakes, and current information is limited about the conditions under which stocking is effective. Potential long-term impacts include the likely shift from a grass-dominated system to one dominated by algae. A bad history of biological control and limited knowledge about potential impacts suggest caution in moving forward with Grass Carp stocking.

Overall costs vary depending on the size of the problem, though thousands of dollars are being spent in affected states each year. Costs can become extreme. According to the WA Dept. of Ecology, control programs in Silver Lake, WA, cost over $1 million. Because all control methods are costly, preventing the introduction of the weed and providing early detection monitoring of a water body are important. Educating the public and encouraging them to examine and clean boats, trailers, and other gear are critical.

Life History and Species Overview

The native range of Brazilian Elodea includes the Minas Garaes region of Brazil and coastal Argentina and Uruguay. Within this area, male plants predominate in a 6:1 ratio. Outside its native range, only male plants are found. Because of this, reproduction occurs only through the rooting of plant fragments. An analysis of Brazilian Elodea's life cycle conducted in Lake Marion, SC, revealed periods of rapid growth and blooming in spring and summer, significant defoliation of older plant material in the late summer, and a period of renewed growth and blooming in the fall. During winter, the plant grew very little, existing in a green state along the bottom of the lake. In coastal lakes of the PNW, plants commonly have reduced, but continuous, growth throughout the winter.

History of Invasiveness

Brazilian Elodea, like many plant invaders, is spread by humans. It first became popular as an "oxygenator" in fishponds (a misleading

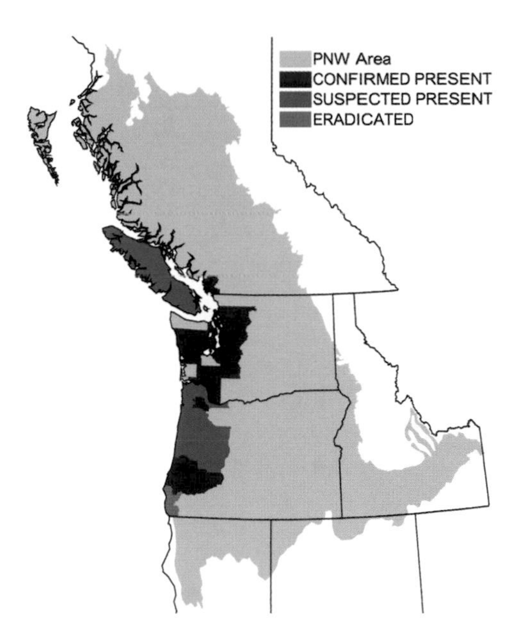

characterization because as dense stands respire, nighttime dissolved oxygen is reduced). Brazilian Elodea was also thought to be important in reducing mosquito larvae (thus reducing malaria). It became popular in aquatic gardens and aquariums, with US sales beginning in 1915 and continuing today, despite national and local laws that prohibit the transport and sale of noxious species. It has also been widely used as a laboratory specimen. Most introductions are likely the result of careless dumping from aquariums, though it can also spread from an infested to an uninfested area by attaching to boats, fishing poles, and trailers. The first known case of Brazilian Elodea in the US was in Millneck, Long I., NY, in 1893. Currently, it has been reported in water bodies in at least seven countries (Australia, Chile, Denmark, France, Germany, Japan, New Zealand, and the US).

Other Sources of Information: 60
References: 117
Author: Allison A. Van

Common Reed *Phragmites australis*

Species Description and Current Range

The invasion of the Common Reed required a bit of detective work to explain. The species is actually native to the US and has probably been present in the PNW for several thousand years. But during the last 200 years, the plant began to spread aggressively on the Atlantic coast and westward. At first, human disturbance was entirely blamed for the expansion, but recent research using genetic sequencing showed that the invader was an exotic strain from Europe. Experts suspect that a broad ecological tolerance and aggressive growth enable the invasive strain to take over areas where the native strain historically grew and to expand into new regions. The exotic strain is now found throughout the continental US and southern Canada and is spreading rapidly in the PNW.

Common Reed is a member of the grass family (Poaceae) and can grow up to 6 m tall. It forms distinctive fluffy seed heads at the top of long stalks. The native and invasive strains look fairly similar but can be distinguished through close examination. The stems of the native plants (visible after the leaf sheath has been removed) are generally smooth and shiny, while the invasive strain has rough stems. The native strain has a red or purple color to the stems in spring and summer, while the invasive stems are tan. The invasive strain also holds onto its leaf sheaths longer and more tenaciously, a diagnostic character for plants from the previous season. The current distribution of the invasive strain is not very well known, but most stands in the PNW are believed to be the invasive strain. Only genetic tests will tell for sure.

The introduction of an invasive strain that looks similar to the native plant is called a "cryptic invasion" and provides an additional set of challenges to invasive species biologists and managers. If the different strains are hard to tell apart by looking at them, it can be difficult to detect a new invasion. Similarly, control efforts must be carefully targeted at the invasive plants without damaging native strains. Managers must also contend with difficult questions about what level of biodiversity they should try to protect. Does it matter that a native strain could be wiped out if the perpetrator is of the same species?

Impact on Communities and Native Species

The biggest problem with the invasive strain of Common Reed arises from its aggressive, competitive nature. Once the plant starts growing in a new area, it commonly increases in density until it is the only plant in the immediate area—a monoculture. Where the invasion is severe, it displaces plant species that provide food for wildlife. The native strain more typically grows in mixed stands with several species of plants coexisting. The invasive Common Reed also causes more insidious changes in the ecosystem, altering nutrient cycles and hydrological regimes. All of these changes lower the quality of the wetland as habitat for waterfowl and the migratory birds that depend on the wetland community. Overall, the invasion causes a net loss of biodiversity.

Control Methods and Management

Because the plant can multiply from small pieces of rhizome (the underground, horizontal stem), typical methods of mechanical control such as mowing or disking can help spread the plant. Although dredging, flooding, draining, burning, or grazing may beat the invader back, the most successful strategy involves the use of an herbicide, glyphosate. The treatment costs about $250/ha and requires follow-up work and retreatment in subsequent years. Unfortunately, the herbicide can have the undesirable side effect of killing other plants, so appropriate permits are needed before use. Biological controls are under investigation, but none is recommended at this time. Monitoring is an important step to ensure that the invasion does not recur.

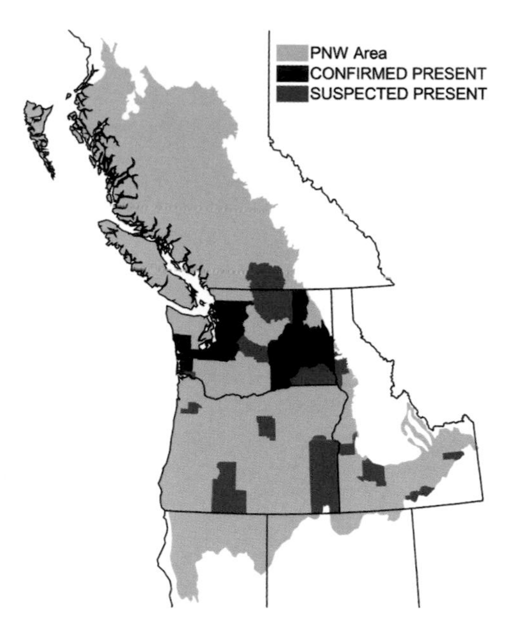

Life History and Species Overview

Common Reed spreads primarily through vegetative growth, forming bigger and bigger clumps along the underground stems (with up to 200 vertical stems/m^2). Even just a small chunk of rhizome can start a new colony of this invasive plant if transported to a new location. Common Reed does flower and set seeds each year, but most new stands are thought to originate from rhizome fragments. As a perennial, the plant survives from year to year and just keeps growing. It can grow in a wide range of wet habitats, most commonly in marshes, on the borders of lakes, ponds, and rivers, in tidal estuaries, and in ditches along highways. Its spread is facilitated by the restriction of tidal circulation, pollution, and destabilization of soil. The species has a global presence, with various strains found in Asia, Africa, the Americas, Australia, and Europe.

History of Invasiveness

The invasive strain probably arrived in the US in the 1800s, growing where ship ballast was dumped at Atlantic coastal ports. Because of its similarity to the native strain, the invader was not recognized and spread rapidly along new railroads and roadways constructed in the late 1800s and early 1900s. Initially, the spread was blamed on human activities causing habitat disturbance, and the plant is still known as an indicator of wetland disturbance. Recent research shows that a key factor in the invasion's success was the introduction of the European strain, which is now found throughout the continental US.

The native strain has almost disappeared in New England, while the invader continues to expand to new areas, especially inland wetlands in the Midwest. The invader has already gained some ground in the PNW. Extrapolating from the rapid spread that has occurred in the last century, the story is far from over. The expansion will likely continue to accelerate through the middle of North America and on the West Coast, with a magnitude as great as that of infamous invaders like Purple Loosestrife (*Lythrum salicaria*) and Saltcedar (*Tamarix ramosissima*). Unfortunately, Common Reed has already expanded into some sites where successful control of other invasives, such as Purple Loosestrife, has made habitat available.

Other Sources of Information: 29, 72, 92, 120
References: 278, 279, 336
Author: Elizabeth A. Skewes

Eurasian Watermilfoil *Myriophyllum spicatum*

Species Description and Current Range

Eurasian Watermilfoil, an aquatic plant with featherlike underwater leaves and emergent flower spikes, is found in lakes, ponds, reservoirs, rivers, and streams. It can often be confused with native, noninvasive milfoils, but because it is highly noxious, it is important to correctly identify it. It can be distinguished from other species by the arrangement of its longish leaves, 2–4 cm long, in whorls of four around reddish or olive-green (in summer) stems. Leaves typically are square-tipped and have more leaflet pairs per leaf (14–21) than the four milfoil species native to the PNW. Flower spikes emerge above the water's surface and have small flowers, with one tiny, shorter leaf beneath each flower.

Eurasian Watermilfoil was once sold as an attractive aquarium plant and now infests water bodies throughout the US. First found in the PNW in 1965, Eurasian Watermilfoil is now established in central and southern BC, throughout WA, and in parts of OR, ID, and CA. In WA, it has invaded over 100 water bodies. It still has not reached its full potential distribution, however, as many susceptible areas are not yet invaded. In WA, it is considered the most problematic aquatic invader because of its wide distribution and resistance to control.

Impact on Communities and Native Species

The introduction of Watermilfoil dramatically alters the ecology of a water body. Eurasian Watermilfoil reproduces rapidly and can infest an entire lake within two years of introduction, forming dense mats on the water's surface. These mats interfere with recreational activities, such as swimming and boating, and also displace beneficial native vegetation, cause flooding, and clog water intakes, such as those used for irrigation and power generation.

Watermilfoil can reproduce from fragments, which is what makes it such a successful invader. In the spring and summer, activity such as boating and swimming, or even removal efforts, can create fragments, starting new plants. Because it establishes early in the season, it is the first aquatic plant to develop and shades out native vegetation, causing declines in plant diversity and displacing native plants and animals. Mats of Watermilfoil provide poor habitat for waterfowl, fish, and other aquatic organisms. Some waterfowl may eat Eurasian Watermilfoil, but it is a low quality food source and usually has replaced higher quality food plants.

Control Methods and Management

Many eradication and control strategies are currently employed in efforts to manage Eurasian Watermilfoil. Eradication efforts include hand pulling, cutting, bottom barriers to shade, herbicides, dredging, and introducing non-native Grass Carp, which are voracious herbivores. Grass Carp, however, eliminate native vegetation as well as invasive milfoil and cause other problems. There are several possible methods available for homeowners who are interested in local control, including hand pulling, cutting (with special cutting tools), raking, bottom barriers, weed rollers, and spot treatment with herbicides.

Some of these methods may not be suitable, as small fragments of milfoil increase the spread. Before using herbicides, other methods should be considered, and neighbors and other interested parties should be consulted. Extreme caution must be exercised in the use of herbicides, as drift may cause accidental contamination of nearby waters.

Life History and Species Overview

Eurasian Watermilfoil is native to Europe, Asia, and northern Africa. It is an extremely adaptable plant, tolerating and thriving in a variety of environmental conditions. It grows in water depths up to 10 m and can survive even under ice. Eurasian Watermilfoil tolerates a range of salinity, pH levels, and temperature, and is found across a gradient of eutrophic conditions. In sites where water evaporates slowly and plants are stranded, Eurasian Watermilfoil can temporarily develop into a land plant. These forms have smaller, stiffer leaves and can survive until they are submerged again. However, milfoil does not survive long out of water, and drying boats and trailers for five days is one method for preventing its spread.

Eurasian Watermilfoil roots underwater and dies back every winter. Once rooted, it grows upward toward the surface of the water. Tiny pink flowers occur several cm above the water on tall spikes and produce seeds, but these are not thought to be important in reproduction. Instead, the plant's extremely rapid vegetative reproduction accounts for its success.

History of Invasiveness

Eurasian Watermilfoil originates from Europe and Asia, but was introduced to North America perhaps as early as the late 1800s, likely when someone dumped their aquarium contents. It may have been introduced to North America at Chesapeake Bay in the 1880s, but there is evidence that the first collection was made in Washington, DC, in 1942. By 1985, Eurasian Watermilfoil was found in 33 states and 3 Canadian provinces.

The first herbarium specimen of milfoil in WA was found in Lake Meridian in 1965. It became established in central BC around the same time, and is currently found throughout the PNW. It is now widely accepted that the spread of Eurasian Watermilfoil has been caused predominantly by boat trailers, which transport the plant from infected water bodies to uninfected ones. It is also still sold by some stores as an ornamental aquarium plant. This human-dispersed invader has resulted in several million dollars worth of control efforts every year. Because of these high costs, there are now laws in WA prohibiting the spread of invasive weeds by boat trailers.

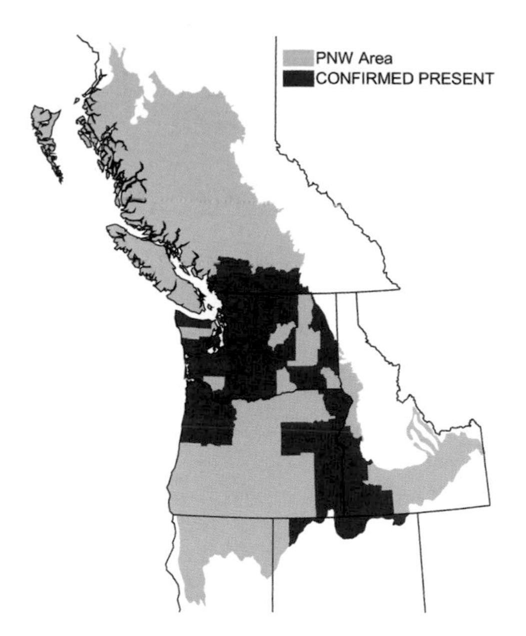

Other Sources of Information: 6, 93, 111, 117, 121

References: 120, 203, 249, 319

Author: Tessa B. Francis

Fanwort *Cabomba caroliniana*

HA

Species Description and Current Range

Fanwort, a mostly submersed plant, has fan-like underwater leaves, as its common name suggests. Because of its attractive foliage, Fanwort is sold worldwide as an aquarium plant, although it is now illegal to sell Fanwort in WA. Fanwort has fan-shaped, deep green or reddish, divided, opposite underwater leaves on stalks. The submersed leaves and stems sometimes have a gelatinous coating. The color of the shoots is strongly influenced by light. In low light the plant is green, and in high light it is reddish brown. Most of the plant is underwater, but oblong leaves sometimes float on the surface, usually when the plant is flowering. The small flowers float on the water's surface and flower color ranges from white to pinkish-purple and yellow, depending on the variety. Fanwort can be confused with other submersed plants with finely divided leaves such as White Water-buttercup (*Ranunculus longirostris*), which has alternate leaves; coontails (*Ceratophyllum* spp.), which lack roots and have whorled leaves; or the milfoils (*Myriophyllum* spp.), which have whorled, feather-shaped leaves.

Fanwort has been introduced to a slough off the Columbia River near Longview, WA, and is found in Cullaby Lake, OR. Fanwort is considered native to the subtropic-temperate regions of eastern North and South America. Fanwort is common in the southeastern US, occurring from TX to FL, north to MA, and west to KS. It is adventive northeast of VI. In parts of the southeastern US, Fanwort plants with purple-tinged flowers have been treated as *Cabomba carolinina* var. *pulcherrima*. South American Fanwort with yellow flowers has been called *Cabomba carolinina* var. *flavida*. Because of its disjunct distribution, there is speculation that Fanwort could be naturalized in North America.

Impact on Communities and Native Species

Fanwort thrives in a variety of climates, from almost equatorial (northern Queensland, Australia) to temperate (New England states and ON, Canada). Once established, Fanwort is extremely persistent and chokes out native plants and animals. An infestation in Queensland reduced numbers of Platypus and Water Rats. Fanwort's dense stands reduce species diversity, adversely impact water quality, interfere with recreational uses, and affect aesthetics. The innate invasive potential of Fanwort appears to be high where it has been introduced, although at the present it does not seem to have spread from its one location in WA. However, based on Fanwort's ability to invade and flourish in a variety of climates and continents, it has the potential to become a major problem plant in the PNW.

Control Methods and Management

As with many aquatic plants, controlling Fanwort is difficult and expensive. No attempts have been made to manage the WA population of Fanwort. However, management efforts are under way in some US states and in Australia. Diver removal and installation of sediment covers are suitable only for small infestations. Mechanical controls such as harvesting are short-term and expensive since the plant rapidly regrows. In drinking water reservoirs, Australia reports some success in draining the water body to completely expose and dry the plants. However, unless the plants dry completely, they can regrow. Grass Carp feed on many submersed plants and have controlled Fanwort in AR and FL. Although at present there are no known classical biological control agents for Fanwort, Australia is currently looking for potential biological control agents in Fanwort's native range. Australia had some success using the systemic and selective herbicide 2,4-D to control Fanwort. Fanwort is also considered susceptible to fluridone, a systemic, broad-spectrum herbicide. Contact herbicides such as endothall and diquat should also knock down Fanwort for a season's control. As with many invasive species, the best control method is to prevent introduction.

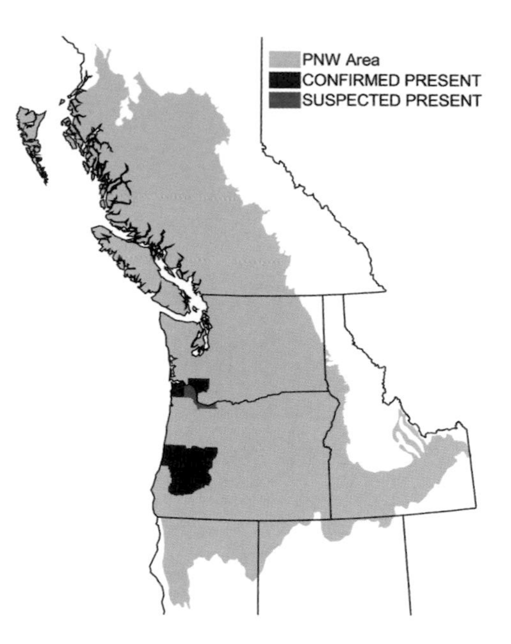

Life History and Species Overview

Fanwort is found in freshwater ponds, lakes, ditches, and slowly moving water. It has been reported growing in water depths of 10 m, but it more typically grows in depths to 3 m. Fanwort prefers silty sediments over hard substrates, but it can also survive freely floating for 6 to 8 weeks in deep water. This species favors nutrient-rich water bodies with low pH, and in more alkaline waters it tends to lose its leaves. Its growth is also inhibited by high calcium levels. Unlike many aquatic plants, Fanwort can grow well in turbid water. In temperate climates, the plants die back in the winter and regrow in spring. The plants flower from May to September, but in tropical areas the plants may grow and flower continuously. Growth rates up to 5 cm/day have been reported from Queensland, Australia. Although in its native range Fanwort can self-pollinate and the seeds readily germinate, it is not known how viable the seeds are when Fanwort is introduced outside its native range. For instance, in Australia, where Fanwort has been widely introduced, the seeds are not viable. However, like many aquatic plants, Fanwort spreads readily by stem fragments or rhizomes. The rhizomes readily break.

History of Invasiveness

Fanwort is native in southeastern North America, but has been introduced outside of its North American range in ON, Canada, and in several states including WA, OR, MI, NY, PA, and some New England states. Fanwort is listed as a noxious weed in ME, VT, WA, and CA (pending final determination). Fanwort is regarded as one of the worst weeds in Australia and is considered a weed of national significance. It is choking waterways along Australia's east coast.

Fanwort was introduced into China in the 1980s and has become a serious pest throughout the Yangzi River (Changjiang) delta in East China. It has been introduced to other areas, including Malaysia, India, Japan, and New Guinea via discarded aquarium plants. Japan also considers Fanwort to be a noxious weed.

The primary invasion route for Fanwort is via the aquarium trade. It is a popular live aquarium plant with widespread commercial availability, especially though aquarium Web sites. To close this introduction pathway, six states have prohibited the sale of Fanwort. Once introduced into a water body, Fanwort can be easily spread by plant fragments on boats, trailers, and fishing gear. Transporting any aquatic plants on boat trailers is also illegal in WA.

Other Sources of Information: 93
Author: Kathy Hamel

Hydrilla *Hydrilla verticillata*

Species Description and Current Range

Hydrilla verticillata is a freshwater perennial herb that closely resembles the native American Elodea (*Elodea canadensis*), and the introduced Brazilian Elodea (*Egeria densa*), which also occurs in WA lakes. Sometimes called Florida Elodea, Water Thyme, Wasserquirl, or Indian Star-vine, Hydrilla can be distinguished by its 3 to 20 mm pointed leaves with serrated margins. The leaves are attached in whorls of usually five around the stem. Hydrilla forms small white to yellowish tubers (potato-like structures) on the roots, buds on the stem, and white floating flowers. In contrast, the American and Brazilian Elodea do not produce tubers; they have larger flowers that do not float free but remain attached to the parent, the leaves occur in whorls of usually four, and their texture is less rough. As Hydrilla can grow 2 cm/day, it spreads rapidly. It can be found in steady or running water in lakes, ponds, streams, rivers, canals, and reservoirs (major habitat types). It is usually rooted on the bottom and forms monospecific stands that cover extensive areas. It was found in WA in 1995 in the closed system of Pipe Lake and Lucerne Lake in King County.

Impact on Communities and Native Species

Hydrilla grows in high densities, covering the water surface and rapidly overgrowing and replacing native species. By blocking the sunlight, it restricts access to light for other plants such as the American Eelgrass (*Vallisneria americana*) and the American Waterweed (*Elodea canadensis*). Hydrilla also reduces the oxygen concentration, making the habitat unsuitable for fish and other aquatic animals. Hydrilla impacts humans as it forms dense mats that clog machinery and impede water flow, as shown by the $170,000 losses at the Gunstersville Dam and Wheeler Dam in NC. These dense mats also create problems in

navigation, lower the quality of recreational activities such as swimming and fishing, and affect flood control as the natural flow of water is blocked.

Control Methods and Management

Hydrilla is classified as a Federal Noxious Weed in the US, and is listed on WA's Wetland and Aquatic Weed Quarantine list. Although Hydrilla is difficult and expensive to control (CA has spent over $15 million since 1976 and TX $2 million since 1984), extensive management programs have produced some good results in FL, CA, and NC. Methods of control include the use of herbicides such as copper sulfate, diquat, endothall, dichlobenil, and fluridone, although these herbicides, especially copper, might be extremely toxic to fish and other aquatic animals. This sort of control costs around $120,000 per km^2, which makes it expensive for treatment over extensive areas. Biological control via herbivorous fish, such as the triploid sterile Grass Carp (*Ctenopharyngodon edella*), has been used in many states in the US and has worked well in small lakes and ponds. Insects like the Australian Leaf-mining Fly (*Hydrellia balciunasi*) have been used in FL and have also shown promising results. Mechanical methods of control, such as cutting or harvesting, are not recommended, as the vegetative reproduction of Hydrilla allows small fragments of the plant to grow and establish again, increasing the rate of spread. In some areas of FL, specially designed aquatic-plant harvesting programs costing between $74,000 and $98,000 per km^2 have been used. Since its introduction in WA in 1995, the state and local governments have begun a control program that includes the use of fluridone, Grass Carp, and hand-pulling removal by divers.

Life History and Species Overview

Though Hydrilla is native to Asia, Africa, and Australia, it is found all over the world, attesting to its success as an invader. Hydrilla can tolerate a wide range of temperatures, acidic conditions, light, and carbon dioxide concentrations, allowing it to photosynthesize and grow in conditions where other plants cannot. Hydrilla can spread by seeds (the monoecious variety only), stem fragments, tubers (subterranean vegative propagules), and turions (compact overwintering buds produced in the leaf axis or in the stem). All of these reproductive methods ensure that Hydrilla grows and spreads rapidly even in unfavorable conditions as tubers may remain dormant for up to 10 years.

Hydrilla can be monoecious, with male and female flowers on the same plant, or dioecious, with male and female flowers on different plants. Female and male flowers differ in their structures. While female flowers have translucent petals 1–5 mm long that float on the water surface, male flowers have three white to red narrow petals about 2 mm long that float to the water surface but break free of the plant when they are mature. In WA and in the northern US, the monoecious type is present. Dioecious Hydrilla (all females) tends to dominate in the southern US. Dioecious and monoecious Hydrilla are also present in Europe, and dioecious in New Zealand (all males). Dioecious plants tend to branch at the surface, forming dense mats, while in contrast monoecious plants tend to branch at the bottom. As such, the dioecious type might have the biggest impact on both humans and native species.

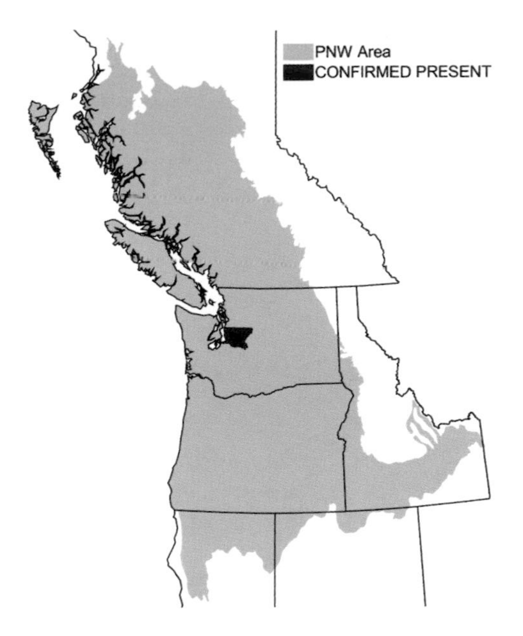

History of Invasiveness

Hydrilla has invaded Europe, New Zealand, and North America and is considered the worst and most successful invasive aquatic plant in the US. It was introduced in the mid-1950s via the aquarium trade and has spread via boats, fishing, deliberate planting, and waterfowl. In WA, when Hydrilla was discovered, the USDA laboratory confirmed that the specimens of Hydrilla were monoecious; further DNA research indicated that these plants are of Korean origin. Hydrilla may have been introduced to King County as hitchhikers on purple-flowered exotic water-lily rhizomes, as introduced water lilies are common in Pipe Lake and Lucerne Lake, and Hydrilla has certainly been introduced in this way in CA. In the PNW the control methods seem to have been effective in preventing the spread of Hydrilla; yet, because of its highly invasive capabilities and the serious problems that this species can produce, OR, WA, and BC remain on alert.

Other Sources of Information: 24, 49, 100, 111, 121
References: 97, 119, 125, 300, 325
Author: Fernanda X. Oyarzun

Reed Canarygrass *Phalaris arundinacea*

Species Description and Current Range

Reed Canarygrass is a perennial, cool-season, mat-forming rhizomatous plant in the grass family (Poaceae). It is widespread across the northern third of the US and throughout much of Canada and is an especially insidious invader in the PNW, the northern Midwest, and in the northeastern states. In the PNW, Reed Canarygrass occurs from the coast inland to MT and WY, and from BC south into CA. It tends to form the largest infestations in low-elevation wetlands west of the Cascades, in both fresh and brackish (somewhat salty) water. Reed Canarygrass is also frequent in river floodplains and wet ditches, along roadsides, and in old pastures where it has been planted for forage. Consuming too much mature Reed Canarygrass plants with high levels of alkaloids can lead to a nervous disorder in grazing animals called "Phalaris staggers." Also, certain alkaloids within Reed Canarygrass can be extracted and are rumored to produce LSD-like hallucinogenic effects when smoked.

In large infestations, Reed Canarygrass often forms a thick sod layer with upright stems up to 2 m tall. It has flat leaf blades, and from a distance can be easily identified, when it is flowering, by its flower color and inflorescence shape. Its flower clusters are compact and resemble spikes when immature, but become open and spreading once flowers are mature. When in full bloom (May–June in the PNW), the inflorescences change in color from pale green to dark purplish, becoming straw colored and compact once again when seeds have developed and dispersed.

Impact on Communities and Native Species

Reed Canarygrass can create large (100+ ha), dense, single-species stands that compromise wetland biodiversity since it can effectively compete with and exclude almost all other plant species. By eliminating native vegetation, Reed Canarygrass displaces wildlife because it provides little food or suitable habitat. Reed Canarygrass also has the ability to alter ecosystem processes and functions by forming thick sod-layers (sometimes up to 1 m deep) that elevate the surface of the wetland, and it therefore increases rates of sedimentation, changes hydrology, and alters nutrient cycling.

Control Methods and Management

A variety of management options are available for the successful control and local elimination of Reed Canarygrass. The best treatment method depends on site characteristics, the size and scope of the infestation, and what resources are available. Successful treatment methods include: digging out plants by hand, covering the plants with clear or black plastic for over a year, mowing at least five times/yr for several years, mowing at the onset of flowering followed by a herbicide application in fall, repeated herbicide applications alone, or repeated tillage combined with prolonged flooding (6+ mo). All treatments should include follow-up monitoring and treatments for 3–5 years, and reseeding or planting of desired plants to create a dense herbaceous canopy to prevent reinvasion.

Life History and Species Overview

Reed Canarygrass is a sod-forming perennial plant and can reproduce vegetatively (clonal spread) by rhizomes and rhizome fragments, or sexually by seed. Each flowering stalk can produce several hundred seeds, but Reed Canarygrass probably has a low rate of successful establishment in mature stands from seeds. Most plants and recurring populations of Reed Canarygrass are likely to be from rhizomes.

Seeds of Reed Canarygrass can be dispersed in animal fur, on human clothing, or in mud carried on automobiles or equipment. The most common vector for seeds and rhizome fragments, however, is dispersal by water. Reed Canarygrass seeds do not germinate in dense shade and can begin tillering (producing rhizomes) within 8 weeks of germination. Seedlings are susceptible to prolonged flooding and prolonged drought and are not highly competitive with perennial native species. Reed Canarygrass plants may flower and produce seed within one year, but most plants germinate in fall or winter, and then flower the following year. In the Pacific PNW, Reed Canarygrass stems and foliage die back yearly in winter, with new upright stems produced the following spring. In mild winters, some foliage may remain green throughout the entire winter. Established populations of mature Reed Canarygrass plants can tolerate prolonged periods of both flooding and/or drought, and can also survive burning, mowing, cutting, or tilling.

History of Invasiveness

Reed Canarygrass is native to Eurasia and possibly also native to North America. There is some debate as to whether Reed Canarygrass is truly native in the PNW, and the current view is that there are native strains of Reed Canarygrass indigenous to inland Northwest wetlands (MT, northern ID, northern WY). Due to their rapid growth, vigor, and ability to survive in a variety of environmental conditions, however, European

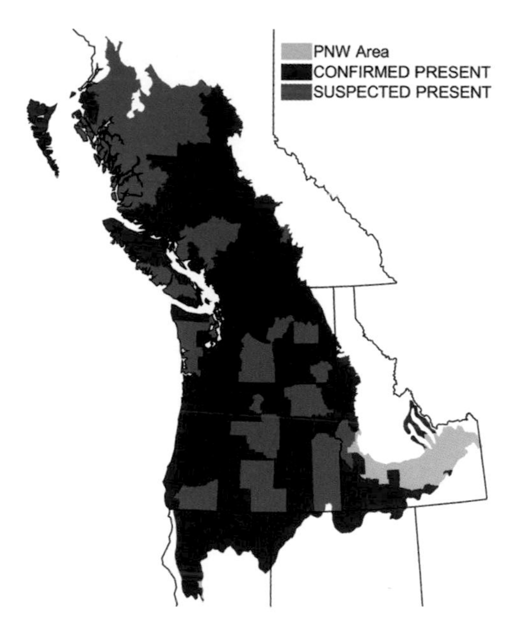

cultivars of Reed Canarygrass were developed and widely planted across the northern third of the US beginning in 1918 for pasture and erosion control. Even though it produces high quantities of good-quality forage (comparable to other grasses) for cattle, horses, and sheep, as the plants age they become less palatable due to increases in alkaloid content. In the PNW, Reed Canarygrass has been cultivated in OR since 1885, and it is likely that what is now abundant throughout many PNW wetlands (especially west of the Cascades), in the northern Midwest, and in northern East Coast states are European cultivars of Reed Canarygrass. No genetic work has been done to identify these different strains, nor have any distinctive characteristics been identified to distinguish the different strains and cultivars from one another.

Other Sources of Information: 91, 125
References: 9, 113, 181, 206, 219
Author: Mandy Tu

Saltcedar *Tamarix ramosissima*

Species Description and Current Range

Tamarix ramosissima is one species in the larger family Tamaricaceae. Of the 90 worldwide Tamaricaceae species, 8 have invaded the US, and all 8 are commonly called Saltcedar or Tamarisk. They are deciduous, generally look like shrubby cedars, and grow in salty soils, hence their name. As most of the 8 species are morphologically similar, some experts argue that 6 of the species, including *T. ramosissima*, be classified as varieties of *T. pentandra* based on their shared five-petal flowers. Of the species, *T. ramosissima* poses the largest threat, and is generally a shrublike plant 1.5–6 m in height with many thin reddish-brown branches extending vertically, covered with small and scalelike green leaves, closely resembling needles, and small rectangular flowers that are pale pink or white. Saltcedar leaves are often covered in sodium deposits, as the plant itself secretes salt.

Saltcedar has become a major problem for much of the US, but it is a major concern in the western states and parts of the PNW mainly because it uses so much water. Around 600,000 ha of riparian habitat, wetlands, and stream banks are infested with Saltcedar in the US, and many native species have been outcompeted. CA has classified *T. ramosissima* as a class A (the highest level) noxious weed, OR and WA have classified it as a class B noxious weed, and NV recognizes it simply as a noxious weed. ID has not added *T. ramosissima* to its noxious weed list, and no documented cases of infestation exist in BC. Saltcedar invasion is a costly problem for the western states, and many states have placed a quarantine on the sale and transport of *T. ramosissima* seeds and plants.

Impact on Communities and Native Species

Saltcedar can outcompete native species and forms monotypic stands of shrubs. Saltcedar is rapidly spreading across native communities at rates of 18,000 ha/yr and displacing native species and altering the hydrology of these areas. Because it has a very deep root system, up to 30 m, it draws salt from lower soil levels and deposits the salt into its leaves. When these leaves are dropped, the salt is transferred to the surface of the soil, and salinity increases. Many native species cannot live in these salty conditions and die out, but Saltcedar is able to thrive in saline conditions and so is aided in forming monotypic stands. Saltcedar creates a large amount of dry plant matter that increases the fire frequency of the area, but Saltcedar is fire adapted and is able to resprout quickly after fires, while native species may take time to recover. Saltcedar also uses large amounts of water. The immense root system finds deep water tables, and a single plant uses up to 750 l/day. Groundwater can be depleted and the possibility of floods can increase because Saltcedar clogs streambeds and underground aquifers. When rain falls, the water cannot travel down the riverbed or be absorbed rapidly, increasing the chance and intensity of floods.

The displacement of native vegetation has a negative effect on the animals of the area. Saltcedar is a useless plant to most animals. The seeds are too small and lack enough protein to be worth foraging for, and the scalelike leaves provide no nourishment. Saltcedar is also not a favorable bird habitat, as the numbers of species per 40 ha dropped

from 154 to 4 between native habitat and monotypic Saltcedar stands.

Control Methods and Management

Control of Saltcedar is difficult as it recovers from most natural disasters such as floods and fires. It sprouts back quickly after fire and can withstand flooding for up to 24 months. However, there are ways to kill it. The major method of mechanical control is deep root plowing, which destroys the root system but at the same time destroys all other vegetation in the area. Removing the crown of the tree with chainsaws is another method of mechanical control, but leaving the root system simply allows the plant to sprout again. Herbicide treatment is effective but is difficult to implement and costs $75–105 per half ha. The most effective treatment presently is a combination of mechanical and herbicide treatments such as poisoning the stump after cutting off the plant at the ground. An enticing option, biological control, involves introduction of predators such as beetles, mealybugs, and moths from Israel or China, where Saltcedar is native. Two species have been approved for limited release in the US, a leaf beetle and a mealybug, and research continues on the others. However, continued monitoring is required to ensure that the reduction of Saltcedar does not further threaten the Federally Endangered southwestern subspecies of the Willow Flycatcher, which has adapted almost completely to use Saltcedar habitat after Saltcedar outcompeted their natural habitat of dense willow and seepwillow species.

Life History and Species Overview

T. ramosissima is able to live in a wide variety of habitats. The only real requirement is that during the initial growing stage the soil surface must be well saturated. Seed germination will begin within 24 hours of moistening. After Saltcedar has established itself, however, it can live in moist areas or very dry areas and grows quickly, up to 30 cm/month in early spring. It is able to survive in dry areas largely because of its extensive root system, which can tap water below dry riverbeds. Saltcedar is also well adapted to living in saline soils, allowing it to invade areas where Saltcedar

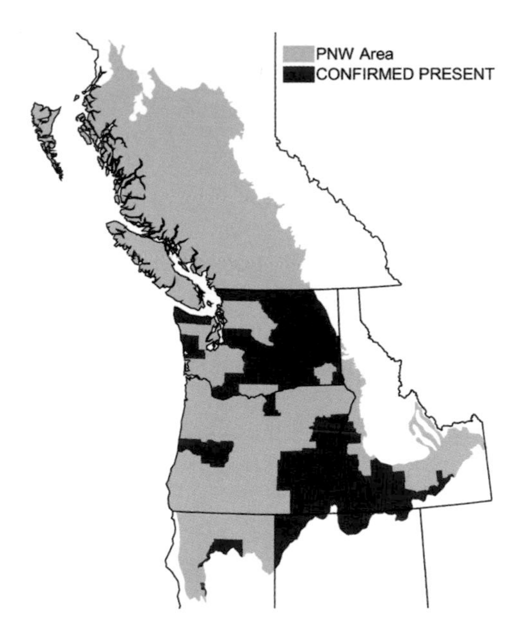

already exists. Once Saltcedar grows to normal size, it is largely there to stay. Fire will burn the above-ground portion of the tree, but it will simply sprout back quickly and reestablish itself after the fire. Floods help the Saltcedars as they are able to grow adventitious roots, and the settling of silt after the flood provides an excellent habitat for seeds to begin growing. When Saltcedar reaches maturity it reproduces sexually and produces around 600,000 tiny seeds (around 1 mm). The seeds are covered with fine hairs that aid in wind and water dispersal, allowing for wide distribution.

History of Invasiveness

Saltcedar was brought from Eurasia in the early 1800s. Its original distribution included Western Europe, the Mediterranean, North Africa, northern China, and India. Saltcedar was introduced as a windbreak, a shade-producing ornamental, and a means of erosion control. The USDA established six species in its DC Arboretum in 1868, but by the 1920s Saltcedar had escaped and invaded drainage sites in arid and semiarid areas, becoming a major problem.

Other Sources of Information: 18, 67, 71, 87, 125
References: 16
Author: Colin A. Olivers

Swollen Bladderwort *Utricularia inflata*

M

Species Description and Current Range

Swollen Bladderwort or Floating Bladderwort is a member of the only group of carnivorous aquatic plants in the US (the bladderworts). Native to the southeastern US, Swollen Bladderwort is found freely floating in fresh waters. The name "utricularia" comes from the Latin *utricularius,* meaning "the master of a raft floated on bladders." The lacy underwater structures that bear the distinctive small round bladders are actually stems, not true leaves. Swollen Bladderwort is the only bladderwort in the PNW with a spoke-like structure that keeps the flowering stalk, topped with yellow, snapdragon-like blossoms, at the surface. Each inflorescence bears 3–15 individual flowers. When not flowering, Swollen Bladderwort looks frustratingly similar to native bladderworts. The best time to distinguish them is in the fall, when *Utricularia macrorhiza*, the native, grows turions (aboveground rootlike structures). Turions are hard to see in the invasive Swollen Bladderwort, but they extend as large, cylindrical structures in the native. Both species produce fruits, with the Swollen Bladderwort bearing a slightly longer fruiting stalk (pedicel 35 mm long) than the native bladderwort (20 mm long).

Swollen Bladderwort is in a pioneering phase in the PNW. The WA Dept. of Ecology monitors WA lakes and has reported Swollen Bladderwort in several western WA counties (Mason, Kitsap, and Thurston). Swollen Bladderwort has not been reported in any other parts of the PNW, though it has the potential to spread into lakes and waterways throughout the region similar to those infested in western WA.

Impact on Communities and Native Species

Swollen Bladderwort forms dense mats, especially when flowering. These mats can cover the entire surface of ponds and lakes and interfere with recreation. In its native range, Swollen Bladderwort provides food for fish, waterfowl, and muskrats, as well as cover for fish and aquatic invertebrates. It is unknown how Swollen Bladderwort affects PNW native plant or animal communities, though the dense mats likely inhibit native plant growth, change water circulation, and alter fish and invertebrate communities. To people, these shifts likely mean changes in the fisherman's catch, the bird-watcher's view, and the family's favorite swimming hole.

Control Methods and Management

Current control methods for Swollen Bladderwort include chemical, mechanical, and manual methods. Treatment with the aquatic herbicide fluridone in Lake Limerick, WA, was effective for about 2 years, although the invasive plant Brazilian Elodea, *Egeria densa*, was the actual target species. Lake Limerick residents switched to manual control, hiring students to remove Swollen Bladderwort plants during the summer. The introduction of carp into Silver Lake (Cowlitz County, WA) appears to have eradicated Swollen Bladderwort from one infested location, but the use of carp is not recommended

in water bodies with anadromous fish, ruling out many regions of the PNW. No research has been directed toward developing or assessing effective, long-term control options for this species. As with many other invasive plants, sale of the plant and its seeds is prohibited in WA. If you sight Swollen Bladderwort, report the location: in BC, to the BC Ministry of Environment, Lands and Parks; in WA, to the WA Dept. of Ecology.

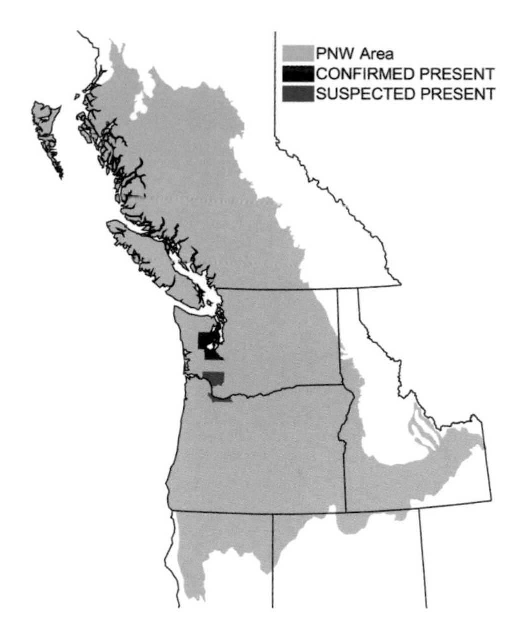

Life History and Species Overview

Swollen Bladderwort spends most of the year in the lower part of the water column of shallow lakes or bogs until it flowers. The plant then floats to the surface, where it flowers. In rare cases, Swollen Bladderwort flowers under the water's surface. Blooms are usually seen in the PNW from May to July. This flowering pattern is very different from the Swollen Bladderwort's displays in its native range. For example, blooms are seen throughout the year in FL. Swollen Bladderwort does produce seeds, but also has an interesting approach to vegetative production that provides some drought tolerance. When the small ponds and marshes that Swollen Bladderwort inhabits dry out, the plants lie exposed on the sediment. Under these conditions, they may grow small, threadlike branches that produce tubers. These tubers remain dormant until the plant is again submerged. Though this is a known mode of growth and reproduction in the native range of Swollen Bladderwort, no tuber production has been observed in the PNW.

The leaflike stems of Swollen Bladderwort bear small, round bladders that act as tiny traps. Each bladder is rimmed with small hairs. When small invertebrates (like crustaceans or insect larvae) disturb the hairs, they trigger a trap door that opens and sucks them in. Once inside, the unsuspecting prey are digested by enzymes, as if in a predator's stomach, and absorbed by the plant. Swollen Bladderwort can also take nutrients directly from the water, but prey trapped by the bladders provides a necessary supplement to the plant in the acidic, low-nutrient waters where it usually grows. Though Swollen Bladderwort can grow in dense mats in its native range, it is considered less competitive there because its primary growing season is in the late winter through early spring and does not coincide with the major summer growing season. The plant is largely inactive during this season and does not inhibit growth of other species. The relatively cooler growing season in the PNW could make Swollen Bladderwort more competitive here. Besides their popularity in the aquarium trade and among carnivorous-plant enthusiasts, other members of the bladderwort group have been used as antibiotics, metabolic stimulants, and a source of iodine.

History of Invasiveness

Native to the eastern US (NJ to FL), Swollen Bladderwort was likely introduced to the PNW through release from the aquarium trade or water-plant market. Someone may have released the plant in hopes that it would trap and eat mosquito larvae, thus providing some respite for the pond owner. It was initially detected in Horseshoe Lake (Kitsap County, WA) in 1980. No sightings have been reported in ID, BC, or OR, though Swollen Bladderwort habitat exists throughout the region. Boaters spread many invasive aquatic plants including the Swollen Bladderwort. This species is also transported among lakes by waterfowl.

Author: Heather M. Tallis

Water Primrose *Ludwigia hexapetala*

Species Description and Current Range

Water Primroses are perennial herbs found in or near lakes and ponds. They can creep along shorelines, float in dense, tangled mats on the water surface, or grow upright. The best way to identify the invasive Water Primrose *Ludwigia hexapetala* is by its bright yellow flowers, which bloom throughout the summer. The flowers have five petals, each 15–30 mm long and 12–25 mm wide; the leaves are slightly hairy and alternately arranged. There is one species of Water Primrose, *Ludwigia palustris*, that is native to the PNW. *L. palustris* is easily distinguished from its introduced relative, especially when flowering—it grows upright and blooms small, green, five-pointed flowers with no petals.

Ludwigia hexapetala has invaded at least two sites in WA: the Longview/Kelso Diking District in Cowlitz County and a nursery pond in King County. The Cowlitz County colony has been established for at least 25 years and covers about 20–25 ha. Water Primrose dominates the plant community despite the presence of several other invasive species. A colony in King Country covering 46 m^2 was discovered in 1997. It is unknown whether any infestations exist in OR or BC.

Impact on Communities and Native Species

Invasive Water Primrose is unusually aggressive. It outcompetes native plants, dominating shoreline vegetation and forming dense, extensive mats. These mats reduce water flow, thereby increasing sedimentation and decreasing oxygen availability for aquatic life. Dense patches block fish passage and prohibit valuable sunlight from reaching life underneath. *L. hexapetala* invasion also alters species diversity. By covering shoreline habitat, mats obstruct important foraging, breeding, and nesting areas sought out by some species (e.g.,

shorebirds, migratory birds), while providing a food source for others (e.g., muskrats, waterfowl, livestock).

Perhaps most disconcerting is the fact that ecological consequences of Water Primrose invasion remain largely unstudied, yet could be quite severe—direct impacts such as altered species diversity and lower water quality could in turn have indirect, unanticipated affects on ecological processes. More research is necessary before potential impacts can be identified and effective management strategies devised.

Control Methods and Management

As is often the case with invasive species, controlling *L. hexapetala* is arduous and expensive. Attempts usually involve mechanical methods such as cutting, which are labor-intensive. These methods create immediate open areas of water, but are analogous to mowing a lawn—plants grow back from intact root systems and must be cut repeatedly. The Longview/Kelso Diking District, WA, spends an estimated $100,000 annually to remove Water Primrose and other invasive plants associated with it. Other strategies such as hand pulling or digging, shading with an opaque material, or application of herbicide may also be effective. In King County,

WA, in 2000, the herbicide glyphosate was applied to Water Primrose (with an added surfactant to cause adhesion to the invasive plant), followed by 2 years of shading. This treatment appears to have been successful, at least temporarily, as no Water Primrose was found in 2002 and 2003. In the Longview/ Kelso Diking District, however, application of glyphosate merely changed the dominant species from Water Primrose to Parrotfeather (*Myriophyllum aquaticum*), another invasive plant.

Life History and Species Overview

L. hexapetala is an herb that grows in areas of shallow water up to about 1 m in depth. It prefers still or slowly moving water, and is usually found along the perimeters of lakes, ponds, and ditches. Its native range includes southern South America and the southeastern US, where it is widely distributed. Though this species is not known to have medicinal value, related species found in India are used to treat skin ailments and dysentery. Scattered populations of introduced Water Primrose now exist in other areas of the US, as well as in France, Spain, and Belgium. A particularly aggressive plant, it is considered a nuisance even within the boundaries of its native range.

Invasive Water Primrose can reproduce either by seeds or vegetatively (by plant fragments). Throughout the summer it blooms bright yellow flowers, usually with five petals, despite a Latin name suggesting six (*hexa* = six, *petala* = petal). During early stages of growth, *L. hexapetala* forms rosette-like clusters of rounded leaves on the water surface. Once the plant reaches reproductive maturity the leaves lengthen to a more elliptical shape, similar to that of willow leaves. The stems also lengthen at this point, growing upright up to a meter above the water surface. It bears a fruit that is a cylindrical capsule.

History of Invasiveness

Invasive Water Primrose has been present in the PNW for decades—a dated herbarium specimen bears proof that the Cowlitz County, WA, infestation was present in 1956. The mechanism of introduction, however, remains

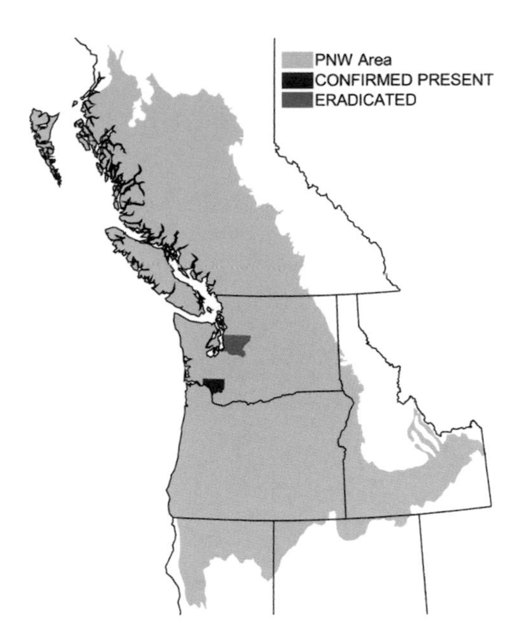

unknown. Fortunately, neither the Cowlitz nor King County, WA, populations appear to be spreading.

Showy flowers have made *L. hexapetala* a desired ornamental plant, and the horticulture industry is largely blamed for its spread outside of the native range. Water Primrose enters water bodies when deliberately planted or when plants or seeds are discarded improperly. Other methods of dispersal include water currents and hitchhiking on feet or feathers of waterfowl. Controlling the spread of this invasive plant will likely depend on informed horticultural practices that curtail its distribution. The outlook is hopeful: as of January 2001, *L. hexapetala* has been banned from sale in WA.

Acknowledgments: Jenifer Parsons (WA Dept. of Ecology) provided valuable information included in this account.

Other Sources of Information: 12, 72, 121, 125

Author: Jennifer M. Moslemi

Yellow Flag Iris *Iris pseudacorus*

H

Species Description and Current Range

Yellow Flag Iris is often planted in garden ponds for showy spring flowers that vary from pale yellow to almost orange and are faintly etched in brown or purple. Cultivars of the plant have been produced with yellow-striped leaves. The 8–10 cm diameter flowers have three large drooping petal-like sepals or falls and three smaller erect petals. The seed-pods are a triangular, elliptical shape and 4 to 8 cm long. The plant is tall and may stand 1 to 1.5 m in height. The stalkless dark green leaves, which are long and swordlike, are bound together in a sheath and arise fanlike from the soil. The species is perennial and can spread from seeds and rhizomes (rootlike stems that grow along the ground sending out both roots and shoots). The plant grows along the banks of lakes, ponds, rivers, and streams and also in wetlands. Yellow Flag Iris is found throughout the PNW at elevations up to 1,000 m. WA lists Yellow Flag Iris as a class C noxious weed because it is widespread in the state. The PNW Exotic Pest Plant List categorizes the plant as A-2: "highly to moderately invasive but still with a potential to spread."

Impact on Communities and Native Species

Yellow Flag Iris can colonize large areas and quickly form dense monotypic stands that outcompete native vegetation. Along the Merced River, CA, the native marsh plant, *Typha*, or Cattails, has been completely excluded. Various insects feed on all parts of the plant in its native range. Within the US, insects may feed on it as well but do not deter the plant's spread. The effect of this species on native fauna is not known, but bees and hummingbirds feed on its nectar, which could reduce the pollination of native species if they are not visited or if pollination is less effective. Yellow Flag Iris is poisonous to some animal species and causes gastro-enteritis in cattle.

Control Methods and Management

The plant is extremely difficult to control once an infestation with extensive rhizomes has been established. Cutting followed by herbicide treatment such as glyphosate may be the best control method. The plant is susceptible to many registered herbicides but is resistant to terbutryn. When hand pulling the species, care should be taken because the leaves are sharp, and resinous substances in the leaves can irritate the skin. For an extremely large infestation, mechanical harvesters may be a last resort. This method could spread the rhizomes if they are not chopped finely enough, and it is not recommended for sensitive areas such as shallow streambeds because the disturbance may lead to the establishment of other unwanted plants. Burning is not recommended because of the iris's tendency to resprout from below-ground rhizomes. Fire has also been found to be ineffective in suppressing seedling establishment.

Yellow Flag Iris continues to be sold in garden centers and on the Internet. The plant should be removed from gardens to prevent further spread. Cities should follow the example of Eugene, OR, which prohibits planting Yellow Flag Iris on city property or in city projects.

Life History and Species Overview

The genus *Iris* is named for the Greek goddess of the rainbow. *Pseudacorus* refers to the similarity to the European Sweet Sedge (*Acorus calamus*). American Sedge (*Acorus americanus*), which is found in northern WA and ID and in BC, resembles Yellow Flag Iris when not in flower and grows in the same habitat. The leaves of the American Sedge smell sweet or citrus-like when crushed, while Yellow Flag Iris is odorless. The name Yellow Flag is thought to derive from the fluttering of the flower in the wind, which suggested the waving of a flag. In Britain it is also called Segg, the Anglo-Saxon word for "small sword," an allusion to the leaf shape.

The plant is native to all of Europe except Iceland and also occurs in North Africa, Western Asia, and Siberia. The flowers can be used to produce yellow ink or dye and the rhizomes a good black dye. Historically the Yellow Flag Iris rhizomes had several medicinal uses for coughs, diarrhea, and toothaches. The plant is also useful for erosion control and sewage treatment and is known to remove metals from wastewaters.

Groups of clones, which can measure 1 m across, are formed from the rhizomes and can produce several dozen to several hundred flowering shoots. Unless the winter is mild, the shoots die back and grow back from the rhizome in the spring. Yellow Flag Iris blooms in April/May in the south and June/July in the north. The flowers are pollinated by bumblebees (*Bombus*), Honeybees (*Apis mellifera*), and the Long-tongued Hoverfly (*Rhingia rostrata*). The seedpods contain from 32 to 46 viable seeds that are brown and flattened and will float if they land on water. Seeds have a germination rate of 48–62% even after being kept in seawater for 31 days, which makes dispersal to nearby islands possible. Yellow Flag Iris grows best in moist spots with shade or full sun and is found in silty and pebbly areas along lakes, streams, and rivers. The plant has a high nitrogen requirement but tolerates high soil acidity, brackish water, and drought. Excavated rhizomes can grow even after three months without water.

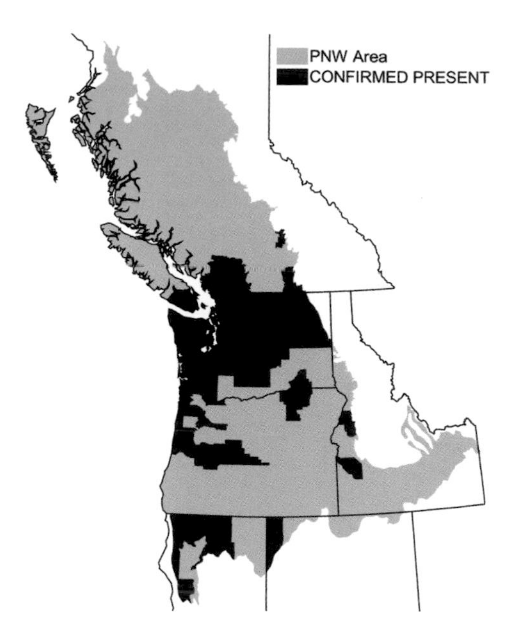

History of Invasiveness

Yellow Flag Iris was brought to the US and Canada in the early 1900s as an ornamental plant. The spread is believed to have started by rhizomes being thrown out of gardens or from low-lying gardens where spring floods carried away the rhizomes and/or seeds. The first Yellow Flag Iris growing wild in North America was found in 1911 in NF. By 1931 it was established in BC and was found in Skagit County, WA, in 1948. The distribution was reported from NF to MN in 1950, and records from CA and MT date to the late 1950s. In Canada the Iris was so plentiful in marshes by 1961 that it had the appearance of a native plant. By the 1970s the plant was common in many parts of the PNW. Today Yellow Flag Iris is found throughout most of the US and Canada except for the Rocky Mountains and parts of the Midwest and Southwest. The species is also considered a weed in New Zealand.

Other Sources of Information: 24
References: 62, 299
Author: Patricia A. Townsend

Yellow Floatingheart *Nymphoides peltata*

Species Description and Current Range

Yellow Floatingheart is a perennial water lily–like plant in the Buckbean family (Menyanthaceae) that carpets the water surface with heart-shaped leaves. It has bright yellow flowers 2–4 cm in diameter with a distinctive fringe along the edges of the five petals. The flowering stalks are held several cm above the water's surface. Two to five flowers emerge from each flowering stalk.The plant has round, floating leaves with slightly wavy margins and purplish undersides. The species name for Yellow Floatingheart, *peltata*, means "shieldlike," referring to the leaf shape. Yellow Floatingheart may easily be confused with the native yellow-flowered species Yellow Pond-lily (*Nuphar luteum* ssp. *polysepalum*). Yellow Pond-lily differs in having larger leaves and cuplike flowers without fringed petals. Another similar species, Water-shield (*Brasenia schreberi*), has white flowers and leaf petioles that attach to the center of the floating leaves.

In the PNW, Yellow Floatingheart grows at only two sites, both in WA. One is an extensive infestation in the Long Lake reservoir in eastern WA; the other is in a small artificial pond in western WA. Though it has a limited distribution now and is not currently widespread in the PNW, the species tends to form dense mats many hectares in size in the locations where it has been introduced, making it a threat to freshwater systems.

Impact on Communities and Native Species

Yellow Floatingheart is an obligate wetland species that forms dense mats in the shallow portions of lakes and ponds. These mats are often many hectares in size and typically compete with and almost completely displace existing native plant communities and animal species. In addition, large infestations diminish water quality by creating stagnant areas with low oxygen levels beneath the floating

mats. Large Yellow Floatingheart mats interfere with boating, fishing, and swimming.

The innate invasive potential of Yellow Floatingheart appears to be high where introduced, although currently it does not seem to spread from one habitat patch to another. It has been growing at the Long Lake site since the 1920s and has not spread to suitable nearby habitat. A significant concern for this species is that it might be a "sleeper" species, whose invasibility potential may suddenly explode.

Control Methods and Management

Little is known about control methods and management of Yellow Floatingheart. It is unknown if there have been serious attempts to control or eradicate this species in North America. It often grows with other invasive plant species such as Eurasian Watermilfoil and Fragrant Water-lily. Due to their similarity, methods effective for the control of Fragrant Water-lily may prove to be useful for Yellow Floatingheart.

There has been some experimentation with chemical control of Yellow Floatingheart in the United Kingdom and New Zealand. Glyphosate is effective, but when eradication is desired, repeated applications with chemical controls are usually necessary. Secondary effects although unknown probably include impacts on native species. Mechanical control methods like hand clearing or covering with opaque bottom-barrier materials may be effective for controlling smaller populations, but preventing introduction is the best control. Mechanical cutting and harvesting may temporarily control larger infestations. If plants are cut, all pieces should be disposed of properly because they spread via plant fragments. There are no known biological control methods for Yellow Floatingheart.

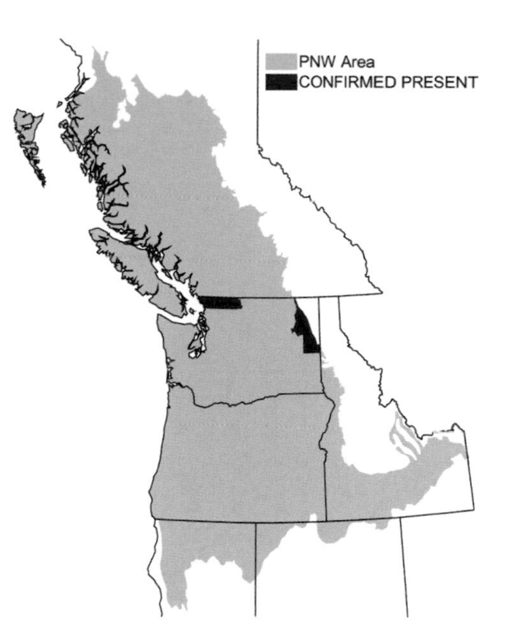

Life History and Species Overview

Yellow Floatingheart is an herbaceous, aquatic perennial whose short- to long-branching horizontal stems allow it to grow in habitats ranging from damp mud to water up to 2.5 m deep. Leaves die back in the winter, but grow up to the water surface via stolons in the spring. Plants typically flower from June to October and produce fruit from July to November. It is a day-flowering species, with each flower withering within a single day. The large, showy, yellow flowers are pollinated by insect visitors, including bees, flies, moths, and butterflies. Viable seeds are produced abundantly and germinate readily. Prominent seed hairs help the seeds float and potentially aid in attachment to waterfowl for dispersal. The seeds initially float at the water surface in dense concentrations, but sink when the frost period begins. The length of seed viability is unknown. Yellow Floatingheart reproduces and spreads by underwater stolons, seeds, and from plant fragments with attached portions of the stem. Its preferred habitat is sunny locations in shallow portions of still, or slow-moving, freshwater reservoirs, lakes, and ponds.

History of Invasiveness

Yellow Floatingheart was first reported from the US in the 1880s, where it was introduced as an ornamental in water gardens. The earliest herbarium collection from the PNW is from 1930 from Long Lake, a large reservoir in eastern WA. The native range of Yellow Floatingheart includes temperate regions of Europe, Eurasia, and the Mediterranean area. It has been documented in 24 states, several of which, including WA, consider it a noxious weed.

Currently, the primary invasion route is via ornamental horticulture and subsequent intentional introductions to the wild. Yellow Floatingheart is a popular and easily grown water-garden species with widespread commercial availability, especially though Web sites specializing in water-garden plants. Plants can be purchased on the Web for as little as $1.99. Several states, including WA, bar the sale and importation of Yellow Floatingheart. Several other species of Floatingheart are also widely sold as ornamental plants for water gardening. These may also have the potential to be invasive if introduced into appropriate natural habitats. The increase in popularity of water gardening will probably increase the potential for more Yellow Floatingheart infestations as well-meaning but not-well-informed individuals transfer plants from managed water gardens into the wild.

Other Sources of Information: 93, 111, 25
References: 56, 70, 169, 321, 324
Author: Kathryn Beck

Cordgrasses *Spartina alterniflora* *S. anglica* *S. densiflora* *S. patens*

Species Description and Current Range

Cordgrass is the general name for several related perennial marsh grasses. It grows most commonly in estuaries and intertidal areas. It spreads rapidly, reproducing both by seeds and underground stems (rhizomes). Cordgrass is 0.6–1.5 m tall, with leaf blades between 0.6 and 1.5 cm wide. The hollow, hairless stems act as a self-supporting structure. Similar native plants, bulrushes (*Scirpus* spp.) and sedges (*Carex* spp.), grow farther inland in marshes. Several species of cordgrass invade coastal and estuarine areas in the PNW. In some cases invasion results in dense cordgrass "meadows," as is the case in Willapa Bay, Grays Harbor, and Puget Sound, WA, and the Strait of Juan de Fuca. Cordgrass has also colonized Cox I. and the Siuslaw estuary in OR, and Vancouver I., BC.

Impact on Communities and Native Species

One of the most devastating impacts of cordgrass establishment is alteration of shoreline topography. The dense stems and roots of stands trap sediments. Over time, bare and gently sloping mudflats become elevated cordgrass meadows with steep slopes to tidal channels. This change in relief alters the speed and direction of water flow. The stands can also reduce the rate of water flow from terrestrial areas, causing increased flooding. Cordgrass outcompetes native flora, including species of economic importance. For example, Eelgrass (*Zostera marina*), a native plant that provides much-needed refuge and food for fish, crabs, shellfish, and other marine life, cannot tolerate the shade produced by cordgrass stands. Unlike Eelgrass, invasive cordgrass appears to be of little to no food value for aquatic animals. Cordgrass invasion also alters the density and diversity of invertebrate communities in mudflats, the consequences of which remain largely unknown. For example, the density of native clams (*Macoma balthica*) generally decreases in areas invaded by cordgrass, whereas densities of introduced Eastern Clams (*Mya arenaria*) and Asian Littleneck Clams (*Venerupis philippinarum*) generally increase. Shorebirds lose foraging areas with the spread of cordgrass, as unobstructed mudflats become impenetrable cordgrass meadows too dense to forage within.

Control Methods and Management

Controlling cordgrass is not easy. Efforts usually consist of a combination of four methods: cutting, hand pulling and digging (or in particularly bad cases, tilling up the infested area), covering, and chemical treatment. Multiple cuttings serve only to slow spread, as root systems are left intact. Individual seedlings or isolated clumps can be eradicated by pulling or digging, but all plant material must be carefully removed—one dropped seed or rhizome fragment can start a new infestation. Small colonies can be covered with an opaque fabric or plastic for at least a year to shade out the unwanted plants. Chemical treatments may involve the application of glyphosate, which is applied with a surfactant to adhere the herbicide to the cordgrass.

Biocontrol is another option; a predator from the native range of cordgrass could be released in invaded areas. Although biocontrol is an attractive alternative—it is cost-effective, long-term, nontoxic, and requires little labor—it can have devastating effects on the native community if the biocontrol agent attacks non-target species, spreads disease, or survives beyond eradication of the problem species. Thorough research prior to release is necessary to determine viability and minimize risks. In Willapa Bay, WA, an insect (*Prokelisia marginata*) was introduced to consume cordgrass. Preliminary studies suggest that the risk of *P. marginata* consuming native plants is low, but time will tell if this approach succeeds.

Controlling cordgrass is expensive. In WA, $718,000 was allocated for the task in 2000 and 2001, and the rate of spread still far exceeds control progress. Not surprisingly, the best method of control is to prevent establishment, or if the species is already established, to eradicate it while populations are still small. If you encounter what may be an undiscovered population of cordgrass, contact your state Department of Agriculture or provincial Ministry of Agriculture, Food, and Fisheries.

Life History and Species Overview

Cordgrass, a member of the Poaceae, derives its name from its historic use as a material with which to make cord or rope (*spate* is Greek for "rope"). The genus *Spartina* includes 17 species that are indigenous to Europe, North Africa, and the Americas. One species, *S. foliosa*, is native to the southwest coast of the US. *S. foliosa* is smaller, spreads at a slower rate, and produces fewer seeds than its invasive counterparts.

Cordgrass reproduces by seeds or vegetatively (via rhizomes). Vegetative reproduction produces clones of the parent plant that can spread rapidly, usually forming circular patches. Small root fragments can grow into new plants, making complete eradication difficult. Cordgrass flowers create seeds by

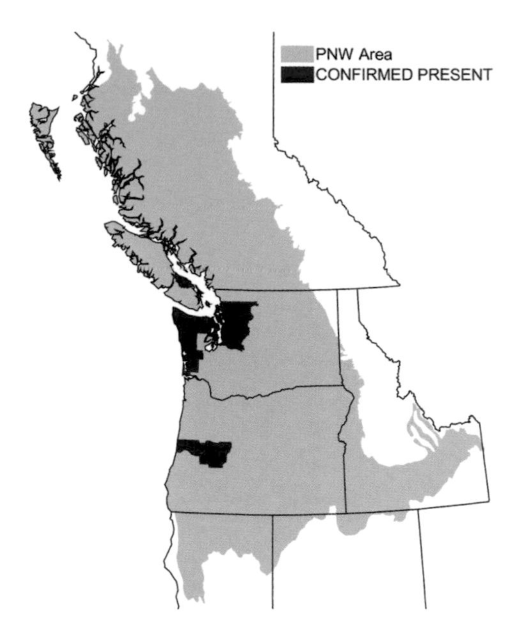

selfing (pollination within one flower) or outcrossing (pollination between flowers). A cool, wet period of 3–4 months is necessary for seeds to germinate—roughly the length of a PNW winter. Seeds can germinate in areas with low oxygen and/or high salinity.

History of Invasiveness

Cordgrass was deliberately introduced into some areas where it is currently a nuisance. For example, in Puget Sound in 1961, an agronomist planted cordgrass at Port Susan Bay for dike stabilization and cattle forage. By 1997, 1,340 ha in 73 sites had been invaded. Other introductions were unintentional. In Willapa Bay, transplantation of eastern US oysters at the turn of the 19th century likely caused the invasion. Fragments of cordgrass (and probably seeds) from the Atlantic coast were used to pack oysters for shipment and, upon arrival, were discarded into the bay. Willapa Bay is now the largest cordgrass colony in the PNW—the estimated number of ha invaded grew from 162 in 1982 to 6,070 in 2002. Other ways cordgrass may spread include accidental transport on boats or in ballast water, via ocean currents, or by hitchhiking on the feet or feathers of waterfowl.

Other Sources of Information: 12, 72, 118, 119, 125
References: 87, 110, 132, 136, 154
Author: Jennifer M. Moslemi

Japanese Eelgrass *Zostera japonica*

H

Species Description and Current Range

Japanese Eelgrass has grasslike, green, ribbony blades with two veins running their length. The venation pattern helps distinguish it from the native species, *Zostera marina*, which has 3–11 veins per blade. Japanese Eelgrass is found in scattered clumps or extensive meadows high on intertidal mud or sand flats. It is commonly called Dwarf, Japanese, or Asian Eelgrass or Duck Grass. Though it is one of the 60 marine flowering plants, and all of its common names include "grass," it is not truly a grass. The native species is commonly found lower in the intertidal zone than the invasive, but they are occasionally found together. Adult plants of the two species can vary substantially in size and can be difficult to tell apart. In general, the native species is larger, with blades 3–12 mm wide and up to 3 m long. Japanese Eelgrass has blades 0.75–1.5 mm wide and 6–30 cm long.

Native eelgrass provides food and shelter (from sunlight and predation) to small invertebrates that are then eaten by fish, birds, or crabs. Some fish, like herring, lay eggs on the eelgrass blades, and when hatched, the young herring use the structure provided by eelgrass blades as protection. Since herring and several other species that live in eelgrass beds are important food for salmon, eelgrass is considered critical habitat for salmon under the Endangered Species Act. There is some debate among governing bodies in the PNW as to whether Japanese Eelgrass plays the same role for native fishes. Besides possibly fostering populations of tasty (and profitable) crab and fish, Japanese Eelgrass provides many other services to humans in its native range. It has been used for upholstery stuffing, fuel, packing or roofing material, fertilizer; as a salt or soda source or fiber substitute, basket material, bedding, sound insulation, dike material, and material for cigars, children's toys, and nets. The range of Japanese Eelgrass in the PNW presently extends from bays and inlets

on Vancouver I. and Cortes I. (about 160 km north of Vancouver) in BC south through WA and OR. The species was reported in Humboldt Bay, CA, but was eradicated.

Impact on Communities and Native Species

Japanese Eelgrass can invade areas that are naturally barren of plant growth and hence is a pioneering species, as is the native eelgrass. By moving into barren areas, the invasive alters habitat structure, changing water flow and sediment deposition, making the sediments finer grained and richer in organics. Establishment of Japanese Eelgrass on mudflats in the PNW is associated with increased species richness and greater abundance of dominant species including several types of worms and amphipods. In the PNW, grazing invertebrates like isopods (a common isopod is the Pill Bug or Potato Bug) and gastropods (snails) that normally feed on the native eelgrass also graze on Japanese Eelgrass, showing that this invasive can play a similar ecological role as the native, providing a food source (the invertebrates) to forage fish. In Boundary Bay, BC, several species of migratory waterfowl also dine heavily on Japanese Eelgrass. Wigeons (*Anas americana*) ate substantially more Japanese Eelgrass than native eelgrass, as did Mallards (*A. platyrhynchus*), Green

Winged Teal (*A. crecca*), and Brant (*Branta bernicla*). The reason for this preference is still mysterious, since analyses showed that the two species of eelgrass do not differ in energy content or in carbohydrate, fiber, protein, lipid, or phenol content. Japanese Eelgrass, located higher in the intertidal than native eelgrass, is exposed and thus available to waterfowl for grazing for more of the day, and the larger leaves of the native species may be harder to crop or chew.

Changes in important ecosystem processes like nutrient cycling and decomposition have also been associated with Japanese Eelgrass. In Yaquina Bay, OR, stands of Japanese Eelgrass were net sinks for nitrogen and phosphorus, meaning that they continually pull nutrients out of the water column. Through this mechanism, Japanese Eelgrass changes the movement of nutrients between the water column and sediments and can reduce nutrients available to other organisms. This invasive can change decomposition rates and the sediment's microbial community. The jury is still out on whether Japanese Eelgrass should be considered a boon or a bane to PNW intertidal communities.

Control Methods and Management

Management strategies for Japanese Eelgrass in the PNW are varied. WA protects it under the same "no net loss" policy applied to native eelgrass (under the WA Hydraulics Code). WA regulators based protection on studies suggesting that Japanese Eelgrass provides habitat and food for birds, invertebrate grazers, and fish. However, concerns about the effects of Japanese Eelgrass in Humboldt Bay, CA, resulted in a successful eradication project.

Life History and Species Overview

Most species of eelgrass are perennial (though annual populations of native eelgrass exist in CA) and flower in midsummer (in the PNW). Reproductive shoots look different from vegetative shoots, having a stemlike base with broad leaves at the end. White, smooth, small (2 mm long) seeds grow in casings (actually the spadix) near the tips of reproductive shoots. These seeds can stay attached to leaves after the leaves die and break off. These dead, seed-carrying leaves can be transported by currents over long distances, providing a likely mode of spread for Japanese Eelgrass. Eelgrass can also spread clonally via rhizomes, rootlike structures that grow beneath the surface.

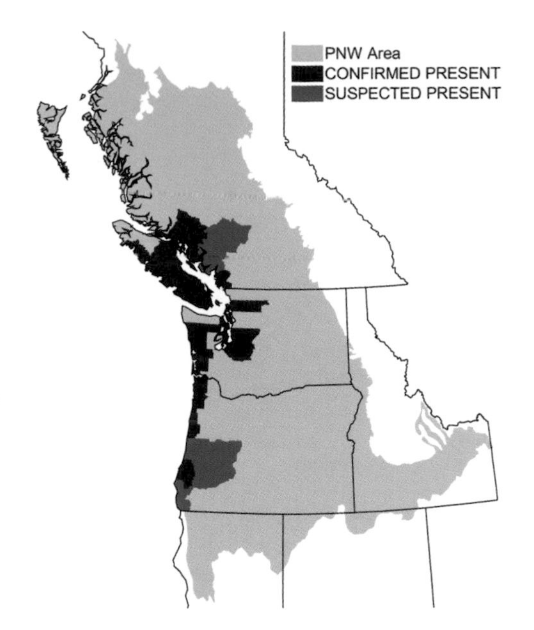

History of Invasiveness

Native to Japan and the West Pacific, Japanese Eelgrass was first reported on the West Coast of the US in 1957. One theory is that it was used as live packing material for a shipment of Japanese oysters imported for aquaculture. It has since spread through the PNW, arriving in Puget Sound in 1974 likely also via aquaculture. The continued success of Japanese Eelgrass in the PNW may be assisted by another invasion; in Padilla Bay, WA, Japanese Eelgrass reaches higher densities when the Asian Mud Snail, *Batellaria attramentaria*, is present. Cover of Japanese Eelgrass is currently expanding in most bays where it has invaded, and monitoring efforts are tracking its spread. If you see Japanese Eelgrass, report the location: in BC, to Precision Identification; in WA, to the University of WA or WA Department of Natural Resources; in CA, to the University of CA Sea Grant Program.

Author: Heather M. Tallis

Sargassum *Sargassum muticum*

Species Description and Current Range

Sargassum (sometimes called Wireweed), a large brown alga covered with little round balls that hold air, is a common introduced species around the world. Native to Japan and Asia, its precise native range is unclear because of its long association with humans. The name Sargassum comes from the Sargasso Sea, an algae-filled gyre in the central North Atlantic infamous in maritime history for foundered ships and confounded sailors. Sargassum is found in rocky intertidal (between high and low tide marks) or subtidal (below low tide mark) habitats down to 10 m but is most abundant in shallow habitats 2–5 m deep. A circular, quarter-sized holdfast attaches adults to rocks so that up to 30 fronds (branches) can extend vertically up into the water column. Fronds are 3–5 mm in diameter and have hundreds of tiny leaflets (1–3 cm in length) and many small (2–5 mm in diameter), spherical floats that imbue it with positive buoyancy, making Sargassum a good competitor for light. Sargassum has chemicals called phenolics, which are distasteful and deter herbivory by many invertebrate grazers (snails, urchins, and the like), and which may be a reason why sea-birds like cormorants use it in their nests.

The genus *Sargassum* contains more than 140 species and occurs in every ocean but the Antarctic. In the PNW Sargassum is easily confused with native species in the genus *Cystoseira* (sometimes called Northern Bladder Chain). These two genera are distinguished by the shape of their floats (pneumatocysts). Sargassum's floats are spherical and *Cystoseira*'s floats are elongated and pointy.

Sargassum is a successful invader in many places around the world, including western North America, the British Isles, mainland Western Europe, the Mediterranean, and Scandinavia. On the West Coast of North America, it probably was introduced at multiple locations in WA and BC and subsequently spread up and down the coast. The earliest record of Sargassum in WA is from 1948, but now it is common throughout Puget Sound, the San Juan Is., Strait of Juan de Fuca, and in sheltered bays along the outer coast. Its current distribution on the West Coast is from AK to Baja California.

Impact on Communities and Native Species

Sargassum outcompeted native red algae like *Neorhodomela larix* (Black Pine) in intertidal experiments done in BC. In subtidal marine habitats of the San Juan Is., WA, Sargassum displaces native algae, including commercially and ecologically important kelp species such as *Laminaria bongardiana* and *Agarum fimbriatum*. Consequently, it also has a negative, indirect effect on the native Green Sea Urchin, which depends on native kelps for food and habitat and avoids areas where Sargassum is abundant. Like many algae, each Sargassum provides habitat for hundreds of tiny animals (epibionts), such as snails and amphipods, that use it for both shelter and food. In WA, Sargassum supports a more abundant and species-rich epibiont community than does the native kelp *Laminaria saccharina*. A strange, pencil-shaped fish called a Tube-snout likes to lay its eggs on the fronds of Sargassum in the PNW. Sargassum chokes shallow waterways in some regions, clogging boat propellers and thereby creating a navigational hazard.

Control Methods and Management

Control is rare and is complicated because Sargassum can regrow if a fragment of the holdfast is left intact. The largest Sargassum eradication effort was in southern England in the 1970s and was a failure, despite the removal of 450 tons in a single year. Burial, herbicides, cutting, trawling, and suction methods aren't successful for permanent eradication. Removal from small areas by hand is time-consuming and must be repeated as the cleared areas are recolonized. If you remove Sargassum, dispose of it away from the seashore to prevent spread to new locales.

Life History and Species Overview

Sargassum has a simple life cycle. In the late summer, each individual grows hundreds of small, cigar-shaped reproductive structures called "receptacles" that make both eggs and sperm. Released eggs stick to the surface of the receptacle, where they are fertilized (this species can fertilize its own eggs, one of many traits that have made it a very successful invader). Fertilized embryos stay attached to receptacles until they grow into small germlings and develop tiny, adhesive rhizoids (rootlike structures), at which point they are released and drift to the bottom within a few meters of the adult. Long-distance dispersal probably occurs when adults that are brooding embryos become detached from the bottom and drift at the surface, dropping juveniles as they float. Although the importance of this latter dispersal method for Sargassum's spread is unknown, other algal species are known to employ similar strategies to good effect.

History of Invasiveness

In the PNW, Sargassum was likely introduced accidentally when the Pacific Oyster was imported for aquaculture in the early 1900s. In most regions that Sargassum has invaded, the introduction was an unintended side effect of oyster aquaculture activities. Once established, Sargassum readily spreads. Human disturbance and accidental transport of reproductive Sargassum (e.g., by boats) may facilitate its spread.

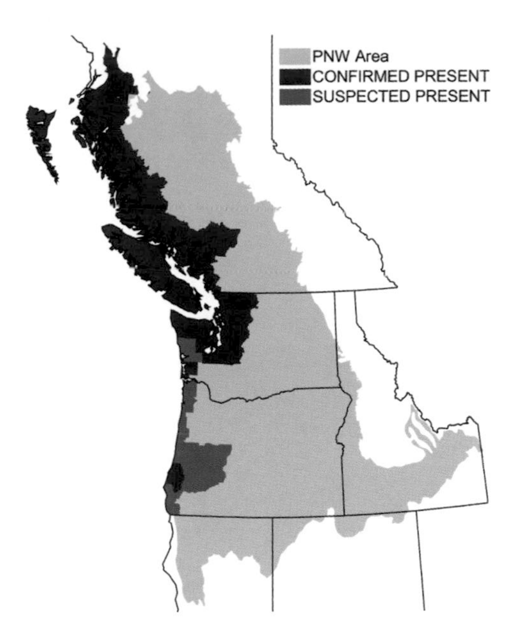

References: 35, 89, 122, 234, 281
Author: Kevin Britton-Simmons

Evergreen Blackberry *Rubus laciniatus*
Himalayan Blackberry *Rubus armeniacus*

Species Description and Current Range

The most widely picked wild-growing berries in the PNW, Himalayan and Evergreen Blackberries are often mistaken for natives. Naturalized in urban, agricultural, and wild areas throughout the region, they grow from sea level to 1500 m. Both species are robust, essentially evergreen shrubs forming dense, impenetrable thickets. Their canes are covered with vicious, slightly curved thorns and grow upward to 5 m, then arch out and downward. Both invaders bear pinkish/white, five-petaled flowers in clusters and shiny, purple, inch-long berries. Leaves are palmately compound with usually five large, oval leaflets. The primary distinguishing characteristic of the Himalayan and the Evergreen Blackberry is the leaflets: Himalayan leaflets have coarsely serrate margins and are dark green on the upper side with grayish-green undersides; Evergreen leaflets are sharply toothed, greenish above and paler below. The native blackberry (*Rubus ursinus*) is not invasive and so should be encouraged. It is easily distinguished from both non-native species by its smaller, straighter, thinner thorns and leaves with three leaflets of a similar color on both sides.

R. laciniatus

The seeds of both invasive species may be "agamospermic" (produce viable fruit without pollination). Because of this property, plants in different places that are the same species may appear dissimilar because they arose from different genetic lines. This has led to much taxonomic confusion. Currently, Himalayan Blackberry is considered to be *Rubus armeniacus*, and Evergreen Blackberry is *Rubus laciniatus*.

Impact on Communities and Native Species

Many people are familiar with the fruit, which is enjoyed fresh or made into jam, wine, and pies. Invasive blackberries are also an important source of nectar in the PNW for another introduced species, the Honeybee, and therefore a major component of honey production. WA produced 3.25 million pounds of honey in 2003.

Unfortunately people are often not as familiar with the negative effects these plants have on the ecosystem and native species. Invasive blackberries alter ecosystems by shading out and killing smaller native species, thereby reducing native-plant and wildlife diversity and hindering the reestablishment of native species during natural succession. The fruit is a seasonal food source for introduced vermin and birds such as rats, blackbirds, and European Starlings and may help to increase their populations. The plants decrease usable pasture, prevent large animals' access to water, trap young livestock, and shelter small animals such as rats from predation.

Control Methods and Management

The best strategy is to prevent blackberry establishment initially by planting or sowing native species immediately after disturbance.

Rapidly growing shrubs or trees are most effective because blackberries don't tolerate deep shade. Including native berries in the planting scheme provides songbirds with replacement food. If blackberries have already appeared it is important not to procrastinate; a patch can widen by 3 m or more a year, smothering every plant in its path.

Eradicating established patches, though difficult, is not impossible. Seedlings can be hand pulled. Larger plants can be dug out, removing as much of the root as possible to limit resprouting. Such digging is an effective but time-consuming technique most suitable to small infestations and around trees and shrubs. In large areas, cutting and mowing the canes is a solution, but repeated mowings are needed. If canes can be removed only once a year, the optimum time is when the plants begin to flower, when food reserves in the roots have been depleted. An effective method of deterring regrowth is to lay wood chips at least 20 cm deep after cane cutting, reapplying as necessary. Treating cut stems with herbicide is another method but can simultaneously promote growth through lateral roots as well as contaminate groundwater if used near or uphill from wet areas. Blackberry is a preferred food for goats; New Zealand and Australia have used goat grazing as an effective control since the 1920s and experiments in CA are in progress. Goats have proved to be inexpensive and efficient at controlling small stands.

Life History and Species Overview

Blackberries grow rampantly on a wide range of barren, infertile soils, tolerating a variety of soil textures and pH but requiring adequate soil moisture. They are usually found on waste ground, disturbed areas, and oak woodlands and along streams, rivers, ditches, ponds, and marshy areas. First-year canes produce no fruit, but may root where the tip of the cane touches the ground, producing a new plant. This process is called "tiprooting," and the new plants at the tip of the canes are the "daughter plants." Second-year canes produce fruit and die; the remaining root system is perennial and continues to produce new canes. Seed production is also prolific; an Australian study estimated 10,000 seeds/m^2 of thicket. Seeds can remain viable in the soil for years.

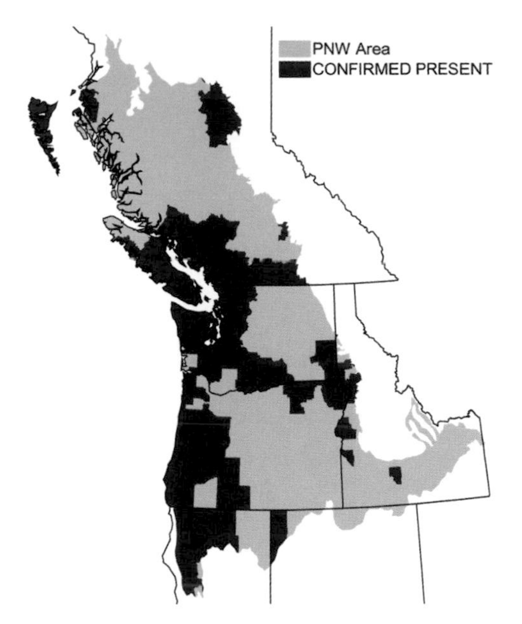

History of Invasiveness

Himalayan Blackberry, contrary to its name, is native to western Europe. It was introduced into America in 1885, possibly by Luther Burbank, and was grown as a crop by 1890. By 1945 it had become naturalized along the West Coast. Evergreen Blackberry was also brought in as a cultivated plant and is of Old World origin. Both species are currently listed as noxious weeds in OR. Neither are listed in WA, where they are so widespread that effective regional control, the purpose of the Noxious Weed Law, is impossible. As recently as 1974, Himalayan Blackberry was officially touted as a beneficial provider of food and habitat to wildlife and for erosion control. In addition, the fruits attract wildlife and are then very effectively dispersed by berry-eating birds and mammals. Passage of the seeds through the bird's gut also improves seed germination. The delicious fruit creates another major obstacle to control: the reluctance to treat these two species as vicious invaders. Even people bent on eradication often keep "just a small patch" to provide themselves with berries for eating.

Other Sources of Information: 49, 61, 63, 64, 129
References: 33, 39, 162
Author: Margo Murphy

Gorse *Ulex europaeus*
Scotch Broom *Cytisus scoparius*

Species Description and Current Range

Scotch Broom (also called Scot's Broom) and Gorse, shrubs in the pea family, generally grow 1–2 m in height and have green stems, small leaves, and numerous large flowers that cover the plant. The flowers are usually yellow, but some cultivated varieties of Scotch Broom can have red to white or purple flowers. Their showy flowers are in a shape characteristic to the pea family, with a "wing" petal to each side of an overarching upper "banner" petal and a lower "keel" petal, curved like a boat's underside. Gorse is distinguished from Scotch Broom by its long sharp spines and the coconut smell of its flowers. Most of the photosynthesis in these two species is carried out in the green stems or spines rather than the leaves. Scotch Broom got its name because it has been commonly used in Scotland to make brooms.

Both Gorse and Scotch Broom are most common near the coast. Gorse forms dense patches in scattered areas along the coast from CA to southern BC. Scotch Broom is widespread throughout western WA, OR, and BC, and is found in scattered locations in ID and eastern WA and OR. Both species are nitrogen fixers and are usually found in infertile disturbed sites including roadsides, dunes, grasslands, and clear-cuts.

Impact on Communities and Native Species

Gorse and Scotch Broom grow rapidly and form dense stands that can shade and thereby outcompete other species. Scotch Broom and Gorse can prevent forests from regenerating by outcompeting tree seedlings. Like other legumes they form relationships with bacteria in their roots that are able to convert nitrogen from the atmosphere into a form that can be used by plants, allowing them to rapidly colonize infertile sites. This converted

Cytisus scoparius

nitrogen increases soil fertility and thereby alters native plant communities. Many native species are adapted to infertile conditions and compete poorly when nitrogen levels increase. The addition of nitrogen improves conditions for other invasive species that are well adapted to fertile conditions.

Gorse has high concentrations of oils in its stems and is highly flammable. This can increase the frequency of fires in the plant community and can also be dangerous for humans. The town of Bandon, OR, is surrounded by dense thickets of Gorse, and in 1936 most of the town burned down when fire raged through the Gorse thickets.

Control Methods and Management

Effective control of both species requires removal of the dense above-ground biomass. Physical removal by hand is labor intensive and only practical for small populations. Burning or repeated mowing is also used to remove the plants. Following fire or physical removal, Scotch Broom and Gorse may sprout from stumps and dormant seeds will germinate, so native species should be planted immediately following removal. Chemical herbicides can prevent resprouting.

Continued monitoring and control may be necessary as viable seeds remain in the soil for decades.

Several species of moths and weevils have been introduced from Europe to control Scotch Broom and Gorse. These insects damage individual plants but have had only limited success in reducing the abundance of either Gorse or Scotch Broom. Many other species of insects feed on Scotch Broom and Gorse in their native range in Europe (35 species of insects were found consuming Scotch Broom at one site in England) but have not yet been introduced to North America.

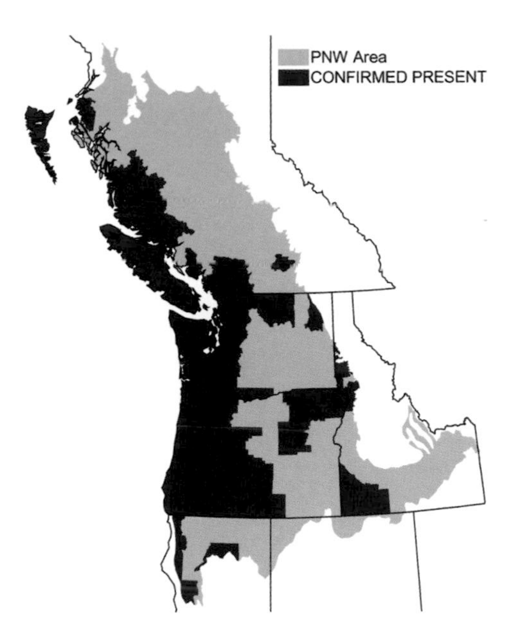

Life History and Species Overview

Gorse and Scotch Broom are woody shrubs that grow very rapidly, gaining up to 1 m in height in the first year. Scotch Broom plants live for 15–20 years while Gorse can live up to 45 years. Both species can produce flowers in their second year, with peak flowering in April–June. Gorse and Scotch Broom produce fruits similar to pea pods that contain 3–8 seeds. Seeds are explosively launched up to 3 m when the pods dry and are also transported by birds and water. Seeds of both species can be viable in the soil for 30–70 years. Thousands of viable seeds can accumulate in the soil underneath dense stands of adult plants and may germinate after fire or other disturbances.

Scotch Broom and Gorse are native to central and western Europe and the British Isles, and Gorse is also native in northern Africa. Gorse has spines that make thickets difficult for humans and large animals to penetrate, a characteristic making it good for hedgerows or natural fences and also a haven for introduced species such as rats, cats, and rabbits. Gorse leaf buds have been used to make tea, and its flowers produce a yellow dye. Lectins from Gorse are currently used in biomedical research and for blood typing. In addition to its use in making brooms, Scotch Broom has been used to make cloth, as a diuretic, and as a substitute for coffee. However, it contains toxic alkaloids that can depress the heart and nervous system and cause death in extreme cases and therefore should not be consumed.

Gorse and Scotch Broom seeds germinate in open areas, such as grasslands, fields, along roads, and in clear-cuts and other disturbed areas. Gorse is often found near the ocean and forms dense thickets in coastal grasslands and dunes. Neither species is able to tolerate cold winters; both species are generally restricted to lower elevations and are more common near the coast where the climate is more moderate.

History of Invasiveness

Gorse and Scotch Broom have invaded coastal areas of North and South America, New Zealand, and Australia. Scotch Broom has also invaded areas in India, Iran, and South Africa. Both species were introduced into the PNW from the British Isles in the late 1800s to remind settlers of their native lands. Gorse was also used to form hedges, but quickly escaped from cultivation. Scotch Broom was widely used in OR and WA to stabilize soil along highways. Both species continue to spread as more land is disturbed.

Other Sources of Information: 71, 92, 125
References: 64, 241, 245
Author: Chad C. Jones

Butterfly Bush *Buddleja davidii*

Species Description and Current Range

Butterfly Bush is beloved by gardeners because it is the lilac of the summer. It is easy to grow, is generally pest-free, and attracts butterflies, bees, and hummingbirds. The plant's small, fragrant flowers are clustered together in groups that are 10–25 cm long. The commercial market has developed flowers in shades of white to deep purple with lilac, which is the most common color in the wild. The inside of the flower is orange-yellow in color, which gives the multistemmed shrub another common name, Orange Eye. The species is perennial and ranges up to 5–6 m in height. Unfortunately, Butterfly Bush has invaded natural riparian and seminatural areas on the West Coast from Los Angeles, CA, to Vancouver, BC.

Impact on Communities and Native Species

Butterfly Bush rapidly colonizes riversides and disturbed areas, forming dense, shrubby thickets that exclude all other plants. In New Zealand, the shrub has invaded national parks and obscured scenic vistas. A study there found a one-year-old population to have several million plants in a space roughly the equivalent of a football field. By the time the plants were 15 years old, they had declined in density enough to finally allow establishment of some native vegetation. The resulting forest, however, remains much different than if the bare ground had been colonized solely by native plants.

Along Salmon Creek and Lake Oswego, OR, Butterfly Bush has displaced native willows, the host plant for the Western Tiger Swallowtail (*Papilio rutulus*), Lorquin's Admiral (*Limenitis lorquini*), and Mourning Cloak (*Nymphalis antiopa*) caterpillars. The shrub provides a nectar source for many species during mid and late summer, which may result in lower pollination for some native plants. Butterflies will not lay eggs on Butterfly Bush leaves because they offer no nutrition for hatchling caterpillars. The species likely has a negative effect on butterfly populations, most of which are limited by host plants and not nectar plants. It is not known how other pollinators are affected.

Control Methods and Management

Shrubs that are merely cut will vigorously resprout and may grow up to 2 m in the following season. The roots must be dug out or the stumps treated with herbicide such as glyphosate. Young plants may be hand weeded, but care must be taken because disturbance encourages seed germination. Reestablishment can be prevented by planting a rapid-growing ground cover of a noninvasive plant. A biological control program has begun in New Zealand to control the spread of this species into forestry plantations. In HI the shrub's popularity in landscaping and widespread distribution likely make it impossible to eradicate. Because it is likely to become more widespread and costlier to control in the future, gardeners should refrain from buying this species and remove it from their gardens, replacing it preferably with natives or the noninvasive species *B. globosa* or *B. macrostachya*. Butterflies are attracted to several natives including Mock Orange (*Philadelphus lewisii*), wild lilac (*Ceanothus* spp.), and Ocean Spray (*Holodiscus discolor*). Several natives for hummingbirds include Trumpet Honeysuckle (*Lonicera ciliosa*), *Penstemon* spp., and Red-flowering Currant (*Ribes sanguineum*).

Life History and Species Overview

Buddleja is named after Reverend Adam Buddle, an English botanist, and *davidii* after a French missionary and naturalist, Father Armand David, who discovered the species in 1869. There are about 100 species of *Buddleja* worldwide. The species is primarily pollinated by butterflies in both its native and naturalized range. The fruits from a single flower cluster can produce 40,000 seeds, which equals approximately 3 million seeds per average plant. Cultivars differ widely in production of viable seeds. "Potter's Purple," for example, produces 20 times more seeds than "Summer Rose."

Seedlings are drought tolerant and will quickly develop an extensive root system. Butterfly Bush reaches reproductive maturity in less than 1 year, grows rapidly at a rate of approximately 0.5 m/yr for 10 years, and then declines. The species is relatively short lived, with records of plants living up to 37 yrs. The shrub prefers full sun and good drainage, and though it may not grow well in acidic and poorly drained soils, it thrives in most others, including low-nitrogen and alkaline soils. Butterfly Bush even grows in the crack of a concrete wall at the Ballard Locks in Seattle, WA. The plant tolerates a wide range of climatic conditions, including oceanic, continental, Mediterranean, and even urban pollution, but does not favor wet soils, which may deter it in parts of the PNW.

History of Invasiveness

The shrub is native to central and southwestern China from Tibet to Hubei at elevations up to 2600 m, where it grows in disturbed areas such as roadsides and river banks. Areas are typically colonized within a year or two of disturbance if there is a nearby seed source.

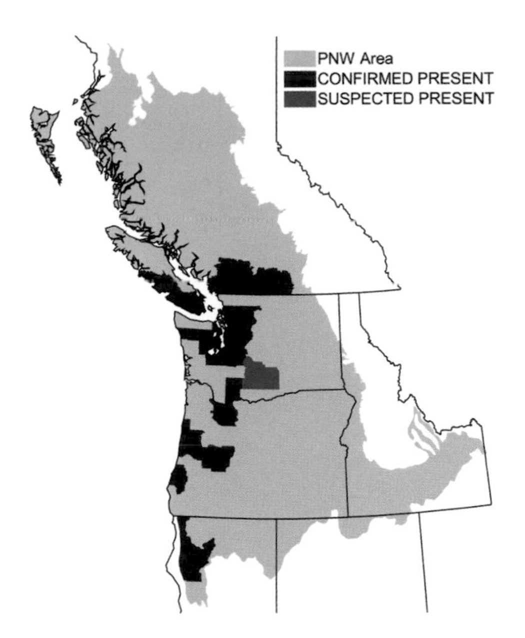

The winged seeds are small and light and easily dispersed by wind or water and sometimes by cars. The seeds have a deep dormancy and consequently may remain in the soil for many years. These characteristics as well as its rapid growth and high seed production have enabled Butterfly Bush to become invasive in many parts of the world.

Butterfly Bush is listed among the top 20 invasive weeds in England. It was introduced in the 1890s as an ornamental plant and by the mid 20th century had naturalized throughout southern England in disturbed areas such as abandoned railway yards and urban areas destroyed by World War II bombs. A similar pattern is occurring in New Zealand, and the plant is also invasive in Australia, Fiji, Europe, and other parts of the US. A survey of garden centers in the Seattle area found a range, depending on the center, of five to hundreds of plants sold per year, which enhances the spread of the species. With a notorious invasive history in climates similar to the PNW the species seems well positioned to invade similar areas. However, the full invasive potential within the US is not yet known.

References: 8, 23, 290, 297
Author: Patricia A. Townsend

Cheatgrass *Bromus tectorum*

Species Description and Current Range

Cheatgrass, an annual grass, grows quickly in the spring, often going to seed before most other grasses. As the grass matures in early summer, it turns a purplish color. It generally grows to 5–60 cm in height, depending on available moisture. Cheatgrass got its name because pioneer wheat farmers felt cheated when the grass invaded their fields and reduced wheat yields. Cheatgrass is also called Downy Brome, because of the soft hairs on its leaves. The seeds have long, stiff hairs (awns) that get caught in socks or other clothing. The roots are highly branched and primarily grow in the upper 30 cm of the soil. Cheatgrass is very common in Sagebrush steppes and grasslands in eastern WA and OR and southeastern BC and is the dominant species on at least 50 million ha in the western US and Canada. It is present throughout the PNW but does not occur in dense forests or at high elevations.

Impact on Communities and Native Species

Cheatgrass invades and becomes very dense, out competing as many as 10–15 species of native grasses. There can be as many as 13,000 Cheatgrass plants/m^2. Common species that are displaced by Cheatgrass include Sagebrush (*Artemisia tridentata*), Bitterbrush (*Purshia tridentata*), and Bluebunch Wheatgrass (*Pseudorogneria spicata*). Cheatgrass is very efficient at using the moisture in shallow layers of soil, preventing other species from obtaining sufficient water. It dramatically impacts the plant communities that it invades by changing the frequency of wildfires. Because it goes to seed and dries out early in the summer, it burns easily in summer or fall lightning strikes. Wildfires naturally occur every 60–100 yrs in Sagebrush grasslands, but invasion by Cheatgrass can provide fuel for burns every 3–5 years. Frequent fires promote the growth of Cheatgrass and other short-lived invasive grasses, while making it difficult for longer-lived native plants, like Sagebrush and bunchgrasses, to survive. In areas dominated by Cheatgrass, fires are not only more frequent but also occur earlier in the summer before other species have produced seeds. Because of these changes in wildfires, Cheatgrass can completely alter the vegetation and prevent many native plants from coming back. Invasion by Cheatgrass can promote other invasive species, such as knapweeds (*Centaurea* spp.), that are also adapted to frequent fires. Cheatgrass also invades wheat fields and dramatically reduces wheat yields. It is estimated that Cheatgrass costs wheat farmers $350–370 million/yr in lost yields and control costs.

Control Methods and Management

Control of Cheatgrass usually involves spraying herbicides early in the spring, before the native plants have emerged. It is important to plant seeds of the native species following the herbicides or new Cheatgrass plants will become dominant again. Several years of treatment are often required for successful control. Pulling Cheatgrass by hand is effective only if there are few plants. Disking and tilling are not effective in controlling Cheatgrass and no biological control is available. It is important to remove all individuals where possible, because when the density of individual plants decreases, each plant will produce more seeds, maintaining a high level of seed production. Since humans often disperse Cheatgrass in socks and other clothing, careful removal of the seeds from clothing before leaving a site can help reduce its spread.

Life History and Species Overview

Cheatgrass germinates in the fall and winter. The shoots grow slowly through the winter but the roots are able to grow at colder temperatures than many other plants. Thus the roots can grow throughout the winter, allowing the plants to grow rapidly in the spring. Cheatgrass plants produce large numbers of seeds (up to 5,000/plant) and die by June. This life cycle allows it to avoid competition from many other plants that grow later in the year. With its seeds safely in the soil, Cheatgrass avoids the negative effects of summer wildfires by completing its life cycle early in the summer. Seeds can be dormant in the soil for 2–3 years but the majority of seeds germinate in the fall of the year they were produced.

Cheatgrass originated in relatively dry communities around the Mediterranean and in southwestern Asia but has now spread to grasslands and shrub grasslands in all 50 states and every Canadian province except NF. It is also found in open Ponderosa Pine (*Pinus ponderosa*) woodlands and less commonly in desert shrublands. It is generally found in areas with 15–55 cm of precipitation/yr and between 600–1800 m in elevation. Cheatgrass is most common in abandoned fields and overgrazed rangelands, along the edges of highways, and in similar disturbed sites, but can also invade undisturbed grasslands. Cheatgrass is an important forage crop for livestock in many areas, but only for the short period in the spring while it is still green and high in nutrients. During most of the summer, after it has dried out, it does not provide a good source of nutrients for domestic livestock or wildlife.

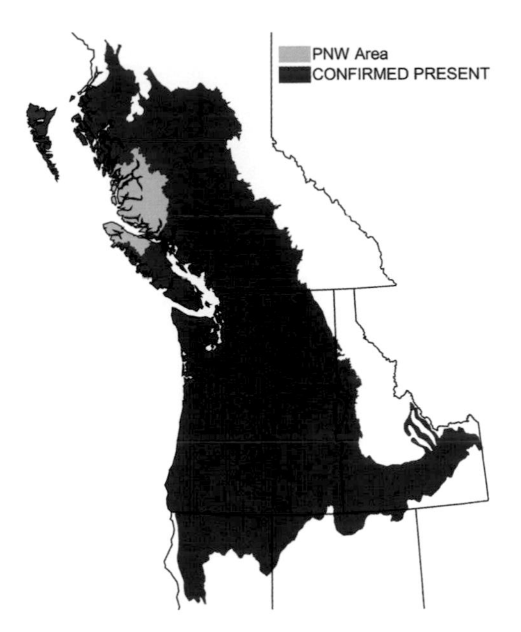

History of Invasiveness

Since 1880 Cheatgrass has spread from its native range to North America, Australia, New Zealand, South Africa, Greenland, and Japan. It was originally transported from Eurasia to North America as seed in soil that was used as ship ballast. It was first brought to the PNW as a contaminant in wheat seed in the late 1800s. The earliest records of Cheatgrass in the PNW are from Spence's Bridge, BC, in 1889 and Ritzville, WA, in 1893. It was deliberately introduced in Pullman, WA, in 1898 in an attempt to find new forage grasses. It was relatively uncommon in the area for several decades, but in the 1920s and 1930s, it spread rapidly and began to dominate natural communities. Cheatgrass most easily invades following disturbances such as fire or grazing by livestock. However, invasion in grasslands and shrub grasslands can occur even without disturbance. Cheatgrass continues to spread as a contaminant in wheat seed and by transport in the fur of animals and in clothing.

Other Sources of Information: 92
References: 88, 200, 226, 316
Author: Chad C. Jones

Common Velvet-grass *Holcus lanatus*

Species Description and Current Range

Common Velvet-grass, a perennial, is green-gray with flat, velvety leaves 4–9 mm wide and 5–10 cm long and stands 0.5–1 m tall. Each plant produces several panicles, soft and densely packed flower clusters that range from 7 to 15 cm long and 4 to 9 cm wide. These panicles are purple-gray in spring and progressively bleach to the color of straw by midsummer. Velvet-grass is abundant across the PNW but is found mostly west of the Cascades, particularly in wetlands, wet meadows, fields, or pastures. Disturbance stimulates growth so Velvet-grass is not surprisingly often one of the first and most successful colonizers of construction sites, roadsides, untended gardens, and other altered lands.

Impact on Communities and Native Species

Velvet-grass is tolerant of water variations and is a hearty competitor of native species in wet areas. In floodplains, deeply rooted native shrubs contribute to soil preservation while Velvet-grass roots rarely penetrate more than 8 cm. Once Velvet-grass is established, it outcompetes many larger native plants and provides little protection against soil erosion in the event of a flood.

Velvet-grass is more productive when growing with other plants and may alter species composition. When grown with plants of similar size and character, it produces three times more above-ground biomass then when it grows alone. As Velvet-grass populations increase, the combined biomass of other species is reduced fourfold. Velvet-grass can impede growth in large plant species. For example, eucalyptus trees grow more slowly with competition from weed mixtures containing Velvet-grass and are 48–60% smaller in diameter than trees grown without weeds. Competition from Velvet-grass continues until the tree is 18 m tall, greatly reducing the economic viability of such cash crops.

Velvet-grass is a common food for herbivores in both wildlife and agricultural settings. Its low tannin concentrations contribute to digestibility in the rumen, although it is less digestible than clover, willow leaves, poplar, and lupin. Herbivores eat green material and require a variety of species with different life cycles so that as one plant type dies off another remains an available source of live nutrients. When Velvet-grass dominates a landscape, its seasonal cycle creates periods of extreme high or low availability of edible biota. Animals eating mainly Velvet-grass have lower reproductive success and gain weight more slowly than when they eat a greater variety of forage plants.

Control Methods and Management

Traditionally, seasonal fires and floods in the PNW caused natural allocation of land among competing plant species while stimulating regrowth of native grasses. Urban and agricultural developments disrupt these natural control mechanisms, changing species growth rates and, consequently, community composition. Prescribed burning promotes competition and often favors non-native species, such as Velvet-grass, that dominate and suppress native regeneration. Mowing of fields before the grass flowers, followed by the removal of cut materials, suppresses the spread of Velvet-grass and is a valuable management tool. However, the growth tissue located at the base of the leaf retains its regenerative ability and the cropped plant will survive. Hand pulling is the recommended method for controlling Velvet-grass populations most effectively. Removing rhizomes and seeds dramatically lessens successful regeneration. Long-term control requires monitoring and continued removal for successful suppression.

Chemical and biological treatments, including soil manipulation and introduction of insects, fungi, and disease, often have negative impacts on native species and can foster Velvet-grass dominance. While herbicides can effectively control annual grasses (91–98% biomass reduction), they are less effective on perennials (45–87% biomass reduction). Velvet-grass is vulnerable to plant-feeding nematodes, but they harm other plants and should not be used.

Life History and Species Overview

Velvet-grass survives in much of the PNW from 0–1,500 m but requires a minimum of 120 frost-free days. It is intolerant to anaerobic, saline soils, and severe frost is lethal. It grows and persists best on poorly drained and nutrient-rich soils of medium to coarse grain texture, like those found in fen-meadow habitats under mildly acidic conditions. Tolerant of full sunlight or shade, this grass colonizes wetlands, shrublands, grasslands, and riparian zones.

History of Invasiveness

Velvet-grass originates from the Iberian Peninsula in southwestern Europe and is considered native to England, a likely explanation for its other common name, Yorkshire Fog. Following introduction to California as a seed mixture component for meadows and hay fields, seeds escaped and spread pervasively throughout the PNW. Following World War I, accelerated logging promoted its spread across BC. This grass displaces native plants and is considered one of western Canada's 41 most troublesome species.

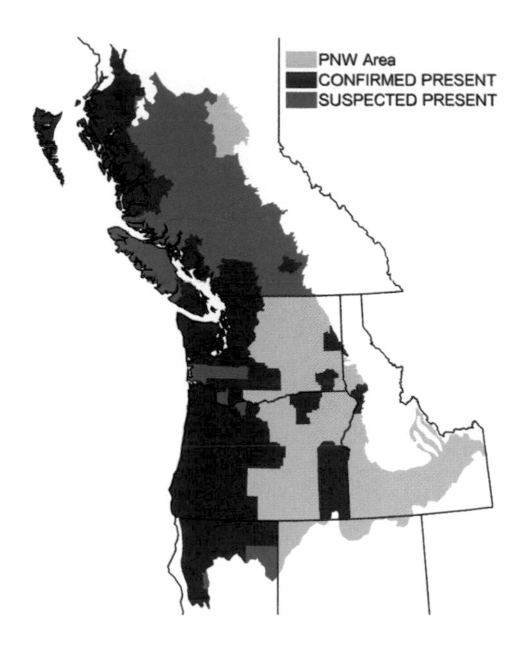

Velvet-grass averages 3,353,124 seeds per kg biomass, yet seed spread rate is low. Each plant produces up to 10 adults in a season. Reproduction occurs by rhizome shoots or seed dispersal by wind and animal movement. In high-water habitats, rhizomes break off and float through the water column, giving Velvet-grass a competitive advantage for colonizing new soil before other species are able to seed. Velvet-grass seeds remain viable in the soil and can begin growth immediately following a disturbance. The soil systems of many disturbed sites contain contaminants to which Velvet-grass is tolerant. Arsenic depositions from mining, and heavy metal accumulation from runoff, inhibit the growth of other species and encourage community dominance by Velvet-grass.

Other Sources of Information: 92
References: 3, 22, 61, 128, 327
Author: Delia R. Kelly

Dalmation Toadflax *Linaria dalmatica* ssp. *dalmatica*

Species Description and Current Range

Dalmatian Toadflax is a member of the snap-dragon family, the Scrophulariaceae. It is an aggressive and highly competitive weed that is native to central Europe and east into central Asia. This herbaceous, short-lived (3–5 yrs) perennial plant grows up to 1.5 m tall, with creeping roots, broad to lanceolate clasping leaves (2–5 cm long), and showy, irregular, yellow flowers. Both the stem and leaves are green and waxy with a bluish or whitish cast. The flowers are bright yellow with orange markings and elongated spurs borne on simple, elongate, terminal racemes. This species was first introduced into eastern North America in the late 1800s as an ornamental plant. It is currently found in 22 US states and 7 Canadian provinces. It can be found in all US states and Canadian provinces of the PNW and has occurrences in nearly 75% of the counties and districts of the region.

There are two close relatives of Dalmatian Toadflax that also occur in North America and have a more limited distribution in the PNW. Yellow Toadflax (*Linaria vulgaris*) is common in the eastern US and Canada and infrequent in the PNW. Broom-leaf Toadflax (*Linaria genistifolia*) is similar in appearance but has narrower leaves and smaller flowers. It also has limited distribution in the PNW. All three toadflax species are invasive exotics that pose similar management challenges.

Impact on Communities and Native Species

Dalmatian Toadflax is a highly aggressive exotic that invades both disturbed areas and intact native habitats. It displaces existing native plants and can dramatically alter the composition and structure of native plant communities. Dalmatian Toadflax will rapidly colonize disturbed sites and is a strong competitor with native plants, spreading through native grasslands, shrub-steppe, and dry coniferous forests. It will also colonize aspen, oak, and riparian forests as well as other shrub- and herb-dominated communities.

The alteration or displacement of native habitats can cause a loss of forage to native ungulates. Deer have been observed to browse Dalmatian Toadflax, but heavy consumption has not been recorded. Some bird and rodent species consume small amounts of Dalmatian Toadflax seed, but the minute seed size limits its use by most animals. Competition with native plant species may result in the elimination of the food source for some animals with diets restricted to a few native species. Insects and birds that are obligate feeders on the native plants displaced by Toadflax are the most susceptible. On sites where Toadflax replaces sod-forming native grasses, surface runoff and soil erosion may increase.

Control Methods and Management

Dalmatian Toadflax has high genetic variability, and successful control and management requires the integration of multiple strategies that exploit the vulnerabilities of individual strains. The goal is to eliminate seed production and vegetative spread. Prevention of infestation is the highest priority due to the difficulty of controlling established populations. Hand pulling can be effective with small populations. Removal of roots is important. Mowing can be an effective method of eliminating seed production, but timing is key, and it must be repeated frequently until the population dies out. Tilling is an effective control where feasible, but must be repeated 4–5 times each summer for at least 2 years for effective control. Toadflax has sufficient palatability that grazing by sheep and goats can effectively suppress stands of it and reduce seed production. But grazing by cattle is counterproductive, as they preferentially select native grasses and herbs, leaving the Toadflax behind. Burning is also counterproductive.

Four beetles (*Brachypterolus pulicarius*, *Gymnaetron antirrhini*, *G. netum*, and *Mecinus janthinus*) and two moths (*Eteobalea intermediella* and *Calophasia lunula*) have been used as biological controls. These insects attack the roots, stems, leaves, and reproductive organs. Use of multiple organisms will be most effective. Effectiveness of herbicides is highly variable. The waxy leaf surface forms a barrier to uptake. Repeated applications (for 3–4 yrs) of picloram or combinations with 2,4-D, 2,4-DB, MCPA, MCPB, or mecoprop have been effective. Surfactants may be necessary to increase uptake. Herbicide levels that are effective in Toadflax control will also kill most native vegetation (including trees), therefore chemical control measures may have undesirable effects if applied in native plant communities. Effectiveness of any of the above control measures depends on rapid subsequent establishment of competitive desirable vegetation cover.

Life History and Species Overview

Flowering occurs from May to October, and seeds mature from July to November. A mature Dalmatian Toadflax can produce up to 500,000 seeds annually. Seedlings emerge in early March to May on most sites. During the first few weeks after emergence, they are exceptionally vulnerable to dehydration and competition. Some seed production can occur the first year. Dalmatian Toadflax plants usually produce a matlike rosette in the early autumn and then regenerate in early spring from vegetative buds on the rootstock. The roots of a mature plant often reach depths of 1–2 m (3 m on occasion) and lateral roots can extend 3 m or more. This extensive root system makes Dalmatian Toadflax an extremely effective competitor with most native plants. It is particularly competitive at this stage and rapidly forms floral stems. Dalmatian Toadflax is self-incompatible and relies upon insects for pollination. Once established, high seed production and the capability to reproduce vegetatively allow for rapid spread and high persistence. Toadflax plants live 3–5 years. This relatively short life span facilitates control measures that target elimination of seed production.

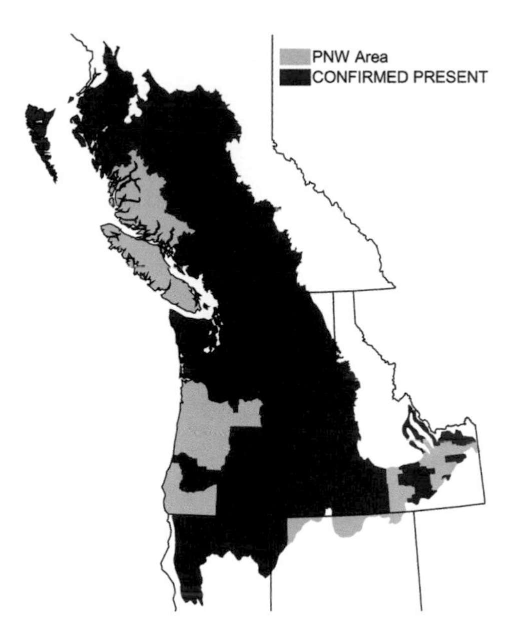

History of Invasiveness

Dalmatian Toadflax has been cultivated as an ornamental for at least 400 years and was introduced into North America in the late 1800s. By the 1920s, it had escaped cultivation. It spread rapidly to infest pastures, farmland, roadsides, rangeland, forests, and riparian zones in much of the US and Canada.

Other Sources of Information: 25, 54, 55, 91
References: 53, 186, 341
Author: Peter Morrison

English Holly *Ilex aquifolium*

Species Description and Current Range

English Holly, an evergreen tree, can reach up to 15 m in height and 0.3 m in the trunk's diameter. The bark is gray and smooth or nearly so. It is well known as a Christmas decoration, with its shiny, dark green, wavy-edged, spiny leaves and clusters of bright red, poisonous berries. The tradition of this yuletide decoration dates back to before Christianity, to the time of the Druids, who hoped that the indoor decoration of English Holly would be a refuge for fairies and other woodland spirits during the harsh winters. In late spring, small white flowers that have four petals bloom on year-old growth. Male and female flowers are on separate plants and the fruit matures (on the female only) in autumn and remains attached through winter.

This common ornamental is found throughout the US, doing best in mild coastal climates and riparian zones. Sustainable populations have been found in the PNW and New England coastal areas.

Impact on Communities and Native Species

English Holly's dense foliage is a refuge for many birds. Blackbirds and thrushes eat the berry. In England and Germany, it is considered bad luck to step on a Holly berry, the favorite food of the robin. Other native bird species, such as Cedar Waxwings, Mourning Doves, finches, and chickadees feed on the berry, as does the non-native House Sparrow. The seeds are dispersed by birds that forage on the berries in autumn and winter. It can take up to 3 years for the seeds to germinate, and seedlings can be slow growers their first few years. English Holly makes up for its slow beginning by living for 250–300 years, and females can produce hundreds of berries each year. Most reproduction is done through seeds, but English Holly will also sucker from roots or, more rarely, spread by layering of branches (both forms of vegetative growth).

In the PNW, English Holly invades edges of wetland areas and forests and gap or interior habitats. It is shade tolerant, and highly competitive with other native, understory plants. Because it creates a different environment, English Holly changes plant and animal species composition of a forest. English Holly regenerates vigorously from well-established rootstock and so is less vulnerable to fires that may kill larger trees. Despite its potential to alter native habitats, English Holly has yet to be proposed for addition to any noxious weed lists in the PNW.

Control Methods and Management

Only physical means of removal can control invading plants. If immediate eradication of the plant is not possible, remove the berries. The next level is removal of female plants (to avoid seed dispersal). For complete expulsion of the species, the plant should be cut down, and an herbicide brushed on the stump. English Holly is resilient, so reapplying the herbicide or cutting back the new growths repeatedly is necessary. Dig up the stump and roots and then continually watch the area for any regrowth.

Removal is much easier when the plant is small and manageable. Because manpower is currently the only way to eradicate this plant, complete removal is difficult and expensive.

People could help reduce the spread of English Holly by not planting it and removing established plants from their property.

Life History and Species Overview

English Holly is native to England, France, Germany, southern Europe, northern Africa, and Asia. It does not grow well in hot, humid areas nor is it very frost tolerant. Holly does best in temperate climates. Able to grow in any soil, English Holly prefers well-drained soils with a sandy base or gravelly loam. It doesn't get to full berry-bearing age until it is 15–20 years old but may produce berries at 2 years of age. The seeds are preyed upon by rodents and dispersed by birds.

English Holly has played a role in European culture since before the Romans. It was considered a holy plant, and some experts believe that "holly" is a corruption of the word "holy." Druids thought that the sun would never forsake the plant. Romans used Holly as a decoration during Saturnalia, a winter solstice festival. In Germany, the plant is called Christdorn, and the Pennsylvania Dutch thought its white flowers represented the purity of Mary and the berries represented the blood of Christ's wounds. Its flowers were used to break fevers, and the leaves were used as a tea to cure colds, bronchitis, and rheumatism. English Holly berries made into a juice were taken for jaundice. However, use was rare, since the berry is highly toxic.

History of Invasiveness

English Holly was brought to all of Great Britain's colonies and is an established invader in North America, Australia, and New Zealand. English Holly invaded both the East and West Coasts of North America's temperate zones. It was first introduced to the PNW in 1869 as an ornamental plant for hedges in landscaping. By the turn of the 19th century, commercial orchards grew Holly for Christmas decorations. English Holly has established populations and is expanding at a moderate to high rate in the PNW, where conditions are ideal for its growth and survival.

English Holly is probably more widespread than is documented. It is also a cash crop for the PNW (although the market is declining)

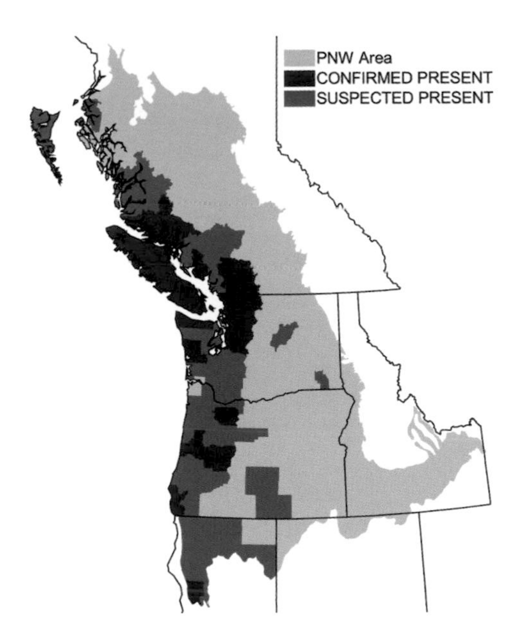

and an extremely common ornamental in yards. Holly orchards in BC produce all of the English Holly that is purchased in Canada. Because of orchards in WA, OR, and BC, and its commercial value, English Holly is not listed as an invasive in these states and provinces but it is considered a species of interest/concern. Only CA lists this species as an invasive.

Other Sources of Information: 12, 76, 93, 95
References: 114
Author: Dawn Olmsted

English Ivy *Hedera helix*
Irish Ivy *Hedera hibernica*

Species Description and Current Range

"English Ivy" is the misused common name for a few different species and many cultivated varieties in the genus *Hedera*. Genetic studies in the PNW have determined that 83% of the invading plants are actually *H. hibernica*, which is more appropriately considered to be Irish Ivy. The remaining 17% were mostly two cultivated varieties of true English Ivy, *H. helix* "Pittsburg" and *H. helix* "Star." In all cases, however, the species were evergreen vines or shrubs with the unusual characteristic of having different forms of leaves on reproductive and nonreproductive plants. The juvenile, nonreproductive, plants are woody vines with palmately lobed leaves. The plants may be a decade or more old before they produce flowers and fruits. Vines that become reproductive have ovate instead of lobed leaves and may either grow on vertical surfaces or become woody shrubs. Small roots along the stem adhere to surfaces and allow the vines to climb. The plant has white flowers in tight clusters in the fall, followed by purple berries in the spring. Because the flowers are usually on vertical surfaces, they may be high in a tree canopy. All forms of ivy have very similar flowers and small, purple, marble-size fruits.

English Ivy is found throughout the PNW, from northern CA into BC, but it is most invasive on the west side of the Cascade mountains, where its evergreen leaves and the mild temperatures and ample rainfall allow it to grow throughout the winter. In colder areas it grows more slowly and is less invasive. It grows fastest in open deciduous forests, but may also be found under more closed canopies. The many forms of ivy with variegated leaves rarely become invasive because of their slower growth rate. The three taxa above are listed as noxious weeds in OR and WA.

H. hibernica (adult form)

Impact on Communities and Native Species

English Ivy is most damaging in forests, where it can form "ivy deserts" on the forest floor. In other habitats it invades, it replaces virtually all herbaceous and most woody species. By preventing regeneration of woody species, it alters the long-term persistence of the forests. The ivy deserts are also reported to have less diversity of birds, mammals, and amphibians. The mechanism that excludes native species is not understood, but it is probably either competition for water and nutrients or the poisoning of other plants through a process known as allelopathy. It also grows up tree trunks and branches, especially using deciduous trees that lose their leaves in the winter; light then reaches the ivy, allowing it to grow throughout the winter. The additional weight of the evergreen leaves may increase storm damage. It is estimated that one acre of a heavy infestation weighs 10 tons. Ivy sap can cause dermatitis in some people. Cosmetic companies promote the extract as a treatment for cellulite in women, but this has not been proven effective. The fruits and leaves are toxic if ingested by humans. Robins, and to a lesser extent starlings, appear to be the primary dispersers of the seeds because they eat the fruit in spring, when there are not many other fruits available. Ivy also provides excellent cover for rats, another harmful invasive species.

Control Methods and Management

Because it has a very waxy leaf, most herbicides do not work well on ivies. Some control is possible if the leaves are scarified with a string trimmer prior to herbicide application. Glyphosate or triclopyr with a surfactant to help it stick to leaves applied on a sunny warm day in winter helps control ivy, but the results may not be apparent for several months. The best control method is probably hand removal, pulling and rolling up the vines as you move along. Be careful not to pull up any remaining native plants and try not to disturb the soil. Pulling is best done in the winter to reduce the impacts to native plants and animals. Because only the shrubs or vines on vertical surfaces will produce fruit, the shrubs should be eliminated and vines on trees cut to sever the connection with the ground. The vines above the cuts will die. Repeated treatments are needed to remove vines missed or those that have begun to regrow up tree trunks. Ivy removal can be done very inexpensively using volunteers, as the No-Ivy League in Portland, OR, and the Ivy-OUT (Off Urban Trees) program in Seattle, WA, have done.

Life History and Species Overview

English and Irish ivies may be 10 or more years old before they flower, but when they do they produce many small flowers in the fall, attracting bees. Seeds ripen the following spring and are eaten by birds, especially robins, which scarify the hard seed coat as it moves through their systems. This can increase the germination, with up to 70% of the seeds germinating, but the main contribution of the birds is very efficient seed dispersal. Ivy can tolerate shade, but grows best at about 65–68% of full sunlight. It persists where established; there are reports of a plant that was dated by annual rings to be 433 years old.

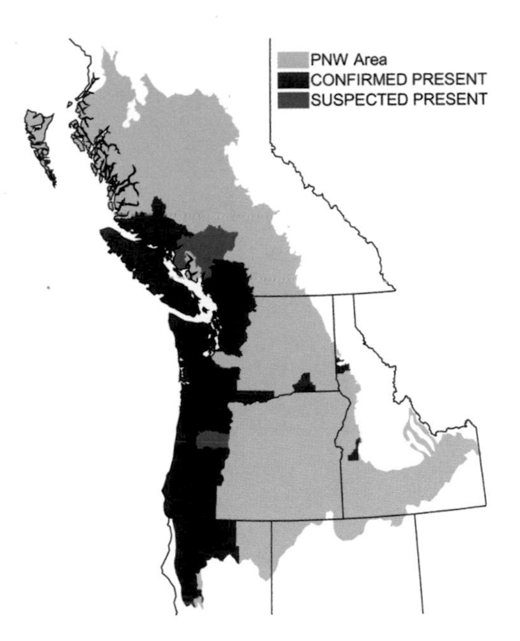

History of Invasiveness

The invasive ivies are native to England, Ireland, the Mediterranean region, and northern Europe west to the Eurasian Caucasus Mountains. In its native range it is widespread and commonly found in forests, where it is also sometimes considered a pest. It was introduced into North America in early colonial times as an ornamental and has been planted in many places to control unstable soil. While it does somewhat prevent surface erosion, it is far too shallowly rooted to stabilize slopes. It has become invasive in the Middle Atlantic states and is one of the most serious invaders of forests in the PNW. It was probably first introduced in this region in the 1890s. While it is currently most prevalent in urban forests, birds are now dispersing it outward into surrounding forests. As eradication efforts mount, it is helpful if nurseries stop selling, and gardeners stop buying, these species and cultivated varieties.

References: 260
Author: Sarah H. Reichard

European Beachgrass *Ammophila arenaria*

Species Description and Current Range

European Beachgrass is an erect perennial with flowering stems up to 1 m tall. The stems are tufted, rising from scaly, tough, rhizomes that spread horizontally and vertically. The leaf blades are rough, fibrous, and in-rolled to a sharp point. They are 2–4 mm wide when flattened, and 10–15 mm broad, without auricles (ear-shaped appendages that occur in pairs at the leaf sheath that envelops the stem). Ligules (a transparent sheath that projects up from the inside of a leaf blade where it joins the stem) are 10–33 mm. The outer surface of the blade is smooth and grayish-green without distinct ribs; the inner surface is whitish and closely ribbed. The flower heads or spikelets are single-flowered and 10–15 mm long, and the lemma (the lower of the two bracts immediately enclosing the flower) of the single flower is 1–3 mm shorter. European Beachgrass is distinguished from the native dunegrass species, American Dunegrass (*Leymus mollis*), which has flat, wide, more bluish-green blades, narrower ligules, and a spike inflorescence.

European Beachgrass has been introduced to North America, Australia, and New Zealand. On the Pacific coast of North America, European Beachgrass dominates coastal dune habitats from the central CA coast to northern BC and AK. Its genus name, *Ammophila*, means "sand lover," because this species is so highly adapted to colonize shifting sand. It is present in all coastal and Puget Sound counties throughout its range. The northern extent of its range, in BC and AK, is limited not by climate but by the lack of suitable dune habitat where clifflike fjords meet the shoreline.

Impact on Communities and Native Species

European Beachgrass is the most pervasive exotic plant species currently threatening coastal dunes along the coast of the western US. The plant is capable of trapping windblown sands more effectively than the native species. The accumulation of sand at the plant crown triggers aggressive growth. European Beachgrass is almost perfectly adapted to a very precise set of habitat conditions. Within that habitat it has no competitors and essentially no predators or diseases. Environmental conditions along the northwest coast of North America are almost perfect for its establishment and spread. Impacts of European Beachgrass to coastal sand-dune ecosystems include its ability to displace entire native plant communities and its significant interference with the natural dynamics of dune systems. In addition, areas heavily infested with European Beachgrass threaten habitat for the Western Snowy Plover, a federally listed bird species. Dunes that form under cover of native dune plants have low slopes perpendicular to the beach. The foredunes (the first ridge of sand parallel with the beach) are low and rise above the beach with a gentle slope. When European Beachgrass communities are established, more windblown sands accumulate around the plants on the foredune. Thus, dune structure changes by forming a higher foredune resembling a wall of monotypic vegetation. As the dunes increase in height, reducing the normal ocean breeze behind the dunes, a new microclimate develops that is no longer suitable for native plant species, and the inland native coastal plant community is colonized by other exotic plant species. Ultimately the integrity of the entire native dunal ecosystem is threatened.

Control Methods and Management

Control of European Beachgrass is necessary to protect the limited occurrences of viable natural sand-dune systems along our coastlines. Unfortunately, the plant is now so widespread on the West Coast of the US that its complete eradication is not practical, unless a more economic means of control is found. The arsenal of known techniques includes manual, mechanical, and chemical alternatives, alone or in combination. Manual removal has been used with success, but at great expense and effort. The extensive rhizomatous network must be removed at least 15 cm below the surface with a shovel or spade. This is the most effective method on smaller sites. Mechanical control is used extensively on larger project areas. However, mechanical removal is only suitable for sites that are easily accessible, relatively flat, and without significant numbers of native plants. The vegetation is excavated and buried to a depth of 1 m and requires follow up maintenance to control resprouting.

Chemical treatment is likely the most cost-effective method of control. There are, however, problems with this method associated with environmental impacts and application regulations. Since European Beachgrass has a relatively low tolerance for alkaline conditions, irrigation treatments with salt water have also been used with mixed success. Removal of large infestations of European Beachgrass necessitates a coordinated, comprehensive effort, with ample project planning, funding, and public review. This process includes procuring all appropriate permits and an ongoing maintenance, monitoring, and adaptive management program.

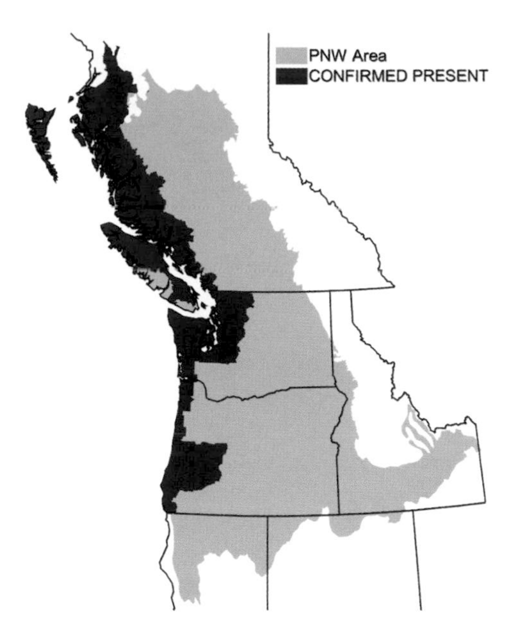

Life History and Species Overview

European Beachgrass is native to the shores of Europe between 30° and 63° north latitude. Flowering occurs from May to August. Mature fruits are dispersed in September. Seeds germinate the following spring, but their viability is low. Reproduction is primarily vegetative, by rhizomes. The plant thrives in areas of active sand movement and most often occupies the windward slopes of exposed dunes. Species adapted to dunes must tolerate wind, sand burial, sand abrasion, salt spray, water deprivation, and salty shifting soils. Only the most tolerant pioneering plants can survive in this specialized habitat.

History of Invasiveness

European Beachgrass was introduced in the early 1900s to stabilize sand dunes. In the 125 years since introduction, it has spread along the entire west coast of North America. Much of its spread can be attributed to planting projects by various government agencies to protect towns, roads and railroads, water supplies, forests, recreation areas, and private property. In addition to intentional cultivation, viable rhizome fragments are spread by wind and water along the PNW coast.

References: 225, 247, 248, 343, 344
Author: Matt Bennett

Fennel *Foeniculum vulgare*

Species Description and Current Range

Fennel, a perennial, anise-scented herb, grows grayish-green, feathery stalks and tiny yellow flowers in umbrella-shaped clusters. Plants establish long taproots and can attain heights of 1.5–2 m after two growing seasons. Invasive Fennel is common on roadsides, construction sites, and other disturbed areas in WA, OR, and CA, and on the Channel Is. in CA. Fennel has invaded much of the coastal US, HI, PR, and the VI.

Impact on Communities and Native Species

Fennel thrives in disturbed areas but generally does not invade established native plant communities. Still, its role in promoting the decline of the now-endangered CA Island Fox (*Urocyon littoralis*) and in changing the life history of the Anise Swallowtail butterfly (*Papilio zelicaon*) warns of its potential to wreak ecological havoc. Ranchers brought plants, including Fennel, and livestock to CA's Channel Is. in the 1850s. When domestic pigs became feral and decimated native plant communities, Fennel took over the disturbed land. Fennel provided pigs with bulbs to eat and protection from hunters; pigs spread Fennel by carrying seeds on their coats and in their feces. Meanwhile, use of DDT as an agricultural pesticide in the 1940s–60s drastically reduced the islands' Bald Eagle (*Haliaeetus leucocephalus*) populations.

With the native eagles largely gone, and a burgeoning population of pigs on which to prey, mainland Golden Eagles (*Aquila chrysaetos*) colonized the islands sometime in the 1990s. Bald Eagles had preyed largely on fish, and had posed no significant threat to Island Foxes. However, each year, after piglets have grown large and difficult to catch, Golden Eagles begin to prey on Island Foxes. About the size of small house cats, Island Foxes are so unaccustomed to predation that researchers can examine them without sedating them, or even using protective gloves. Moreover, Island Foxes hunt during daylight, when they are especially visible to Golden Eagles. By promoting invasion by Golden Eagles, pigs and Fennel have together imperiled the Island Fox. Since 1995, fox populations on San Miguel I. have declined from 2000 to 70 individuals. The Nature Conservancy, National Park Service, and other organizations are working to control Fennel and pig populations and to return Bald Eagles to the islands.

Observations from mainland CA suggest that successful removal of Fennel would not be without complications. The black and yellow Anise Swallowtail Butterfly historically laid its eggs on the Water Hemlock (*Cicuta maculata* var. *bolanderi*) and the Water Dropwort (*Oenanthe* spp.). The species continues to use these plants in rural areas. However, habitat destruction has sharply reduced populations of these native plants, and Anise Swallowtail larvae in urban areas now depend heavily on Fennel and another invader, Bishop's Weed (*Ammi majus*), for food. While Fennel provides good nourishment, Bishop's Weed is toxic for the Anise Swallowtail larvae. Some naturalists now fear that removing Fennel from areas that lack Water Hemlock and Water Dropwort would force the butterflies to lay their eggs on the toxic Bishop's Weed and, ultimately, devastate urban Anise Swallowtail populations.

Control Methods and Management

Fennel should not be used in gardens. Native plants or noninvasive non-native species are better choices for home gardens.

When Fennel is already present, mow or cut the plants before seeds brown and mature in early fall. Pre-empting seed production prevents spread by sexual reproduction.

Remove the entire taproot when digging out the plants, since they can regenerate from root fragments.

Controlled burning can be used in large stands to remove dead vegetation and render regenerating Fennel vulnerable to pesticides. However, fire should be used with caution, since wildfire may be a significant risk in vegetated areas.

Life History and Species Overview

Fennel is a member of the family Apiaceae, which includes carrot, celeriac, parsnip, celery, parsley, dill, caraway, and anise. Though Fennel prefers well-drained, limestone-rich soils, it is robust in soils with pH levels ranging from acid to basic, temperatures of 4–27°C, and precipitation of 0.3–2.6 m annually. Under conditions too dry for germination, seeds remain dormant in soil for up to 7 years. Fennel requires two growing seasons to reach full sexual maturity; younger plants may spread asexually when floodwaters pick up taproot fragments and deposit them elsewhere. Fennel grows most quickly during summer, but can produce leaves year-round if the winter is mild. In the PNW, Fennel typically blooms in July and August and sets its ridged, oblong seeds in the early fall.

Fennel is native to Mediterranean Europe. The wild Fennel common in the PNW generally is a bit less sweet and has a smaller bulb than the cultivated variety available at groceries and farmers' markets. Humans have

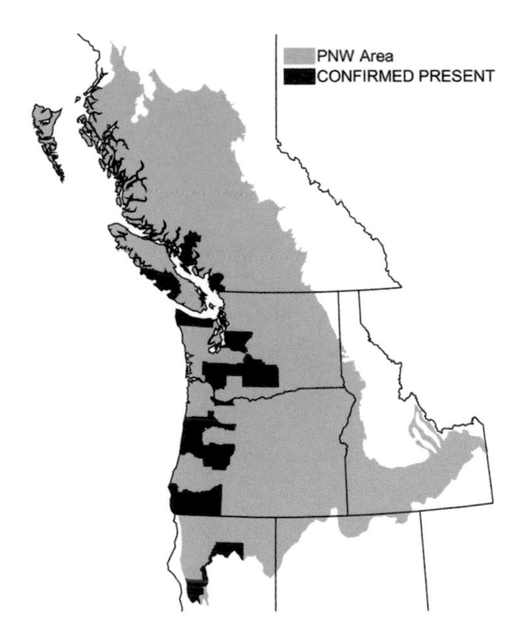

used Fennel for centuries: its leaves as a garnish, its celery-like stalks for cooking, its seeds for their medicinal properties, and the entire plant as a perennial ornamental. Various cultures have credited Fennel with promoting lactation, suppressing appetite, healing snakebites, curing colic, repelling flies and ticks, improving eyesight, relieving nausea, and silencing stomach growling during long church services. Roman gladiators were awarded a Fennel garland when they triumphed in a bout; sweet Fennel seeds are often offered as a breath-freshener in Indian cuisine.

History of Invasiveness

Fennel was separately introduced to the east and west coasts of North America. During the 1600s, Puritan colonists in VA began to cultivate Fennel for use as an appetite suppressant and digestive aid. Spanish priests planted it over two centuries ago in missionary gardens in present-day CA and used its feathery fronds to perfume mission buildings. Fennel escaped Puritan and missionary gardens and began to proliferate in the wild.

Other Sources of Information: 92
References: 32, 69, 195
Author: Diane P. Genereux

Garlic Mustard *Alliaria petiolata*

Species Description and Current Range

Garlic Mustard grows as a single-stalked herb, branching off near the top of the plant to show small, white, four-petalled flowers in the early spring. After the flowers die back, the slender seedpods are visible above the plant. The mature plant is typically not more than 1 m tall, although it can produce flowers when it as short as 10 cm or as tall as 1.9 m, depending on where it grows. The leaf shape is triangular and the leaves are increasingly smaller toward the top of the plant. The leaves grow alternately off of the main stem on short stalks. In early spring the new leaves smell like garlic if you crush them. The garlic smell fades later in the season. The slender taproot usually has a distinctive S- or L-shaped curve just below the soil. During the immature stage, Garlic Mustard forms a basal rosette or low-growing cluster of kidney-shaped leaves. The following spring a tall flowering stem forms on all plants that overwintered, and the basal rosette leaves wither away. The plant dies after flower production.

Garlic Mustard is relatively new to the PNW. It was discovered in 1999 in the Seattle area of King County, with the majority of infestations in Seattle parks, Woodland Park Zoo, and adjacent residential neighborhoods. It was reported from the Vancouver, WA, area in 2003, where it was found in mitigated wetlands throughout a creek drainage. Garlic Mustard is currently reported from the Portland, OR, area and in several locations from BC, including Victoria and Vancouver.

Impact on Communities and Native Species

Garlic Mustard is considered one of the fastest-spreading invasive plants in woodland habitats of North America. It germinates and begins to grow in early spring while native species are still dormant. Mature stands can produce more than 62,000 seeds/m^2 and quickly outcompete local flora, changing the structure of plant communities on the forest floor. Wildlife that depend on local plants are affected when they lose their food source of leaves, fruit, seeds, nectar, pollen, or roots, and when they lose plant material needed for nesting or cover. Garlic Mustard is allelopathic, producing phytotoxic chemicals and two root chemicals (sinigrin and glucotopaeolin). Sinigrin breaks down to AITC, and glucotopaeolin breaks down to BZITC. Both are known to inhibit growth of mycorrhizal fungi. In some areas, Garlic Mustard is considered a population sink for the rare butterfly species, *Pieris napa oleraceae*, and *P. virginiensis*. These are members of the family of butterflies to which the Cabbage White (*P. rapae*), a European immigrant disliked because their larva feed in vegetable gardens, also belongs. When these native white butterflies lay their eggs on Garlic Mustard, the eggs hatch, but the larvae die.

Control Methods and Management

How to control Garlic Mustard depends on quantity and location. For most of the PNW, where it is still absent or relatively new, prevention is the recommended control option. Familiarize yourself with the flower, the plant, and the habitat where it grows. Mature plants can be hand pulled in early spring, or the flowering stalk can be cut to only a few inches above the ground just before flower production. Remove the stems and flowers from the site. Monitor sites regularly and remove plants prior to seed set. Rosettes should not be hand pulled, since they tend to snap off at the roots and the plant will resprout and continue to grow. Herbicides are effective on the rosettes. Garlic Mustard is very widespread in the Midwest and northeast US, and their control options are dramatically different. Natural resource managers sometimes prescribe fire as a control method. Biological control research began in 1998, and research continues.

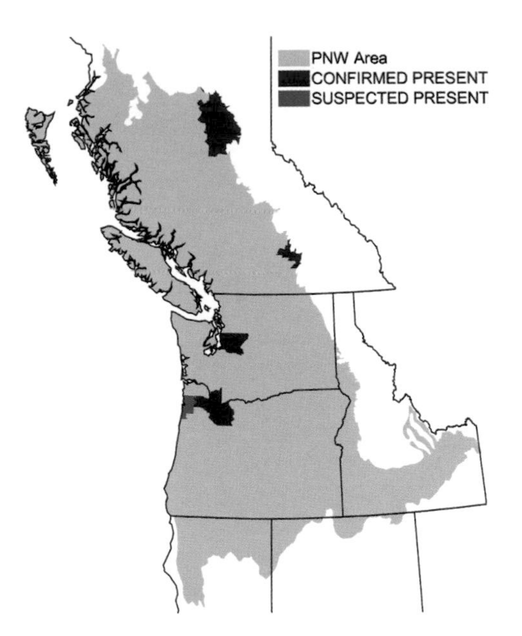

Life History and Species Overview

Garlic Mustard is native to Europe. This biennial herb is edible and was probably passed along among gardeners. However, it does not stay put in the garden, and its impacts in natural areas outweigh any horticultural benefit. Garlic Mustard thrives in the shade of forested parks and natural areas. Often first noticed along paths, trails, or near parking areas, it quickly spreads throughout wooded areas.

This non-native species is difficult to control once it reaches natural areas or parks: it is self-fertile, has a high seed-production rate, is short-lived, germinates in early spring while native plants are still dormant, and can quickly establish in a relatively stable forest understory. In areas of high disturbance, the population can increase from 214% (flood zone) to 1000% (canopy loss in forest windstorm). Average plants produce about 350 seeds, but robust plants can produce up to 8,000 seeds. Seed banks in heavily infested areas can range from an average of 9,533 seeds/m^2 to 107,580 seeds/m^2. North America has no known natural predators for Garlic Mustard. In Europe, 69 insects and 7 fungi are natural predators against Garlic Mustard.

History of Invasiveness

Garlic Mustard has spread to North Africa, India, New Zealand, Canada, and the US. It was first collected in the US from Long I., NY, in 1868 and may have been brought over for food or for medicinal use. The largest North American populations are in New England and the Midwest. Herbarium collections from western states indicate sporadic populations, and early collections of Garlic Mustard are recorded from ID (1892) and Portland, OR (1959).

Garlic Mustard reproduces by seed. It is shade tolerant, and germination can occur in light or under the dark forest canopy. The seeds often fall close to the parent plant and are moved along by pedestrians, animals, and vehicles. The plants can cross-pollinate or self-pollinate, producing individual plants that are genetically similar and interfertile. Most populations self-pollinate before the flowers open and before the stigma is exposed, implying that one plant can infest an area. Because of the biology of this non-native, invasive species, and because of the lack of any biological or abiotic controls (i.e., weather) to keep it in check, once established, Garlic Mustard often becomes a permanent part of the forest ground cover.

Other Sources of Information: 49, 92, 120, 125
References: 116, 236
Author: Bridget Simon

Giant Hogweed *Heracleum mantegazzianum*

Fashionable country gentleman had some
cultivated wild gardens,
In which they innocently planted the Giant
Hogweed throughout the land.
Botanical creature stirs, seeking revenge.
Royal beast did not forget.
Soon they escaped, spreading their seed,
preparing for an onslaught,
threatening the human race.

Genesis, "The Return of the Giant Hogweed," from the album *Nursery Cryme*, 1971.

Species Description and Current Range

Few invasive plant species are the subjects of popular rock songs; however, the danger Giant Hogweed presents to human health and its enormous size and dense populations in the UK inspired the band Genesis. The sap of Giant Hogweed contains a glucoside that makes the skin vulnerable to severe sunburn. When skin or eyes are exposed to its sap and to sunshine, serious burns can result, leading to temporary or permanent scarring and blindness.

A member of the carrot family (Apiaceae), Giant Hogweed is spectacularly large, able to grow more than 4.5 m tall, with 1-m compound leaves and 0.75-m wide inflorescences. The flat-topped umbel inflorescence is composed of many small white flowers. Its stems are large, purple, and hollow, and can make tempting play swords, peashooters, or musical toys for children and fun-loving adults—however, avoid this plant. Giant Hogweed resembles native Cow Parsnip (*Heracleum lanatum*), but is much larger. It prefers moist areas with partial sunlight and will grow in riparian and seminatural areas and in forests. Populations of Giant Hogweed have been reported from Seattle, WA, and Portland, OR, as well as Vancouver I., BC. It is likely that unreported populations occur outside these areas. Although Giant Hogweed populations are currently small, they are steadily growing. Without control, this species has the potential to become a serious problem.

Impact on Communities and Native Species

In continental Europe and the UK, Giant Hogweed forms dense patches that outcompete native species. Furthermore, when the patches die back in winter, the exposed soil is subject to erosion. Similar problems are likely in the PNW if Giant Hogweed populations increase in size and density.

Control Methods and Management

Extreme caution must be used when dealing with this species. Contact with the sap can lead to painful burns, permanent scarring, and blindness. Wear protective clothing, including eye shields. If skin contact occurs, thoroughly wash the exposed area with cold water, avoid sunlight, and seek medical advice.

In the PNW, Giant Hogweed populations are generally small and can be affordably controlled. Giant Hogweed can be dug out mechanically, if the rootstock is removed. If this is done, remember to wear protective clothing. Mowing stimulates growth, but if done consistently may eventually kill the plant by starving the rootstock. Biocontrol methods are not being used at this time; however, cattle and pigs seem unaffected by Giant Hogweed and their feeding and trampling can kill the plants. Chemically, glyphosphate is effective. However, it is non-selective and may kill all plants it contacts. Giant Hogweed may regrow from seed or roots after treatment; thus, treated areas should be monitored for several years and any new growth killed.

Life History and Species Overview

Originally from the Caucasus Mountains and southwestern Asia, Giant Hogweed has spread to North America, continental Europe, and the UK. Giant Hogweed prefers moist areas with partial sunlight. In the PNW, it is most common in disturbed urban areas, though it can also be found along roadsides and riparian corridors.

Giant Hogweed is a perennial plant that takes up to 4 years to flower. Plants die back to the base in early fall, but the tall, dried stem generally remains standing. Plants sprout from seeds or the rootstock in early spring and flower mid-May to July. They produce abundant, long-lived seeds that are dispersed by water and the transport of soil; the species can also reproduce vegetatively. The combination of long-lived seeds, persistent rootstock, and vegetative reproduction has contributed to this species' success as an invader.

History of Invasiveness

Because of its impressively large size, Giant Hogweed was imported around the world for horticultural purposes and has become a problem in continental Europe, the UK, and North America. Giant Hogweed fruits are also used to make the Persian spice golpar, another reason for importation. Giant Hogweed is currently on the US Federal Noxious Weed list, and all imports to the US are banned.

Without monitoring and control, Giant Hogweed has the potential to spread rapidly. Seeds can survive for 7 years and are passively dispersed through water and soil movement. In spite of efforts to control the species in the PNW, its numbers increase every year.

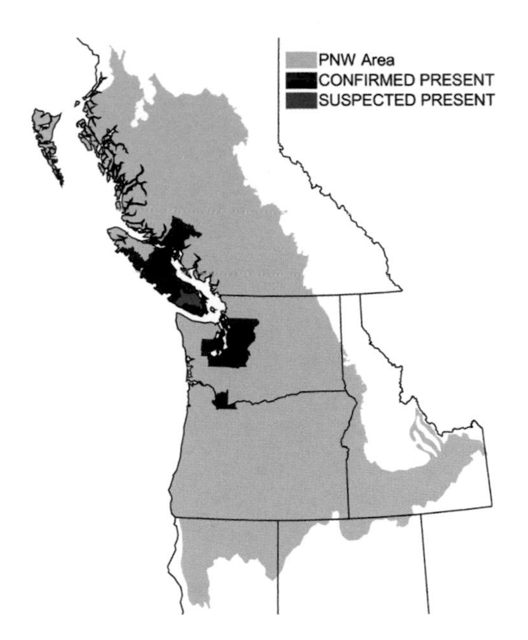

Other Sources of Information: 49, 125
References: 36, 118
Author: Tara S. Fletcher

Hedgehog Dogtail *Cynosurus echinatus*

Species Description and Current Range

Hedgehog Dogtail is a widespread annual grass of dry and often disturbed habitats. It forms a terminal, one-sided inflorescence that is 1–4 cm long, typically ovoid, and easily identifiable at a distance. Hedgehog Dogtail is distinguished by its paired spikelets borne opposite each other; one spikelet is fertile and few-flowered, and the other sterile and many-flowered. Awns on the lemmas and glumes of each extend up to 1 cm and are responsible for the inflorescence's bristly appearance. Hedgehog Dogtail is variable in height depending on local conditions but is usually within the range of 20–50 cm. The generic and common names are thought to be from the Greek *kynos* (of a dog), and *oura* (tail).

Hedgehog Dogtail is established in North America, Europe, Eurasia, South America, South Africa, and Australia. In the PNW, Hedgehog Dogtail is a frequent component of grassland, shrubland, and woodland habitats, especially those that are subject to grazing. In addition, it is common to dominant in many disturbed habitats, such as roadsides, clearings, and developments. Hedgehog Dogtail is most prevalent in western OR, where it is common to dominant in most Willamette Valley prairie and oak woodlands. In North America as a whole, Hedgehog Dogtail's distribution is currently limited to 22 US states and one Canadian province (BC), but it is thought to be expanding.

Impact on Communities and Native Species

Hedgehog Dogtail is one of a suite of annual grasses that have fundamentally altered the nature and composition of grassland and woodland communities over large areas of western North America. These changes occurred so rapidly and extensively that the original nature of some of these habitats (e.g., northern CA grasslands) remains unknown. However, it is clear that the effects of invasions by Hedgehog Dogtail and other annual grasses include a dramatic reduction in native plant diversity, changes to nutrient cycling and water availability regimes, increased soil erosion, and impacts to rare plant species. Hedgehog Dogtail and other annual grasses have been linked to the decline of Blue Oak (*Quercus douglasii*) woodlands in CA, where the establishment and survival of young oaks is increasingly uncommon. In BC, annual grasses including Hedgehog Dogtail have been cited as threats to the rare prairie species White-top Aster (*Aster curtus*) and Golden Paintbrush (*Castilleja levisecta*). In Australia and Tasmania, a fungal pathogen infesting Hedgehog Dogtail is suspected in outbreaks of acute bovine liver disease, a potentially fatal illness in cattle.

Control Methods and Management

Control methods for large infestations of Hedgehog Dogtail and other widespread invasive annual grasses are not well established. As a result, prevention and early eradication are the preferred management methods. However, repeated use of either fire or mowing should be an effective control if performed early and over multiple years, because this treatment prevents formation of a persistent seed bank. Recent tests of prescribed burns in OR prairies did not bring about substantial reductions in Hedgehog Dogtail cover, but data analysis for these efforts is not yet complete. Research in CA suggests that some native plants can be strongly competitive with annual grasses in the absence of other disturbances such as grazing. As a result, restoration of some invaded areas may be possible using repeated applications of native seeds, even without the use of active control methods.

Many herbicides will provide short-term control of Hedgehog Dogtail, and efforts to control other annual grasses (e.g., Cheatgrass [*Bromus tectorum*]) with a combination of herbicides and seeding with native plants have been at least partially successful. However, herbicide resistance has been reported in Hedgehog Dogtail infestations of agricultural land in Chile. No biological agents are known or thought to have the potential to control the species.

Life History and Species Overview

Hedgehog Dogtail is an annual member of the grass family (Poaceae), one of the largest and most economically important plant families. Hedgehog Dogtail seeds typically germinate in fall, overwintering as seedlings. The plants mature and begin flowering in spring of the following year, dispersing their seeds and dying off by midsummer. Since mortality rates for individual seeds are quite high (>80% in one test), and the seeds lack a dormancy mechanism (100% of seeds germinate within 144 hours under controlled conditions), no seed bank is formed. Hedgehog Dogtail can grow in very thin and rocky soils, but is thought to be particularly competitive in areas with high soil nitrogen (such as those already invaded by Scotch Broom [*Cytisus scoparius*], a nitrogen-fixing shrub). No medicinal or traditional uses are known.

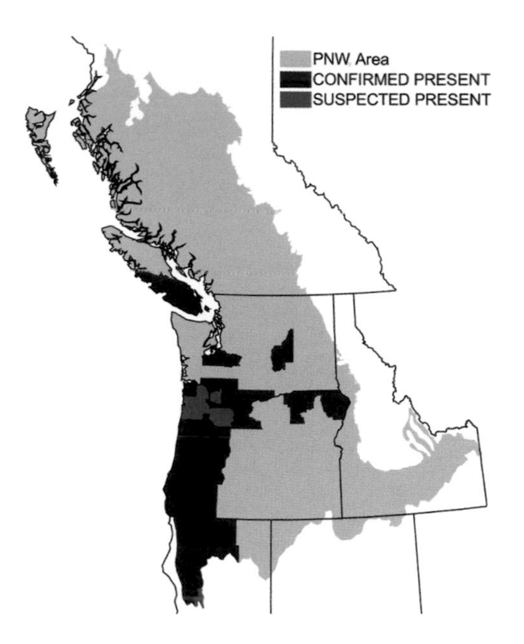

History of Invasiveness

Hedgehog Dogtail is thought to be native to southern Europe or Eurasia, but details of its introduction to North America are unknown. However, long-term data from experimental oak woodlands in Mendocino County, CA, provide a window into its invasiveness. Hedgehog Dogtail is a dominant species in parts of these woodlands, but as recently as 1983 its presence was only occasional or sparse, much as is the case in parts of BC today. (It is unclear whether such expansion by Hedgehog Dogtail represents a loss of native or exotic plant cover.) Despite this history of invasion, Hedgehog Dogtail is currently not listed as a noxious weed in the US or Canada.

Other Sources of Information: 10, 28, 38, 47, 88

References: 60, 159, 262, 285

Author: Devin R. Malkin

Herb Robert *Geranium robertianum*

Species Description and Current Range

Herb Robert belongs to a group of plants known as the hardy or wild geraniums. Wild geraniums are different from the geraniums typically sold at nurseries, which are actually in the genus *Pelargonium*. Wild geraniums tend to be disease resistant and have few pests, traits that make them popular horticulturally. Herb Robert is known historically as a medicinal plant, used in teas and poultices to cure everything from conjunctivitis and toothaches to "afflictions of the kidney," but it is not used much today. Herb Robert is also known as "Stinky Bob," and any question about its identification can be resolved easily by noticing its strong, somewhat unpleasant smell.

Herb Robert is a herbaceous plant with two main forms—a low-growing rosette, and a multibranched reproductive form that can reach 60 cm in height. The leaves, on long petioles, branch out from each node. They are deeply divided to the base into 3–5 lobes, which are also heavily divided. Leaf color ranges from bright to dark green, with the amount of red present in the leaves and stems increasing with more sunlight. Fine white glandular hairs cover the entire plant. The flowers, commonly paired, also arise from the nodes. The flowers have five petals, average 15 mm across, and are typically pink or pink with white stripes. The fruit is a long, tapering, beaklike capsule, which gives the genus *Geranium* its common name of Cranesbill, derived from the Greek word *geranion*, for "crane." The seeds are brown, appear wrinkled, and are 2–3 mm long. The rosette stage of the plant is low growing, with multiple leaves with long petioles arising from a short stem with no further branching and no flowering. Young Herb Robert can be recognized by its kidney-shaped seed leaves and hairy stems. Herb Robert is found primarily west of the Cascades from central BC through central OR. Its habitat ranges from moist,

shady forests and riparian corridors to somewhat drier, rocky areas. It is seen frequently in city parks and can also be found growing in such extreme environments as roof gutters and along cracks in the sidewalk.

Impact on Communities and Native Species

Where Herb Robert occurs it seems to become the competitive dominant species and may be causing a decline in native herbaceous understory species in PNW forests. The mechanisms behind this success are not known, although competition for light and allelopathy are likely. Once Herb Robert arrives at a new area its population rapidly expands, forming a dense covering and reducing space for native plants to grow. Herb Robert's dense growth may reduce the number of native seeds reaching the soil or shade them out. Herb Robert also has the ability to overwinter as a rosette, giving it a head start in growth over many annuals and perennials in the spring.

Control Methods and Management

Removal of Herb Robert before it sets seed is critical. The plant is weakly rooted, so hand pulling is easy. Existing populations can be pulled, mowed, or cut with a string trimmer. Controlling the seed bank is more difficult, requiring long-term, ongoing maintenance to prevent new seedlings from reaching maturity. Areas that contain seeds can be covered with a thick mulch, which may help to suppress germination, although the mulch will have to be reapplied as it breaks down. Seedlings should be pulled as they emerge. Seeds germinate from early spring to late fall and can remain viable in the soil for at least 6 years.

Life History and Species Overview

Herb Robert is native to Europe, the British Isles, North Africa, and southwest Asia. Its small pink flowers are attractive and it is a common plant in the woods and countryside. It is also present in northeastern North America, but it is uncertain whether it is native to that area or was introduced by early colonists. Herb Robert is tolerant of a broad habitat range. It grows well in moist, shady areas, but it can also be successful in drier habitats where it receives full sunlight. Herb Robert is variable in form, ranging from low, compact, or sprawling plants to taller, more upright plants. It has been described as an annual, winter annual, biennial, or short-lived perennial. In the PNW, it appears to be mainly a biennial. The majority of germination occurs in the spring and the fall. The seeds germinate well under a variety of conditions, with a higher germination rate after the first year. Most plants overwinter as rosettes, form the adult reproductive structure the following spring, and die by late fall.

Flowers are produced from mid-spring through late fall. In milder climates flowers can been seen even into the winter. The flowers of Herb Robert display a strong ultraviolet signature that makes them attractive to pollinators, which are primarily flies and bumblebees. Herb Robert is not deterred by an absence of pollinators because it can successfully set fruit from self-pollination over 90% of the time. Five seeds are produced in each capsule. These ripen within 3 weeks and are dispersed explosively when the capsule splits open and the seeds are ejected along a pointed cylinder for up to 6 m—nature's version of a missile launcher. The seeds have a sticky fiber at one end that allows them to attach to adjacent foliage or anything else they contact. A second dispersal event occurs when the seeds attach to a mobile object, such as an animal or a car tire, providing an opportunity for the seed to spread into new, distant areas. Herb Robert is not very attractive to large herbivores and has few pests even within its native range, although leaf miners and aphids can cause some damage.

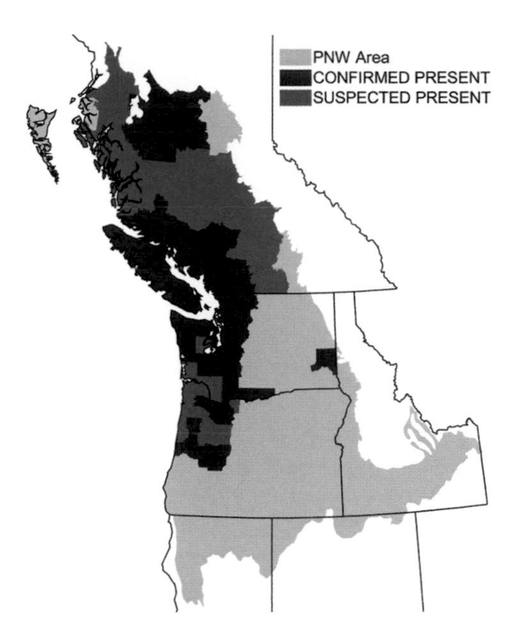

History of Invasiveness

The earliest PNW record of Herb Robert was in 1911 in Klickitat County, WA. Herb Robert was sold as a medicinal as well as an ornamental plant and probably escaped from a garden. Large populations of the plant have been documented in the Olympic Mountains and the foothills of the Cascades in King County, WA. These populations appear to have increased considerably over the last two decades. Currently, Herb Robert is primarily spread by human activity along trails and roads. Its sticky seeds hitch a ride on hiking shoes, camping equipment, car and bicycle tires, or any other moving object. Once transported to the new area, Herb Robert, with its prolific seed set, long lasting seed bank, and widely dispersing seeds, is able to establish new populations and spread rapidly.

Other Sources of Information: 125
References: 254, 352
Author: J. Katie Barndt

Houndstongue *Cynoglossum officinale*

Species Description and Current Range

Houndstongue is an herbaceous biennial or short-lived perennial with two growth stages. The low-lying, nonreproductive rosette stage has large, somewhat hairy, dark green leaves. The leaves are oblong and lance-shaped, resembling a dog's tongue, hence its common name. The reproductive, bolting stage produces single or multiple erect flowering stems (30–120 cm or 1–4 feet) with terminal flowers. Each dull red-purple flower produces up to four flat, hard, teardrop-shaped seeds (nutlets) that are covered in hooked barbs. As nutlets mature, the barbs dry and attach easily to passersby, functioning much like velcro. Large infestations of Houndstongue occur in forest areas cleared for logging or cattle grazing, disturbed pastures, roads, abandoned cropland, and construction areas such as mining operations. In the PNW, Houndstongue is in almost all inland habitats, particularly in dry forests and on rangelands.

Impact on Communities and Native Species

Houndstongue can overgrow many plants because it begins growing early in the spring, allowing it to successfully compete with forage and native species for moisture, light, and nutrients. The large rosette leaves of Houndstongue shade other plants. Houndstongue invades primarily disturbed habitats and is therefore rarely a threat to established native plant communities. However, minor disturbances, such as burrowing by rodents, may provide sites for Houndstongue to germinate and spread to other disturbed areas. Houndstongue produces toxic chemicals, pyrrolizidine alkaloids, that damage the livers of mammals that ingest it. Although deaths of wildlife are not known, Houndstongue reportedly kills cattle and horses. When Houndstongue is green, most ungulates avoid feeding on the foliage because of its pungent odor and unpalatability. However, when dried and a contaminant of hay, Houndstongue is inadvertently consumed. The nutlets attach to passing livestock, wildlife, clothing, or pets and can be irritating when individuals are heavily covered.

Control Methods and Management

A broadleaf phenoxy and sulfonylurea herbicide such as 2,4-D and metsulfuron is an effective control of Houndstongue. Often, the use of herbicides is not feasible because Houndstongue typically grows in rough, inaccessible terrain, making herbicide use costly and impractical. Small infestations of Houndstongue can be controlled by hand pulling rosettes, as long as most of the taproot is removed to prevent regeneration. Houndstongue reproduces only by seed and mowing bolting plants reduces seed production. Stems with seeds should be cut and removed from the site, as even unripe seeds are capable of germinating. Burn cut plant material to prevent seed germination and dispersal. Newly disturbed areas, even ones created by hand pulling weeds, should be seeded immediately with competitive native perennial species. To avoid further seed spread, seeds should be removed from clothing and pets before moving to a new location. A biological control program, initiated in 1988, resulted in the release of two insect biocontrol agents in Canada, but because of concerns about the potential effects the insects may have on native species, release in the US is unlikely. A beetle native to Europe that feeds on young seeds might prove to be an effective control of Hounds-tongue, provided it doesn't damage native species.

Life History and Species Overview

Houndstongue is a member of the Boraginaceae family. Seeds germinate in spring and form a rosette for 1–2 years until the taproot reaches approximately 2 g of dry biomass. The root overwinters and the aboveground leaves die back before the plant produces flowering stems between May and July. Nutlets mature in July and August.

Houndstongue is native to Europe and Asia Minor and grows in small isolated populations primarily along roadsides and sunny open woodlands. It grows in coarse, gravelly to sandy soils and is not well suited to dry sites (those that receive less than 30 cm of precipitation a year). Houndstongue population sizes are highly variable but plants persist for decades.

Historically, Houndstongue has a number of medicinal and household uses including as a bactericide and a treatment for skin problems such as eczema, acne, burns, and corns. Leaves and roots of Houndstongue repel rodents from gardens and stored produce.

History of Invasiveness

In the late 1800s, cereals contaminated with Houndstongue seeds were brought into North America and plants were first reported in Montreal, Canada. Houndstongue has since spread throughout most of North America and is becoming increasingly prevalent in the northwestern US and southwestern Canada. It is declared noxious in NV, OR, WA, and BC (also in MT, CO, AB, WY).

The dispersal of Houndstongue seeds by attaching to hair, wool, feathers, clothing, pets, wildlife, livestock, and vehicles enables this weed to spread quickly. Picking nutlets out of socks or dog fur is a nuisance and helps disperse the plant. Disturbance of the terrain through human activity continues to provide sites for invasion.

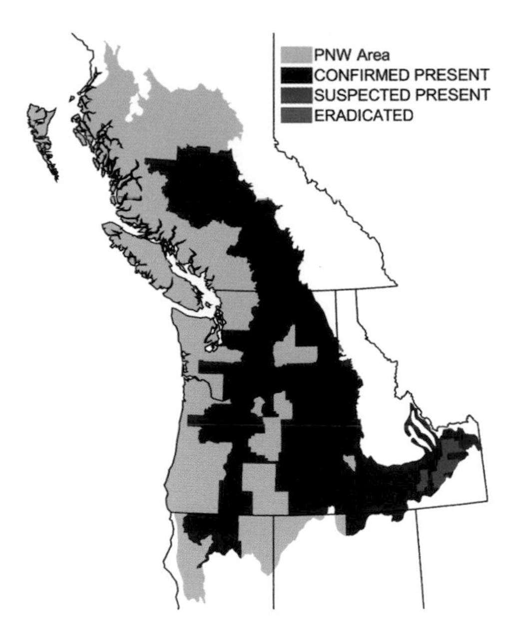

Acknowledgments: Thanks to Linda Wilson and Mark Schwarzlaender for reviewing the text.

Other Sources of Information: 12, 25, 60, 89, 93

References: 85, 86, 183, 314, 315

Author: Jennifer E. Andreas

Diffuse Knapweed *Centaurea diffusa*
Meadow Knapweed *Centaurea pratensis*
Russian Knapweed *Acroptilon repens*

Species Description and Current Range

Knapweeds are plants in the *Centaurea* and *Acroptilon* genus in the Asteraceae (sunflower) family and include many species of invasive, noxious weeds infesting millions of acres in the PNW. Russian Knapweed, Diffuse Knapweed, and Meadow Knapweed are three of the most problematic species and are listed as noxious by the USDA, as well as by most states and provinces in the PNW. All species invade primarily grasslands, pastures, woodlands, and dry forests. Diffuse Knapweed is found in dry grasslands, pastures, and shrublands; Meadow Knapweed prefers moist sites mostly along the Cascades. Russian Knapweed invades mainly disturbed areas such as roadsides, ditches, pastures, clear-cuts, and croplands in semiarid areas throughout the PNW. Knapweeds are aggressive invaders that reduce grazing capacity, crop yields, and diversity and abundance of native species and increase soil erosion.

Knapweeds grow 0.3–1 m tall and have white, pink, or purple flowers resembling small thistles growing at the ends of clustered branches. Knapweed species can be identified by the bracts (the small scales at the base of the flowers). Diffuse Knapweed has a terminal spine on the floral bract. Bracts on Meadow Knapweed have a fringed edge. Russian Knapweed bracts are papery at the tip. Meadow and Diffuse Knapweeds grow individually from basal rosettes and reproduce solely by seed. Diffuse Knapweed has highly divided basal leaves, while Meadow Knapweed has undivided leaves. In addition to spreading by seed, Russian Knapweed can spread via well-developed root systems, which can reach more than 7 m, forming very dense colonies.

Centaurea diffusa

Impact on Communities and Native Species

Knapweeds can quickly invade disturbed plant communities but may also impact relatively pristine sites. Once these plants have a foothold, they are superior competitors against most other plant species and often form dense infestations. Diffuse Knapweed can suppress other vegetation by competing for soil water. Both Diffuse and Russian Knapweeds secrete toxic chemicals from their roots (allelopathy), which reduces the growth of many native and economically important plants. All knapweeds cause significant reductions in desirable pasture grasses, crops, and native plant species. They cost tens of millions of dollars annually in the PNW in reduced forage, crop losses, and costs of control. The spines of Diffuse Knapweed can damage the mouths and digestive tracts of livestock. Russian Knapweed is toxic and can cause severe neurological damage to horses.

Control Methods and Management

The most effective control is locating new infestations quickly and eliminating them before they become well established. A major goal of control efforts is to limit transport of seed (mainly by contaminated hay, vehicles, and livestock) and to educate people to identify and monitor areas where invasions are likely. Currently, the main method for control of knapweeds is herbicide (typically picloram), although timing and method of application are key. Biological control programs are so far ineffective. Nine biological control agents were introduced in the US for Diffuse Knapweed. So far, the most promising is *Larinus obtusus*, a weevil that eats seeds as a larva and eats leaves and shoots as an adult; it has had some success in eastern WA. Two biocontrol species were introduced for Russian Knapweed, with minimal impacts to date. Fire can eliminate adult Diffuse and Meadow Knapweeds, but must be combined with post-burn reseeding and/or herbicide to prevent reestablishment. Fire is likely to be ineffective for Russian Knapweed, as plants can resprout from the root system. Smaller infestations can be controlled by pulling out adult plants, although this is especially difficult for Russian Knapweed, as the entire extensive root system must be removed. Mowing is ineffective; disking or plowing can work when done carefully and combined with reseeding. Russian Knapweed is the most difficult to control; combinations of herbicide, mechanical control, and revegetation are required. Gloves and protective clothing should be worn when working in knapweed, as the plants are rough and can cause itchy rashes in some people.

Life History and Species Overview

Knapweeds have a variety of life history strategies. Diffuse Knapweed grows as a low rosette for one to several years (depending on environmental conditions), then flowers and dies. Meadow Knapweed and Russian Knapweed are perennials. Meadow Knapweed is a fertile hybrid of Black and Brown Knapweeds. Russian Knapweed can rapidly spread horizontally from underground root systems.

Diffuse and Meadow Knapweeds spread only by seed. A single knapweed plant produces hundreds, sometimes thousands, of seeds. Some species (like Diffuse Knapweed) have seeds that can survive in the soil for many years. Diffuse Knapweed is attacked by gallflies (discussed in Spotted Knapweed, p. 62), which can reduce seed production. Meadow and Russian Knapweed seeds naturally disperse only a few meters but can be carried long distances by vehicles, waterways, livestock, and wildlife. All knapweeds are frequently transported in contaminated hay and hayseed. The majority of Diffuse Knapweed seeds remain in the seed head. When the plant dies, the stalk detaches from the root crown and becomes a tumbleweed, which can travel long distances, gradually dispersing its seeds. In some areas masses of tumbleweeds pile up, often causing substantial damage to fences.

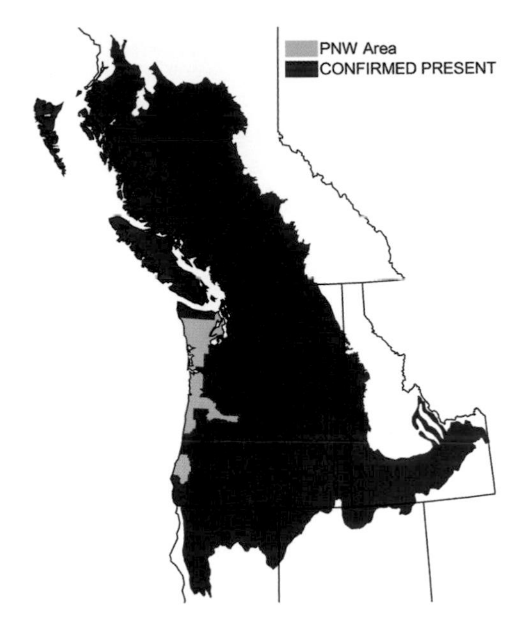

History of Invasiveness

Most knapweed species are native to southern and eastern Europe, southern Russia, and Turkey. Knapweeds are often rare or patchily distributed in the native range. Many knapweed species were introduced to the US and BC in contaminated hay and alfalfa seed around the turn of the century. Knapweeds had limited distribution for a number of years and then seemed to rapidly expand. They now cover millions of acres in the PNW, mostly in semiarid areas like eastern OR and WA and eastern BC.

Other Sources of Information: 45, 91, 93, 125, 126
References: 266, 363
Authors: Amanda G. Stanley and Francis S. Brown

Spotted Knapweed *Centaurea maculosa*

Species Description and Current Range

Spotted Knapweed is one of the worst rangeland weeds in North America. The plant has a long woody taproot and grayish-green, highly dissected leaves growing from the root crown in a rosette form. Flowering plants have one to many stems, with pink to purple flowers on the ends of clustered branches, and are 0.3–1 meter tall. Flowers resemble small thistles and can produce up to 30 seeds each. Spotted Knapweed can be distinguished from other knapweeds by the black spots on the tips of the bracts (the small scales at the base of the flower). It is listed as a noxious weed federally and by many states, including WA, OR, and ID, and by BC. It is found throughout the PNW. Spotted Knapweed infests grasslands, woodlands, dry montane forests, mixed-conifer forests, rural disturbed areas, and rural/managed pastures.

Impact on Communities and Native Species

Spotted Knapweed easily invades disturbed sites, such as roadsides and overgrazed pastures. However, it also invades pristine areas. Once established, Spotted Knapweed tends to form a dense monoculture (50–400 plants/m^2), reducing biodiversity and forage for livestock and native ungulates. Spotted Knapweed is unpalatable to most grazers, except in very young stages. Heavily invaded sites have increased water runoff and decreased water retention capacity. Spotted Knapweed costs millions of dollars each year in reduced forage and costs of control.

Control Methods and Management

Spotted Knapweed can be controlled by herbicides, careful grazing regimes, or mechanical means. While a dozen species of insects have been released for biological control of knapweeds, so far none has proved effective. Most promising is *Cyphocleones achates*, a root-boring weevil that may kill adult plants. However, adult *Cyphocleones* are flightless, and disperse only a few meters. It will take many years for *Cyphocleones* populations to grow and spread. Most land managers currently control knapweeds with herbicide (typically picloram). Small infestations can be controlled by pulling plants in summer after flowering begins but before seeds are set. Pulling must be repeated for several years. Mowing and tilling are usually ineffective. Fire is ineffective unless combined with reseeding and targeted post-burn herbicide. Sheep can eat young knapweed plants in spring, before plants begin to bolt. High-intensity sheep grazing in early spring is currently being tested as a method of Spotted Knapweed control.

Two species of seed-head gallflies were released as biological control agents in the 1970s, *Urophora quadrifasciata* and *U. affinis*. These small flies lay eggs in developing knapweed buds. The eggs hatch into larvae that induce the plant to form a gall; the larvae feed on the gall tissue. By taking up space in the seed head and consuming plant resources, gallfly larvae reduce seed production. Each seed head can contain up to 16 larvae, although the average is usually 2–4. Both species of gallfly have become abundant and widespread but have proved an ineffective control as knapweed is so prolific; even with relatively high rates of gallfly infestation, knapweed can produce 2,000–10,000 seeds/m^2. Gallfly larvae overwinter in the seed head. Many native species eat the gallfly larvae during the winter, including chipmunks, White-tailed Deer, Mule Deer, chickadees, and Deer Mice. In most areas, Deer Mice appear to be the dominant predator, consuming about 40% of the larvae over the winter. Mouse populations are twice as high in areas infested with Spotted Knapweed as compared to those with native vegetation. Mice are the main reservoir for hantavirus, which can cause severe pulmonary disease in humans. New evidence shows that hantavirus prevalence is three times as high in mouse populations living in knapweed infestations compared to mice inhabiting native vegetation. In other words, large Spotted Knapweed populations support large populations of gallfly larvae, which in turn provide plentiful food for Deer Mice. Mouse densities have increased in response to the abundant food, and these mice are more likely to be carrying hantavirus.

Life History and Species Overview

Spotted Knapweed seedlings germinate in spring and fall, depending on rainfall conditions. Seedling survivorship is better in wetter summers. A seedling eventually grows adult leaves and becomes a rosette. Rosettes bolt in spring, and plants begin bud development in early summer, flower in late summer, and set seed in late summer or early fall. Each plant can produce 50 to thousands of seeds depending on plant size, soil moisture, and abundance of biocontrol insects. Spotted Knapweed reproduces only by seed, and some seeds can survive in the soil for up to 9 years. Plants are short-lived perennials and may live up to 9 years. After seed set, stems dry out and the plant dies back to the root crown. These stems remain upright on the plant for 1–2 years. New rosettes and stems can grow from the root crown.

Spotted Knapweed can secrete toxins from its roots (a process known as allelopathy) that reduce the growth or even kill many

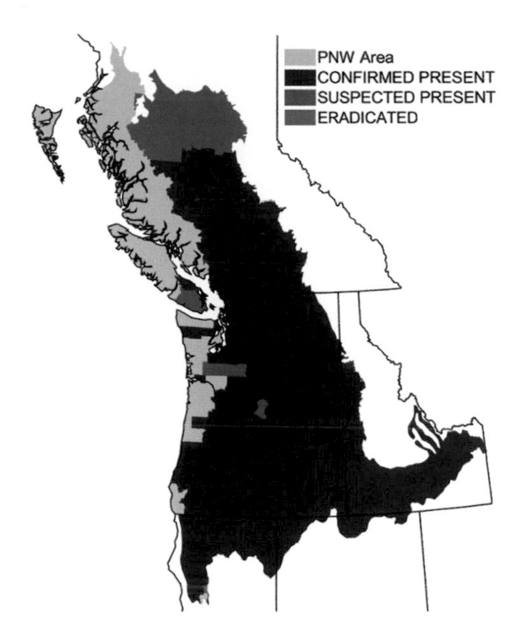

species of native plants, giving it a strong competitive advantage. Spotted Knapweed may also be such a strong competitor because of its associations with soil fungi called mycorrhizae. Mycorrhizae colonize plant roots and form symbiotic associations with host plants, whereby the plant provides sugars to the fungi, and the fungi provides nutrients to the plant. Knapweed may even be able to take nutrients from native species via shared mycorrhizal networks.

History of Invasiveness

Spotted Knapweed likely came from eastern Europe and invaded North America via contaminated hay seed at the turn of the 20th century. It first appeared in BC and the San Juan Is., WA. Seeds are transported in contaminated hay or feed, by vehicles, livestock, wildlife, or people. By 1980 it had spread to 48 counties in the PNW, reaching 326 counties by 1998. It is now present in 45 of the 50 states and in AB and BC, Canada. As of 2000, Spotted Knapweed infested an estimated 7.5 million acres.

Other Sources of Information: 45, 93, 125, 126
References: 49, 243, 363
Author: Amanda G. Stanley

Bohemian Knotweed *Polygonum bohemicum*
Giant Knotweed *Polygonum sachlinense*
Japanese Knotweed *Polygonum cuspidatum*

Species Description and Current Range

Bohemian, Giant, and Japanese Knotweeds are closely related congeners invading riparian, roadside, parkland, and other areas throughout the PNW, the eastern US, and Europe. Referred to by their genus name *Polygonum* in the US and Japan and known as Fallopia in the UK and Europe, until recently all three species were commonly called Japanese Knotweed. However, Giant and Bohemian Knotweeds (the Japanese-Giant hybrid) are invaders with similar appearances, habitats, and ecological impacts. All three species are tall, rhizomatous perennials that form large single-species stands. Japanese Knotweed often grows to a height of 2–3 m, Bohemian Knotweed to 3–4 m, and Giant Knotweed to 3–5 m. Their rhizomes appear "knotty," with a dark brown exterior and bright orange interior. Their aerial stems have a woody, hollow, bamboo-like appearance and are pale green, often with purple speckles. After fall senescence, the dead stems turn reddish-brown and can remain standing for 2 to 3 years. They have arching branches and ovate-shaped leaves with a flat (Japanese), rounded (Giant), or intermediary (Bohemian) base. Knotweeds form clusters of small white or cream-colored flowers in spring that produce large amounts of seed in late summer.

The extensive rhizome systems of Japanese, Giant, and Bohemian Knotweeds make these species very difficult to control. Rhizomes can reach a depth of 2 m and extend up to 7 m from the parent plant. Knotweeds can regenerate from a piece of rhizome as small as 7 g (about the size of a fingernail). The rhizomes and stems are so strong that knotweeds can penetrate concrete sidewalks, driveways, and house foundations. Knotweeds prefer sunny, moist habitats and grow in coastal forests, riparian areas, wetlands, and seminatural areas. In the PNW, these species have invaded many areas in WA and OR and are reported in southern BC, western ID, and northwestern CA.

Impact on Communities and Native Species

Knotweeds are able to quickly colonize large areas and outcompete native vegetation for light and soil resources. Knotweed foliage creates a dense canopy, shading out virtually all other plant species. The accumulation of knotweeds' persistent stem litter creates a thick mat over the ground, further restricting native plant establishment.

Knotweed is of particular concern in the PNW because it invades riparian areas. The regular occurrence of flooding disturbance creates the opportunity for knotweeds to

P. cuspidatum

invade. Once introduced, these species spread quickly throughout a riparian corridor as floodwaters pick up root and stem fragments and deposit them downstream. Knotweeds' displacement of a diverse riparian flora has negative consequences for wildlife and humans. Riparian trees and shrubs provide terrestrial wildlife habitat, erosion control, shade, large woody debris for rivers and streams, and litter inputs that are an important source of food and habitat for stream organisms. The tall, dense growth of knotweeds restricts public access along riverbanks for recreation and fishing. When the vegetative growth has died back in winter, the exposed bare soil may increase erosion, especially on steep riverbanks.

Control Methods and Management

Prevention of new infestations is the best method for control. It is important to remove knotweed populations as soon as they are seen colonizing an area and to take precautions not to spread soil contaminated with knotweed stems and rhizomes. It is extremely difficult to manually control knotweed once it has become established. Digging of established populations is not recommended because rhizome fragments will regenerate into new plants. Repeatedly cutting the stalks during the growing season has had mixed success for control. Cutting at least three times in combination with thick mulch or heavy weed fabric topped with >15 cm of mulch can weaken the plants' rhizome system. The success of repeated cutting is possible on small populations only with a long-term commitment of extensive labor. Repeated cutting of the stems and application of a systemic biodegradable herbicide such as glyphosate is a more effective means of control.

Life History and Species Overview

Japanese Knotweed can thrive in a wide variety of habitats. In Japan, it is commonly found in sunny places on hills and high mountains. It can grow in most soils and is a dominant pioneer on volcanic slopes. In their introduced range, knotweeds are primarily found in moist, unshaded habitats. Knotweeds are dioecious perennials reproducing both vegetatively and by seed and flowering from July to October in Japan and in August and September in the UK and US. They are pollinated by bees and other insects. Seeds appear several weeks after flowering and are wind dispersed. Reproduction is primarily through regeneration of stems and rhizomes. The role of sexual reproduction in knotweed dispersal in the PNW is unknown.

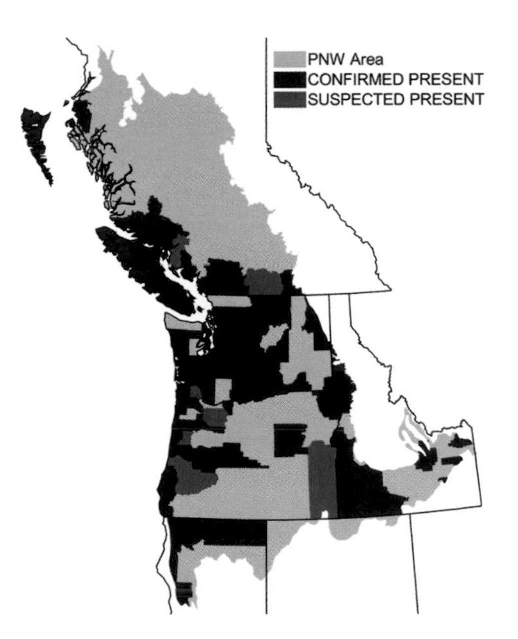

Japanese Knotweed was considered a folk medicine in eastern Asia. Chemicals isolated from Japanese Knotweed have been used as an antibiotic, as antioxidants and antimutagens in cancer research, to promote healing of burns, and to enhance the immune system and cardiac functions.

History of Invasiveness

Japanese and Giant Knotweeds are native to eastern Asia and have invaded Europe, New Zealand, Australia, Canada, and the US. They were introduced from Japan to the UK as an ornamental in 1825 and spread from there to North America in the late 19th century. In North America, they have spread widely along the East Coast and have been found as far north as NS, Canada, and south as far as NC in the US.

Other Sources of Information: 20, 46, 79, 89, 92
References: 19, 59
Author: Lauren S. Urgenson

Kudzu *Pueraria montana var. lobata*

Species Description and Current Range

Kudzu, a vigorous vine that produces dense mats and suspended canopies of rapidly growing foliage, blankets everything it encounters. Leaves have three dark-green leaflets, up to 18 cm long and 12 cm wide. Leaflet edges and undersides are hairy, and leaflet margins can either be rounded or have shallow lobed indentations. Young stems are hairy, and vines may grow 18 m in a single season. Kudzu produces long clusters of fragrant purple flowers that blossom from July to October. Flowers only form on upright vines in sunny areas, and hairy pods containing small, beanlike fruits are produced in late summer or early fall. The leaves are dropped and younger vines may die back after the first frost, while older vines retain their viability and resprout the following spring.

In the continental US, Kudzu occurs east of TX, OK, and NE, and south of MA and IL. Kudzu is one of the most problematic invasive plants in the southeastern US, where its rampant growth has earned it nicknames such as the "the-weed-that-ate-the-south" and "mile-a-minute vine." Kudzu thrives in many habitats, including forests, riparian areas, farmlands, and seminatural areas such as roadsides and abandoned fields. Several small infestations of Kudzu have been controlled in WA and OR.

Impact on Communities and Native Species

Kudzu kills native vegetation by climbing over it and blocking out sunlight. Once established, Kudzu patches can produce thousands of plants/ha because the vines are capable of growing roots at each node. The ability of Kudzu to fix nitrogen alters soil nutrient levels, benefiting other invasive species. Kudzu has detrimental effects on timber production in the southeastern US, because it grows over the trees, destroying hundreds of thousands of dollars' worth of forest products each year; infestations on farmlands are often so extensive and entrenched that affected agricultural lands are removed from productive use.

Control Methods and Management

The key to eradication of Kudzu is repeated defoliation, which eventually kills the plant by exhausting the deep, energy-rich root system. Although repeated mowing and grazing can eliminate Kudzu without the use of herbicides, obstacles concealed in deep tangles of vegetation may make mowing dangerous, and the construction of fencing for grazing requires a substantial investment of labor and money. Herbicides are often used to control Kudzu after cutting or burning above-ground vegetation. The choice of herbicide and application method depends on the age of the Kudzu patch, the soil type, and the surrounding vegetation. Small infestations may be treated by applying a broad-spectrum herbicide, such as glyphosate, to the cut surface of the larger vines near the root crown; this treatment works best if the herbicide is applied immediately after cutting the plant. Successful control requires all root crowns in an infestation to be treated, and repeated treatments are necessary for eradication. Treated infestations should be monitored for several years. If herbicides are used to control

Kudzu, care must be taken to ensure safe application practices and prevent damage to nearby plants; further, chemical control is costly, ranging from \$120–\$400/ha. Research into the use of fungi and insects as biological control agents is currently under way in the US and China.

Life History and Species Overview

Kudzu is a perennial legume, belonging to the pea family (Fabaceae). Seasonal growth begins in May and continues until the first frost. Hairy vines grow from starchy, fibrous roots that can extend more than 3 m into the soil and weigh over 100 kg. In addition to storing energy, in times of drought these roots help Kudzu take advantage of groundwater beyond the reach of many native plants. Most reproduction is vegetative, with new plants developing when runners growing from the root crowns contact the soil and form new roots. Flowering typically does not occur until the third year, and little reproduction is thought to occur by seed. Seed germination rates are low but may be increased by burning.

Kudzu prefers moist to well-drained soils, with a pH of 4.5–7, mild winters (5–15°C), warm summers (25°C or above), and plentiful rainfall (100+ cm annually). Preferred habitats are forest edges, abandoned fields, roadsides, and rights-of-way, but Kudzu can tolerate a variety of nonoptimal conditions, living in nutrient-deficient, sandy, and clayey soils. The species is classified as intermediate in shade tolerance. In temperate Asia, where Kudzu is native, it has been valued for centuries: the leaves prepared as vegetables, the roots serving as cooking starch and medicine, and the fibers used to make cloth. In the southeastern US, Kudzu is currently a forage crop, a craft and basket-making material, and an ingredient in jelly and quiche. It is also used in Asian cuisine, and researchers are investigating its effectiveness in treating alcoholism and in removing heavy metals from contaminated soils.

History of Invasiveness

Kudzu was brought to the US in 1876, as a gift from the Japanese at the Philadelphia Centennial Exposition. It was first used as

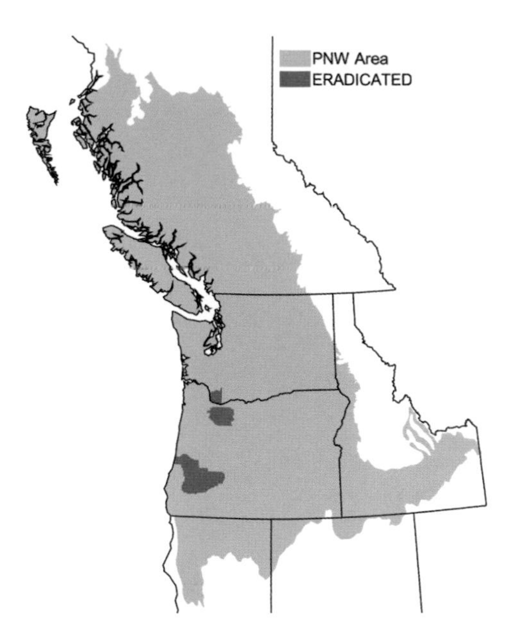

an ornamental plant for porches and arbors, and by the early 1900s, it was promoted as livestock forage through mail-order catalogs. During the Depression, farmers in AL, GA, and MS were paid nearly \$20/ha by the Soil Extension Service to plant Kudzu in tobacco and cotton fields to control erosion and replenish nutrients.

By the 1940s, Kudzu was known as the "savior of the South," and Kudzu clubs were formed to promote its benefits. However, by the 1950s Kudzu was viewed as a pest, and in 1997 it was declared a noxious weed by the federal government. Kudzu is currently estimated to cover 810,000 ha, mostly in AL, GA, and MS. New infestations typically occur through the movement of living plants in soil or plant materials. Although rates of local proliferation can be impressive, long-distance dispersal is not thought to be important in its spread.

Other Sources of Information: 24a, 26a, 45, 72, 85a
References: 107a, 109a, 109b, 141a, 153, 205a, 223b, 232a, 321a, 329a
Author: Nancy LaFleur

Leafy Spurge *Euphorbia esula*

Species Description and Current Range

The name Leafy Spurge comes originally from the Latin word *expurgare*, meaning "to purge," because of its medicinal use as a purgative. Leafy Spurge is a perennial plant with erect stems, smooth-edged bluish-green leaves, and heart-shaped clusters of yellow flowers. The stems can grow to a meter above ground, with a deep underground root system growing up to 7.9 m and laterally spanning 4.6 m or more. It is distinguished from similar-looking species by the milky latex found in all parts of the plant. Leafy Spurge is adapted to many conditions, including disturbed areas, fields, roadsides, woodlands, rangelands, prairies, wastelands, and neglected croplands, and can grow in all soil types. It is especially aggressive in semiarid zones and in areas with coarse-textured soils, and is found throughout the PNW and Canada. Within the PNW, it is most likely to be found in agricultural areas where grazing animals have thinned competing plants.

Impact on Communities and Native Species

Leafy Spurge owes its success as an invasive species to its multifaceted and persistent approach. It gains a competitive advantage by emerging in spring before other plants and is allelopathic, producing a chemical compound in its root system that inhibits other plants' growth. Diterpenes and terpenoids contained in its latex can cause blistering skin irritation and temporary blindness in humans and mouth ulcers and severe diarrhea in cattle. When cows graze on native grasses in lightly infested areas, the grasses recover more slowly because of Leafy Spurge's chemical offense. Cows will not graze in areas where Leafy Spurge comprises more than 1,000 kg/ha, or 10% of the total available plant material, drastically reducing rangeland productivity. In the PNW, ranches infested with Leafy Spurge have dropped in value up to 90%. If left uncontrolled, Leafy Spurge can become the dominant plant within a few years.

Control Methods and Management

Controlling Leafy Spurge is a challenge. Pulling, mowing, or burning the above-ground portion of the plant does little to prevent the spread of its stubborn root system, which holds substantial stores of nutrients. Likewise, the roots can bud even when tilled, so tilling of the root system is ineffective in halting its spread. Chemical controls include applications of pesticides such as picloram, 2,4-D, and dicamba in both spring and fall for several consecutive years.

Goats, and sheep to a lesser extent, are immune to Leafy Spurge's toxins in moderate amounts and have been used to control its above-ground spread. Sheep prefer other plants but will graze on Leafy Spurge if it is the dominant food. Goats, however, eat it readily without adverse effects because enzymes in their saliva and digestive systems allow them to digest the toxins. However, it can be toxic even to sheep and goats in high amounts. Grazing can control the spread of the plant, but generally, infestations are not reduced by grazing alone. Insects from

Europe, where the plant is native, including five species of flea beetles, the Black Dot Leafy Spurge Beetle, the Cypress Leafy Spurge Beetle, two species of black flea beetles, and the Copper Leafy Spurge Beetle, can weaken Leafy Spurge by feeding on leaves and roots. Unfortunately, the habitat requirements of these beetles are sometimes different from the habitat requirements of Leafy Spurge. The gall midge *Spurgia esula*, a fly native to Italy whose larvae eat new shoots and inhibit flowering, is in ND, but the introduced midge may be more difficult to establish elsewhere because it is vulnerable to local parasitoids. The Spurge Hawkmoth, native to both Europe and Asia, was introduced because the larvae consume the plant's leaves, but as Spurge is fairly tolerant of defoliation at least one other biological control is required to reduce an infestation's spread. Other potential biocontrols include a rust fungus and several bacteria. Irrigation of native plants improves their ability to compete and can help in the management of Leafy Spurge. Without disturbance to its surroundings, Leafy Spurge will die out on its own in about 12 years, so for major infestations in uninhabited areas, the best option may be simply to wait.

Life History and Species Overview

Leafy Spurge's flowers produce large amounts of pollen and nectar and are pollinated by bees and other flying insects. Each shoot can produce 100–200 seeds, which are propelled 5 m or more by exploding seed capsules, and seeds can be viable 5–8 years after release. The oblong gray seeds are spread by animals and waterways, and about 80% germinate. Once germinated, seedlings develop buds, making vegetative reproduction possible within 7–10 days of emergence. Each bud can produce a new shoot, and one plant can spread 1 m/yr or more via vegetative reproduction. Seedlings emerge in the spring and flowers appear 1–2 weeks later, with up to 16 branches in each cluster; flowering ends in midsummer.

Leafy Spurge has medicinal value, and was used historically as a topical treatment for warts and calluses and taken in small doses as a powerful emetic. It has also been used as an efficient fuel when processed into pellets and burned; due to its early flower emergence, it provides an early-season food source for bees.

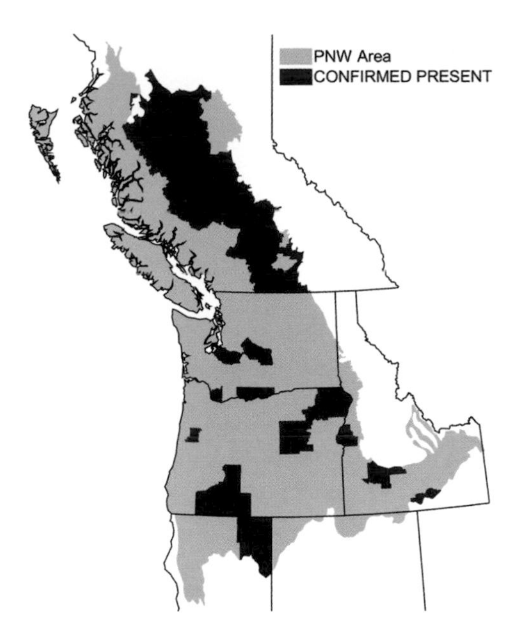

History of Invasiveness

Leafy Spurge was thought to have been introduced to eastern North America in the early 19th century, and to have spread westward to ND within 80 years of its introduction. It is now believed that two or more strains were introduced at different times from their native continents, Europe and Asia, eventually producing an aggregate species. The plants were likely introduced accidentally in shipments of contaminated grass seed and cereal, or in ships' ballast.

It is easy to see how Leafy Spurge gained its foothold once on the continent. With its high seed production and explosive dispersal, its extensive root system with vegetative reproduction, and its chemical offense system, it is a formidable competitor for any native plant.

Other Sources of Information: 78, 125, 126
References: 21, 308
Author: Ericka Kendall

Medusahead *Taeniatherum caput-medusae* ssp. *asperum*

Species Description and Current Range

This annual grass of woodlands, shrublands, and grasslands has seeds with rough terminal bristles up to 10 cm long. The awns twist and spread as they dry, evoking Medusa, the snake-haired monster of Greek mythology who turned anyone who saw her to stone. Medusahead resembles two native perennial grasses with wiry heads: foxtail barley and squirreltail. However, their heads break apart as they mature, unlike those of the Medusahead. Medusahead plants are up to 60 cm tall, branch from the base, and have narrow, rolled leaves. The inflorescence is a spike or head nearly as wide as it is long. Medusahead remains green 2–3 weeks longer than other annual grasses, so it is recognizable by its green color in early summer. As it dries, the foliage bleaches to a distinctive tawny color, and is rough because of its high silica content.

Medusahead has invaded disturbed rangelands from WA to the Interior Valley of CA and east to CO. It is most likely to invade soils with a high clay content but also colonizes well-drained soils where adequate soil moisture is available, especially if the native vegetation has been disturbed.

Impact on Communities and Native Species

When Medusahead arrives at a site, its litter, which decomposes very slowly, creates conditions that exclude most other species. Several centimeters of thatch quickly accumulate, forming a mat that keeps seeds from touching the soil surface. Medusahead germinates readily under these conditions, but the seeds of its competitors do not. Medusahead can reproduce in litter because its seedlings can survive desiccation. They have roots up to 10 cm long, and even if a young root dies because it dries out, a new one will develop when moisture returns.

Medusahead litter ties up nitrogen and burns easily, seriously modifying ecosystems. Nitrogen and other nutrients remain bound up in the residual stems and inflorescences instead of being recycled through decomposition. Medusahead grows well in the resulting nutrient-poor soil, but its competitors suffer. As it prospers, Medusahead produces still more litter, creating a mat of fine, dry, highly flammable fuel and increasing the likelihood of fire. When burned, Medusahead litter forms an insulating layer, allowing Medusahead seeds that are on or in the soil to survive burning, a process that further degrades native plant communities and leaves them vulnerable to still more invasion by Medusahead. This positive feedback cycle is capable of converting sagebrush shrub-steppe, juniper woodlands, chaparral, and Palouse grassland into homogeneous swards of Medusahead with little biological or structural diversity and low value for wildlife or livestock. Over half a million hectares in the western US are degraded where Medusahead has replaced more valuable foods for livestock and wildlife. Livestock will graze Medusahead while it is green, but it becomes less palatable it as it dries out. Seed-eating birds also avoid Medusahead. The nutritional value of Medusahead is low because it is 72–89% ash; furthermore, grazing on dry Medusahead causes injury to mammalian herbivores when its barbed bristles pierce their soft tissues. In habitats such as talus slopes and the margins of vernal pools, Medusahead threatens populations of rare plants such as Malheur Valley Fiddleneck (*Amsinckia carinata*) and Sierra Valley Mouse-tail (*Ivesia aperta* var. *aperta*).

In dry environments, where the sparse native vegetative cover carries fire poorly, the altered disturbance regime that follows Medusahead invasion is especially pronounced and extremely detrimental. Fire destroys the stabilizing soil crust and leads to increased erosion. Erosion of the surface layer exposes underlying clay sediments, which shrink and swell with changes in moisture, setting the stage for further invasion because Medusahead is one of the few species adapted to such conditions. In short, by altering nutrient cycles, disturbance regimes, and soil chemistry, Medusahead changes the physical and biological potential of the sites it invades.

Control Methods and Management

Medusahead can be killed with a combination of chemical and mechanical methods. Because it produces huge seed banks, new Medusahead crops will emerge for several years unless the seed bank is depleted. This can be done with preemergent herbicides. Some authorities recommend burning Medusahead before it sheds its seeds, but because fire has potential to exacerbate infestations, burning is unwise. Regardless of the techniques that are used to remove Medusahead, revegetation with perennials is necessary for effective long-term control. Squirreltail is a native perennial bunchgrass that is quite successful in competition with Medusahead. It grows well in a variety of grassland, woodland, and shrubland habitats and is tolerant of fire. Minimizing disturbance to native vegetation helps to prevent Medusahead from getting a foothold.

Life History and Species Overview

Medusahead seeds typically germinate in autumn. Plants continue to grow throughout winter and spring, flower and set seed by early summer, and dry up by midsummer. The roots of Medusahead are able to grow at low temperatures and use available moisture in winter. By summer, they develop thick cell walls that allow them to survive even when the soil is hot and dry.

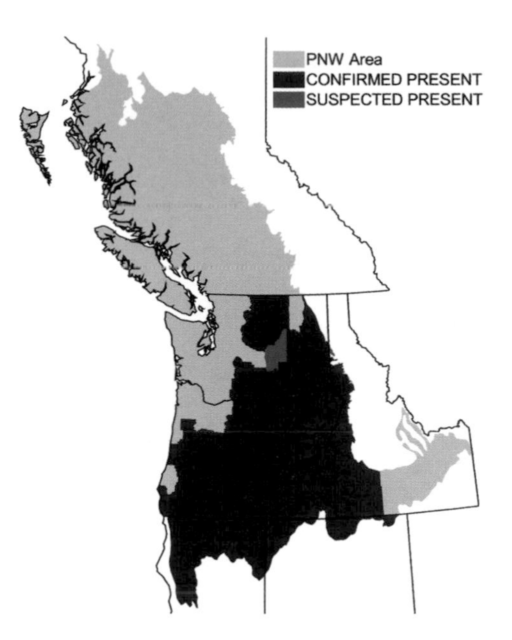

History of Invasiveness

Medusahead arrived from Eurasia, probably by way of imported livestock, in the latter half of the 19th century. It was collected in southwest OR in the 1880s and again in eastern WA in 1901. Spread by livestock, it expanded its range southward into CA, where it thrived in the hot, dry summers and mild wet winters. But this annual grass was not limited to areas with a Mediterranean climate; it also invaded, albeit more slowly, rangelands in ID, NV, UT, and CO.

Other Sources of Information: 15, 91
References: 63, 142, 152, 353, 355
Author: Bertie J. Weddell

Orchard Grass *Dactylis glomerata*

Species Description and Current Range

Orchard Grass is a long-lived, cool-season bunchgrass characterized by a cocksfoot-shaped seed head, or panicle, hence its common name "Cocksfoot" in Canada, Europe, New Zealand, and Australia. The panicle is strongly tufted, clustered on one side of the branch, and reaches 3–15 cm in height. The selection of the genus name *Dactylis*, which derives from the Greek word *dactulos*, meaning "finger," is attributed to the stiff branches of the panicle. Orchard Grass typically reaches a height of 50–120 cm. The base of the leaf enclosing the stem is compressed, giving it a flattened appearance that is diagnostic for identification. The leaves are 2–12 mm wide and typically 10–40 cm long, but can reach 1 m in length. The lower leaf surface is not shiny and has a distinct center ridge. Leaves vary in color depending on variety and soil conditions, but are typically green to blue-green.

Orchard Grass occurs throughout the US and north into southern Canadian provinces. It is well adapted to many eastern states as well as to moist or irrigated areas of the west. Vegetation communities in the PNW particularly suited to Orchard Grass include Sagebrush, Ponderosa Pine, Douglas-fir, Aspen, and Oregon White Oak (*Quercus garryana*, also called Garry Oak) ecosystems. Oregon White Oak/Orchard Grass communities are found in the Umpqua River valley and the open oak woodlands of coastal northern CA. It is also commonly found along roadsides and in pastures, meadows, and waste areas.

Impact on Communities and Native Species

Orchard Grass is quite competitive when ample nutrients and moisture are available. However, it is not considered invasive in most areas. It is a favored forage grass, and consequently animals control its spread in areas with intensive grazing. Where conditions are favorable for growth, it can become a noxious weed because it is long-lived and, once established, can form a dense sod and dominate a community. In the Oregon White Oak ecosystem in BC, Orchard Grass is considered one of the worst exotic-species threats. Non-native grasses make up to 30% of the vegetation in the BC Oregon White Oak ecosystems. Invasive grasses negatively impact other species by competing aggressively for nutrients and water and forming a dense litter layer that blocks light and fosters high-intensity fires. Orchard Grass alters the soil nutrient regime by adding nitrogen from decaying grass litter, which benefits invasive species at the expense of natives.

Control Methods and Management

Mechanical control methods have limited success on Orchard Grass. Repeated mowing may stimulate shoot growth, occasionally making Orchard Grass a problem in lawns. Since it does not aggressively spread vegetatively, manual pulling of young plants in early summer before seeds are set can be effective for small areas. In some locales, Orchard Grass declines after repeated summer burns, while in other areas it persists after a burn. Many herbicides, including Aatrex (atrazine), Arsenal (imazapyr), Bromax (bromacil), Direx (diuron), Kerb (pronaminde), and Proncep or Caliber (simazine), are effective on Orchard Grass; however, they are not specific to it and therefore may not be practical where Orchard Grass persists in a landscape with native species. No known biological control exists.

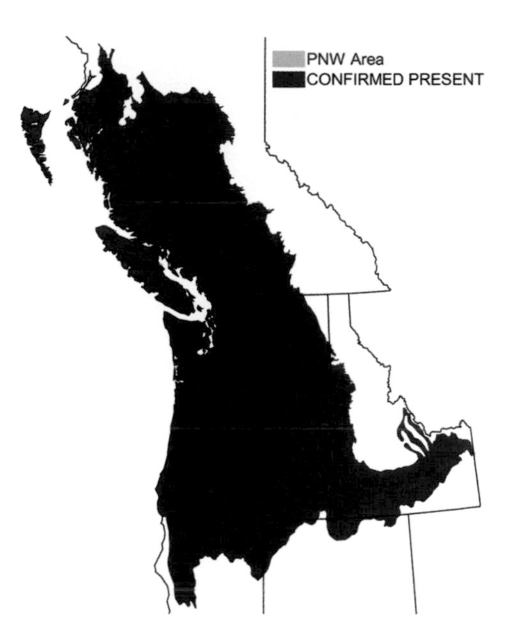

Life History and Species Overview

Orchard Grass is a favored forage crop, producing high-yield and high-quality forage that is palatable to livestock. Its use is particularly favored for cattle, domestic sheep, and goats. It is also one of the earliest grass species to grow in spring, exhibiting prodigious growth during cool conditions. Strong summer growth can occur as long as there is some moisture because the grass is so deeply rooted. It is highly tolerant of shade, which probably earned it the name "Orchard" Grass, but it also grows well in open areas. Orchard Grass prefers well-drained soils but tolerates moderately poor drainage as well as drought conditions, persisting in areas with as little as 250 mm of annual rainfall. Its ability to withstand drought is a result of its extensive root system. In dry areas, it may lie dormant during the summer and in the fall, add new growth, and possibly flower again. It does not grow well in areas of high water tables or on saline soils. It produces a dense sod in the upper 8 cm of soil, but some roots extend 46 cm or more below the surface.

Orchard Grass can reproduce two ways: sexually by seed and asexually by new shoot formation from underground stems and roots. Its seeds remain viable for only 2–3 years, which likely prevents Orchard Grass from becoming invasive in most areas. Germination often occurs in fall and is largely controlled by moisture availability; therefore, Orchard Grass does not tend to build up seed banks in the soil. Seeds can collect on animal hair and disperse for long distances.

Orchard Grass is also used for erosion control and upland wildlife and conservation plantings. Its value for erosion control stems from its tendency to form highly dense root networks. It is also used as a cover crop under orchards and in other disturbed areas, such as after a prescribed and controlled burn. It is also highly palatable for Elk, Bighorn Sheep, White-tailed Deer, and Mule Deer, and the persistent, green basal rosette is used for winter forage. Other animals known to graze on Orchard Grass include wild turkeys, rabbits, and Canada Geese.

History of Invasiveness

Native to western and central Europe, Orchard Grass was introduced to the eastern US around 1760. In the 1830s, its forage value was recognized by settlers in VA and its use spread across the US. Orchard Grass has been used in the PNW since the early 1900s as a hay, pasture, and silage crop.

Other Sources of Information: 33, 41, 52, 86, 89
References: 155, 191, 210, 293, 346
Author: Wendy Gibble

Perennial Pepperweed *Lepidium latifolium*

Species Description and Current Range

Perennial Pepperweed, a member of the mustard family (Brassicaceae), is also known as "Tall Whitetop" because of the dense, round clusters of white flowers perched atop 1–2 m stems. Stems are semiwoody at the base. The toothed leaves are grayish green, lance shaped, and somewhat leathery. Flowering occurs in mid June, with 6–8 small blossoms produced at the end of each stem. Flowers are cross-shaped and have four petals and four sepals. The small round fruits contain tiny reddish-brown seeds. Seedlings have round leaves, which are replaced by the more typical, feather-shaped leaves several weeks later as plants mature and begin growing vertically.

Rapidly spreading across the western US, Perennial Pepperweed has scattered populations throughout the PNW. It is most often found near fresh or salt water at elevations ranging from sea level to above 3,000 m. Common habitats include riparian areas, wetlands, alpine meadows, shrublands, and seminatural areas such as roadsides, pastures, and residential areas.

Impact on Communities and Native Species

Perennial Pepperweed has strong impacts on native plants and animals, soil nutrients, agriculture, livestock, and community biodiversity. It outcompetes native vegetation for moisture, nutrients, and light. It forms dense monocultures, further excluding other species by reducing available light. The dead stalks accumulate to produce a dense layer of woody debris that prohibits the seeds of native species from germinating. Buildup of organic material also affects the soil nutrient content by changing the surface carbon-nitrogen ratios. Roots of Perennial Pepperweed account for 40% of a plant's biomass and can reach 3 m below the surface. Unlike the roots of many native species, Perennial Pepperweed's creeping roots do not form dense clusters, so they promote soil erosion. In addition, Perennial Pepperweed acts like a salt pump, using water with a higher salt content than native species, it pumps salts from deep in the soil and deposits them near the surface. This changes the distribution of minerals in the soil and negatively affects less salt-tolerant species, changing plant community compositions and decreasing diversity.

Perennial Pepperweed affects animals and wildlife by altering the structure of foraging and nesting habitat. It outcompetes native food plants, and the accumulation of semiwoody debris degrades nesting habitat for many birds and animals. Near streams, it prevents the regeneration of cottonwoods and willows, two trees that provide important wildlife habitat in these ecosystems.

Livestock are also negatively affected by Perennial Pepperweed, which has less protein and is less digestible than hay, making it unwanted by ranchers. Woody stem accumulation physically hinders the grazing ability of livestock, and even goats will not consume mature plants. Livestock refuse to eat this species if other forage is present and it may even be toxic. The spread of Perennial Pepperweed to farmland is also an economic threat to the alfalfa and sugar-beet industries.

Control Methods and Management

The best method of control is to prevent spread and establishment by minimizing soil disturbance and overgrazing. Early detection and hand pulling or digging to remove all the roots can control Pepperweed. Once established, control is extremely difficult because the plant is competitive and can spread underground. Mechanical control and burning are not useful because new plants can easily sprout from root fragments. Control through grazing may be helpful but is complicated because of the woody stems and the potential of toxicity to livestock. Several herbicides, including chlorosulfuron, metasulfuron, imazapyr, and 2,4-D, kill established plants, with mortality rates of 50–99%. However, even with vegetative mortality of 98%, resprouting from roots can lead to a monoculture of Perennial Pepperweed in the following year. Herbicide application is further complicated because chemicals may kill plants indiscriminately, pollute the water and soil, and are not approved for wet areas where infestations are most severe. No biological control agent is available, and it is unlikely that one will be approved because there are 11 native species of *Lepidium* in the west (one is endangered, *L. barnebyanum*) and many valuable crop species (e.g., canola, mustard, and cabbage) are closely related to Perennial Pepperweed. Control efforts for Pepperweed could also damage these crops.

Life History and Species Overview

Perennial Pepperweed is native to Africa, Europe, and Asia. The natural range includes the Mediterranean basin, temperate Europe, the Middle East, and the Himalayan region of Asia. Plants are typically found in moist areas and are especially adapted to saturated and salty soils. It is flood tolerant and can quickly change its root and vegetative structures to survive periods of flooding. Pepperweed is unusually tolerant of salt-affected soil, has adaptive physiological responses to low oxygen conditions, and can even tolerate long periods of dessication. These adaptations enable Perennial Pepperweed to thrive in a wide range of habitats and make it a potent invader.

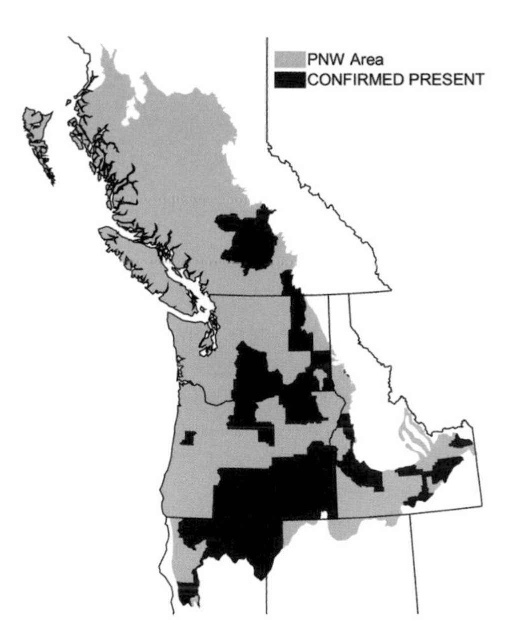

History of Invasiveness

Lepidium latifolium was first introduced to North America around 1900 in contaminated sugar beet seed imported from Eastern Europe. Inadvertent introductions are also known from New England to Mexico. Perennial Pepperweed has invaded every state in the US and has spread rapidly in the western states over the past few decades. Initial establishment is often associated with waterways or agricultural irrigation systems. Once established, it can spread quickly by both seed production and creeping rootstocks. A tremendous number of seeds are produced (over 6 billion per acre of infestation) and germination rates are high. Seeds are often spread long distances by farm or construction equipment and are also transmitted in alfalfa seed, hay, or straw from grain crops. Plants spread underground by producing new shoots along the rootstock. Rootstocks are buoyant and small pieces can float long distances downstream to produce new colonies of plants.

Other Sources of Information: 91, 125
References: 221, 356, 357, 358
Author: Christopher N. Templeton

Purple Loosestrife *Lythrum salicaria*

Species Description and Current Range

Purple Loosestrife can form exceptionally dense stands in wetlands, with seedling densities approaching 10,000–20,000 plants/m^2. As a result of its aggressive nature, it has earned nicknames such as "Beautiful Killer" and "Marsh Monster." This species has small (7–10 mm) reddish-purple flowers arranged in dense, showy spikes and square stems, with opposite leaves that are hairy and lance-shaped. Mature Purple Loosestrife can have more than 30 stems and be 2 m tall and 1.5 m wide. Dense stands of Purple Loosestrife give invaded wetlands a reddish purple color, making it possible to detect this species from a distance. Native Fireweed (*Chamerion angustifolium*) and Spirea (*Spiraea douglasii*) superficially resemble Purple Loosestrife. Fireweed has much larger flowers, alternate leaves, and is not a wetland species. Spirea grows in wetlands but has flowers arranged in clusters and oblong, alternate leaves. The common name "loosestrife" is applied to both native and alien species and refers to species in both the loosestrife family (Lythraceae, which includes Purple Loosestrife) and the primrose family (Primulaceae). Primrose loosestrifes, including the native Tufted Loosestrife, are yellow-flowered and easily distinguished from Purple Loosestrife. Hyssop Loosestrife, an annual native relative of Purple Loosestrife, is distinguished by its shorter stature and white to rose petals. Purple Loosestrife is now common in wetlands across North America; it is present in every Canadian province and all contiguous US states except FL. It is likely found in all PNW counties.

Impact on Communities and Native Species

Purple Loosestrife forms incredibly dense stands that outcompete native plant species for space, light, and pollinators. The changing plant composition alters the structure and function of invaded wetlands and deprives

native animals of the food resources and shelter that native plants provide. Purple Loosestrife clogs waterways and reduces the food quality of livestock forage.

Control Methods and Management

Controlling large, established populations of Purple Loosestrife in natural areas is difficult. Mechanical methods, such as disking, mowing, and applying herbicides, are damaging to native plant and animal species. Biocontrol may offer the best hope for general control in natural areas. Two species of leaf-eating beetles, released as biocontrol agents, have greatly reduced Purple Loosestrife in some areas of Canada and the eastern US and have shown promise in the PNW.

Because it is difficult to control large populations, it is important to monitor for new infestations and attack these while they are small and controllable. Hand pulling is appropriate for young plants; the entire root should be removed. Spot treating with glyphosphate herbicide is best for older plants. Removing the flowers before they produce seeds can control spread into new areas. Ultimately, the best method for controlling Purple Loosestrife is to limit new invasion. Removing Purple Loosestrife plants from gardens will greatly reduce seed sources.

Life History and Species Overview

In its native range of Eurasia, Purple Loosestrife shares wetlands with Cattails, Reed Canarygrass, Bulrushes, willows, sedges, and Horsetails, which are close relatives of species found in North American wetlands. Thus, Purple Loosestrife is particularly well suited to thrive in North American wetlands. However, unlike their Eurasian relatives, North American wetland species have not evolved to compete with Purple Loosestrife, and as a result can be overwhelmed by its invasion. Purple Loosestrife prefers sunny habitats and can invade many wetland types, including freshwater wet meadows, river and stream banks, pond edges, reservoirs, and ditches.

Purple Loosestrife flowers from late June to September and is a prolific seed producer. The average 1-year-old plant produces more than 100,000 seeds, a mature plant more than 2 million. Many seeds sink in the water, germinate, and then rise back to the surface. The small, newly germinated plants float through the water until they find a suitable place to root. It is likely that ungerminated seeds also float to new habitats, and others become stuck to wildlife, humans, and vehicles and are transported to new locations.

Mature Purple Loosestrife produces multiple stems and can cover several square meters. Above-ground stems die back each fall; energy stored in the woody rootstock enables it to regrow each spring. The rootstock also allows it to regrow if the stems are cut or damaged. The combination of woody rootstock and phenomenal seed production contributes to Loosestrife's success as an invader.

History of Invasiveness

Purple Loosestrife was likely first transported to eastern North America from northern Europe with ship ballast in the 1800s. An attractive plant that has a long history as a medicinal herb, Purple Loosestrife was also

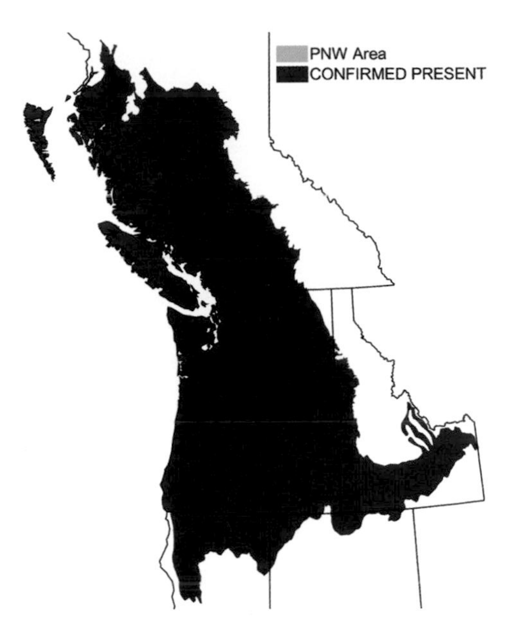

imported for horticulture. More recently, beekeepers have planted the species for its nectar. Initially, Purple Loosestrife was isolated around eastern North American harbors but spread inland as canals and highway systems were constructed. Purple Loosestrife is invasive in nearly every US state and Canadian province. It entered the PNW in the 1940s and has spread widely since that time. Because of its widespread impact, it is illegal to sell Purple Loosestrife in some states and provinces, including WA, OR, and AB. It can still be sold in more than half of the states and provinces, and although it represents a small proportion of nursery sales, its exceptionally high seed production means that any plant purchased and planted can have a large impact on natural areas. Recently, so-called sterile Purple Loosestrife has become available, but it is in fact quite fertile and should never be planted.

Other Sources of Information: 29, 78, 91, 92
References: 38, 148, 289, 305, 307
Author: Tara S. Fletcher

Russian Olive *Elaeagnus angustifolia*

Species Description and Current Range

Russian Olive, a small tree/shrub, can reach 14 m in height and is armed with 2–5 cm-long thorns. It is easily recognized by the combination of narrow leaves that are arranged alternately along the stems, and by the overall silvery appearance produced by tiny scales coating the young twigs, leaves, and flower sepals that reduce moisture loss. Russian Olive produces small (up to 1 cm broad), fragrant, yellow flowers in May and June that are insect-pollinated. Though not a member of the true olive family (Oleaceae), Russian Olive has 1–2 cm fruits bearing a "stone" in the center. Individual plants produce copious amounts of tart fruits that change from green to brownish-red during ripening. Some people use the fruits for jams and jellies. A symbiotic bacterium (*Frankia*) found in the roots of Russian Olive produces supplemental nitrogen that enhances its ability to outcompete other plants. The bacterium is especially beneficial for Russian Olive plants that establish in nutrient-poor soils.

In the late 1800s, Russian Olive was introduced to the PNW as an ornamental from Europe. The US Soil Conservation Service promoted the tree as a wildlife attractant and windbreak. Russian Olive is found throughout the arid interior portions of the PNW along rivers and streams (riparian zones). Seed dispersal continues through the droppings of dozens of bird and mammal species that feed on the fruits of the plants. Russian Olive thrives in a variety of natural habitats that include open fields, stream banks, and marshes. Presently, the worst invasions are occurring in riparian areas where perching birds drop the seeds into moist, rich soils.

Impact on Communities and Native Species

Russian Olive commonly forms dense stands that eliminate established native plant species through competition for light and soil nutrients. In riparian areas, Russian Olive displaces native willows and cottonwoods, important nesting trees for birds. Russian Olive wood is dense, limiting the ability of cavity-nesting birds such as Northern Flickers, woodpeckers, Black-capped Chickadees, and American Kestrels to make nest holes. An overwhelming majority (>90%) of the bird (e.g., Robin, Cedar Waxwing, Northern Flicker) and a majority (>50) of the mammal (e.g., Raccoon, Opossum) species that feed on Russian Olive fruits in North America are native. However, researchers find that bird diversity is greater in areas without Russian Olive. The nitrogen-fixing roots of Russian Olive increase soil fertility in the surrounding soil, which can disrupt native plant communities adapted to nutrient-poor conditions. Russian Olive resprouts from its roots and trunk, enabling its quick reestablishment where native vegetation is managed through burning and mowing. Invasion by Russian Olive is disruptive in agricultural areas as well, where its roots clog irrigation channels.

Control Methods and Management

Currently there are no biological controls for Russian Olive, though it is mildly susceptible to a canker-causing fungus. Russian Olive control is most effective when plants are young. Many states encourage people to hand pull seedlings and saplings in the spring when the entire root systems are more readily removed. Control of mature plants requires cutting at the base and brushing herbicide on the exposed trunks. Herbicide application to the leaves of mature plants is particularly difficult in riparian areas; overspray can contaminate the water and deliver the herbicide to non-targeted aquatic vegetation. Burning and mowing of Russian Olive stands without subsequent herbicide application tend to facilitate regeneration through root sprout production. In most instances, successful control of this species is expensive due to dependence on labor-intensive and/or chemical means.

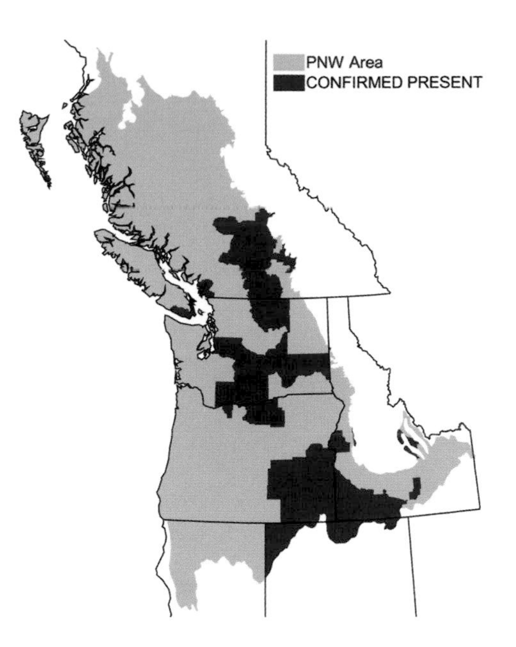

Life History and Species Overview

Russian Olive is native to temperate western Asia and southeastern Europe, and it was first cultivated as an ornamental in Germany in 1736. Ecologically, it is considered a pioneer species because it establishes in areas shortly after the soil or vegetation is disturbed. The seeds of Russian Olive are able to survive ingestion by birds and mammals because the seed coat protects the seed from digestive acid. Additionally, the seed coat contains a coumarin-like germination inhibitor that maintains seed dormancy despite passage through the warm, moist digestive tracts of birds and mammals. Exposure to low temperatures (<5°C) causes the seeds to produce a second compound that degrades the inhibitor and allows seed dormancy to be broken. Russian Olive seeds are viable in the soil for up to 3 years. The seedlings are shade tolerant, annual stem growth on saplings can approach 2 m, and plants can withstand both high (46°C) and low (–34°C) temperature extremes. Flower and fruit production typically begins 3–5 years after seed germination, with mature plants producing up to 4 kg of fruit annually. The combination of these traits helps make Russian Olive such a successful invader.

History of Invasiveness

Escaping from cultivation shortly after its introduction to the central and western US in the late 1800s, Russian Olive became naturalized in NV and UT by the 1920s, and in CO by the 1950s. Today Russian Olive is listed as a noxious weed in CO, NM, and UT and as an invasive weed in CA, NE, WI, and WY. It is particularly problematic in CO, where it has disrupted and displaced native vegetation in hundreds of thousands of hectares of riparian habitat. Its current range includes the entire continental US (except portions of the Southeast), as well the southern portion of Canada from BC to ON. It is most prevalent throughout the Great Basin Desert and in riparian areas of the Great Plains. Russian Olive is found over a wide elevational gradient, growing at sea level in eastern North America and up to 2,500 m in the West. The US Soil Conservation Service promoted Russian Olive up until the 1940s for uses including windbreaks, wildlife attraction, and soil stabilization. Russian Olive cultivars were developed to exploit this plant's ability to thrive under poor soil conditions, such as those found along highways and on soil reclamation projects. CO Cooperative Extension promotes Russian Olive on its gardening advice Web site, and the recommendation of this species by federal and state agencies contributes to its broad distribution.

Other Sources of Information: 22, 60, 92, 93
References: 91, 131, 184, 284
Author: David Giblin

Slender False-brome
Brachypodium sylvaticum

Species Description and Current Range

Slender False-brome is an invasive grass that is rapidly expanding in the PNW. It is distinguished from many other grasses by its hairy leaf margins and lower stems, broad drooping leaves, nodding flower spikes, and a long-lasting bright green color that often persists through fall and at least part of winter. The species is abundant in western OR from sea level to greater than 1,000 m. Infestations have been reported in CA and will likely be found in WA. One population has been observed along the Metolius River east of the Cascades in OR, suggesting that the species can occupy riparian zones in arid lands as well as moist west-side sites.

This perennial grass is invading a wide range of habitats at an alarming rate. Not only can it take over shady environments, such as a forest floor, to the exclusion of most other plant species, but it is also equally capable of dominating sunny sites such as pastures and prairies. Such a broad tolerance to varying light levels makes Slender False-brome unusual among PNW weeds and enables it to invade multiple habitats. Because this species seems capable of growing almost anywhere, its full range of habitats is not yet known.

Impact on Communities and Native Species

Once it invades a site, Slender False-brome dramatically reduces the diversity of native plants and animals. Its presence in a forest can result in a "silent spring" because wildlife find the species unpalatable and it eliminates most wildflowers. Dense growth of False-brome can alter fire regimes, and, especially where it builds up a heavy layer of thatch, will increase the risk and rate of spread of wildfire. The species itself appears to be fire tolerant, resprouting within two weeks of a burn. False-brome can become a serious

pest after timber harvest and can inhibit tree seedling establishment, either by providing cover for hungry rodents or by suppressing tree growth through competitive interactions. It also invades pastures and reduces forage quality for livestock, which dislike the fibrous texture of the grass. It may have negative effects on small and large mammals, native insects, lizards and snakes, and even songbirds as it alters the native understory vegetation these animals depend upon for food and shelter. Additionally, this grass can reduce establishment of planted riparian trees that provide shade and structure to fish-bearing streams. Because of its ability to completely dominate a site and smother other plants under its foliage, Slender False-brome is a serious threat to several endangered species, such as Kincaid's Lupine and Fender's Blue Butterfly. Fender's Blue Butterfly must lay its eggs on lupine so that the caterpillars can feed upon its foliage.

Control Methods and Management

Control of Slender False-brome should focus first on the prevention of its spread by cleaning machinery used in forest management and the boots, clothes, and equipment of forest workers and recreationists and by removing infestations along roadsides and trails. Seeds from roadside patches cling to passing vehicles, people, and wildlife and germinate readily in new areas. Where the species is already established, herbicides (e.g., glyphosate) are an effective control method. Although Slender False-brome continues to grow during the fall, giving it a competitive advantage over many other species, it is more vulnerable to late-season herbicide applications than many native herbs, which are often dormant at this time. Therefore, spraying in October results in effective control of Slender False-brome with good retention of native herbaceous plants. Grass-specific herbicides, such as Fusilade®, can kill Slender False-brome while minimizing impacts to native forbs. Mowing and burning alone appear to be ineffective for controlling the species. Grazing by goats may be effective, and hand removal works in small patches, but care must be taken to remove all root fragments. Follow-up control measures are essential to effective eradication. Seeding with native species, such as Blue Wild-rye, helps establish desirable vegetation and protects the site from reinvasion.

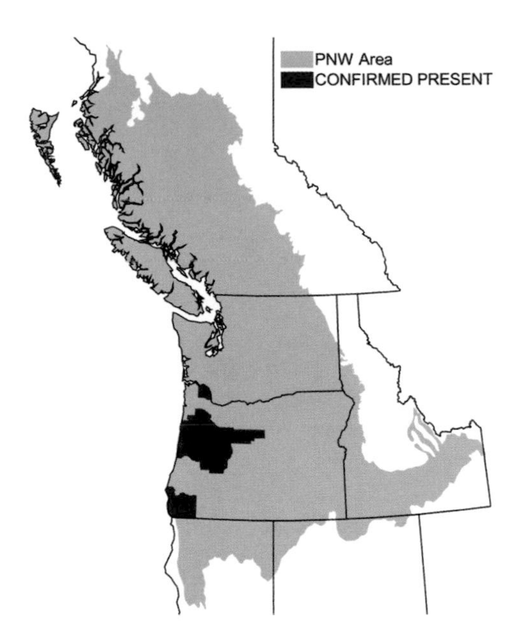

Life History and Species Overview

Slender False-brome has an expansive native range across Europe and Asia and into North Africa. In Europe it is protected from plant-eating insects by a fungus (*Epichloe*) that lives within its leaves. This grass is a long-lived perennial and forms large clumps that eventually coalesce into patches that can reach hectares in size. The species reproduces rapidly from seeds landing close to the parent plant, allowing individual clumps to unite and dominate a local site. The seeds are also transported over long distances via wildlife and humans, giving the species the ability to establish several meters to many kilometers from a source site.

History of Invasiveness

The earliest record of Slender False-brome in North America is a specimen collected in 1939 from near Eugene in Lane County, OR. By 1966, the species was growing in at least two large colonies in the Corvallis-Albany area of Benton County, OR, where it was well established. It was probably grown in test plots near Eugene, OR, in the late 1930s and 1940s at a time when the US Department of Agriculture was evaluating grasses from all over the world for suitability as forage crops and other uses. Seeds from these test-plots likely escaped from cultivation and spread to the thousands of hectares of public and private lands now occupied by the species.

Humans and their equipment act as the primary vectors for False-brome seed dispersal. Initially, the plants establish along roadsides and trails and then rapidly radiate into undisturbed areas or forest clear-cuts by way of continued human traffic or wildlife, especially deer.

Acknowledgments: We wish to thank the False-brome Working Group for assistance in assembling this information and advocating the eradication of the species from North America.

Other Sources of Information: 13, 44, 91, 93

References: 30, 57

Authors: T. N. Kaye and M. Blakeley-Smith

Spurge Laurel *Daphne laureola*

Species Description and Current Range

Spurge Laurel is a compact, multistemmed shrub that is native to Europe. It rarely grows to be taller than 1 m. The glossy evergreen leaves are alternate and spirally arranged along the branches, about 7 cm long, and are wider toward the end of the leaf than at the base. The leaves look somewhat like a small laurel, and the genus is named after the Greek nymph Daphne whose human figure was transformed into a laurel tree. The species name, *laureola*, and the common name Spurge Laurel, reinforce its likeness. Like other Daphnes, it has fragrant flowers that bloom in the winter. In this species the flowers are small and yellow and are found in the axils of the leaves. It has bluish-black fruits in the spring that are poisonous to humans but highly edible to bird species. Spurge Laurel is very shade tolerant and is mostly found invading both open deciduous forests, such as the Oregon White Oak woodlands in BC, and coniferous Douglas-fir forests. In Europe, it is often found growing with Ivy (*Hedera helix* and *H. hibernica*) and English Holly (*Ilex aquifolium*) and it associates with these invaders in the PNW as well. Although it is not commonly sold now, like its cohabitors, Spurge Laurel has been used as a landscape ornamental. It is often used as rootstock for slower-growing Daphnes. When buying another evergreen Daphne, check along the base of the stem to determine if there is a fused area, indicating a graft. If there is, make sure that you remove the roots if the above-ground portion dies after planting.

Impact on Communities and Native Species

Because Spurge Laurel has only recently begun to invade, little is known about its impacts, although it has spread to many parks in the PNW. Horticulturists at the WA Park Arboretum in Seattle, where it was invading parts of both native and ornamental landscapes, reported that when the species was removed, nearby plants began to grow more quickly. This indicates that Spurge Laurel is highly competitive for soil resources such as water and nutrients. Even small plants have extensive and deep roots and large plants may have numerous thick roots. The species has formed thickets on Vancouver I., BC, and appears to be able to invade shaded areas with little disturbance. By changing the light availability and possibly the soil chemistry and pH, Spurge Laurel may affect normal forest regeneration. The waxy leaves of Spurge Laurel form such a solid canopy that tree seedlings, ferns, and all those species that should normally grow appear to be excluded. By shading out native plants it is changing the composition of the understory and may eventually change the overstory of some forests. The ground underneath is often bare, which may contribute to soil erosion in some places.

Control Methods and Management

Although larvae of night-flying moths feed on Spurge Laurel in its native Mediterranean region, the moth is not present in the PNW. Spurge Laurel is not enough of a problem to consider introducing moths as a biological control agent. Moreover, the introduction of the moth might have impacts on other closely related species. In the PNW Spurge Laurel has few pests other than aphids. In BC, a bioherbicide made of a native pathogenic fungi is being tested, but is not yet available. Because the plants's leaves are poisonous, sheep and other grazers do not eat it. Young plants may be uprooted, but removal of larger plants may disrupt the soil too much to be worthwhile. If plants are not too large, a weed wrench can be used to pull them up by the roots. Mature plants can be cut to the ground and, if re-sprouting occurs, the freshly cut surface can be treated with an herbicide such as glyphosate. Resprouting may occur mostly on plants with a stem less than 1 cm in diameter.

Life History and Species Overview

In the PNW, it appears that bird species are the major dispersers of seeds, especially Robins. In Spurge Laurel's native habitat in Spain, seeds are also spread by mice. Mice and other rodents, such as squirrels, may also be spreading it here. The flowers are most powerfully fragrant at night, suggesting that moths may be a major pollinator. Like most members of the Thymelaeaceae, the leaves and seeds are toxic to humans, and this species has long been used in Europe to cause abortions and to stimulate menstrual discharge. It has also been recently investigated as a potential leukemia treatment. Spurge Laurel should be kept away from children and pets. Sensitive adults may find that it causes a painful dermatitis. Care should be taken when handling it; some people have reported shortness of breath when using a brush cutter.

History of Invasiveness

Spurge Laurel was originally introduced as a garden ornamental in the early to mid 1900s, though it is not commonly sold now.

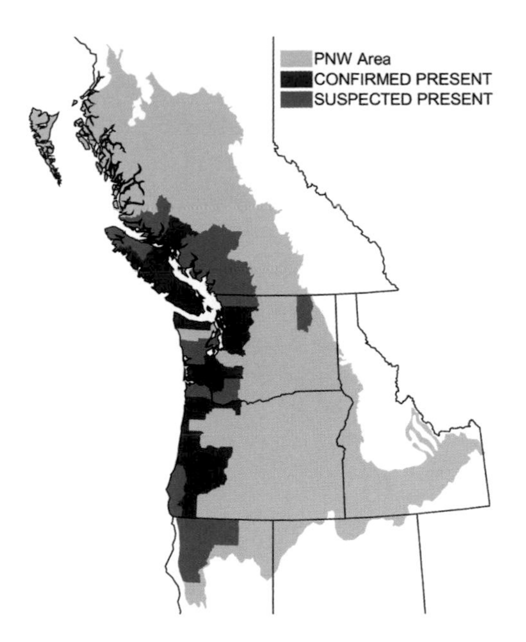

Probably the attractiveness of its leaves and winter flowering made it a desirable horticultural plant. It still may be found on some lists of garden plants that are deer- and drought-resistant. Although it is not currently widespread, it has been found to be rapidly expanding in areas of BC, WA, and OR, and has been described by one biologist as "the next Scotch Broom." It is unknown when the invasion began; Hitchcock and Cronquist do not list it in their 1973 *Flora of the Pacific Northwest.* They included most known invasive plants that were present in the PNW at that time, suggesting that this plant is a recent addition to our flora. It is highly likely that this species will become more prevalent over time. As it spreads, its impact on native forests will likely increase. This is one species that should be removed when it is identified invading natural areas and parks.

Acknowledgments: Two unpublished reports by Joe Percival on the restoration of Witty's Lagoon Regional Park, BC, were enormously helpful in preparing this report. Ray Larson provided additional information about the plant's use in horticulture.

References: 29, 124, 208, 303

Author: Sarah H. Reichard

St. John's Wort *Hypericum perforatum*

Species Description and Current Range

St. John's Wort is a perennial herbaceous plant up to 1 m tall with blunt opposite leaves. The stems are reddish with two distinctive ridges that run along opposite edges. The flowers are bright yellow, 2–3 cm across, with numerous stamens and five separate petals that have small black dots along the edges. The fruit is a three-parted dry capsule that contains many small seeds, each less than 1 mm long. The seeds are mucilaginous when moist and stick to animals, equipment, and vehicles, dropping off when they dry, an effective mechanism of dispersal. A distinctive trait of this species is that the leaves are perforated by tiny transparent "windows." These areas, lacking chlorophyll, are readily visible with a 10X hand lens. St. John's Wort is a common roadside weed, found at elevations up to approximately 1,500 m. It typically invades disturbed roadsides, transmission rights-of-way, and other disturbed sites, particularly in areas with sandy or gravelly soil. However, it may also be found in relatively undisturbed sites away from roads. The current distribution of common St. John's Wort is throughout, or nearly throughout, the PNW, as well as in most of the US except for a few southernmost states. This species extends north in BC to central Vancouver I. and the lower Fraser River Valley.

Impact on Communities and Native Species

St. John's Wort can be a serious pest, aggressively crowding out native species. Because it is not readily eaten by livestock, it typically increases under grazing pressure as other species are preferentially eaten. Although St. John's Wort may most effectively invade and spread in disturbed areas, it is established and persists in relatively undisturbed sites, including natural area preserves. It may have a severe detrimental effect on high-quality native communities.

Control Methods and Management

While St. John's Wort is difficult to control, biological controls are successful in some areas. Klamathweed Beetle (*Chrysolina quadrigemina*) was introduced in the PNW in 1948, and it reduced large populations of the plant. Another leaf-eating beetle, *Chrysolina hyperici*, is used in wetter areas. A root-boring beetle, *Agrilus hyperici*, and a seed-head fly, *Aplocera plagiata*, are used in limited areas. St. John's Wort is not tolerant of frequent cultivation and can be controlled by regular tilling. Frequent mowing to prevent seed maturation may slow the spread of the plant. Herbicides that are specific to broad-leaved plants may be effective in grass pastures but are detrimental to desirable broad-leaved species such as legumes.

Life History and Species Overview

Although it can also propagate by rootstocks, St. John's Wort spreads primarily by seed. Some plants appear to be able to produce seeds asexually, and the species occurs in mixed populations of plants with different numbers of chromosome sets. As with many invasive plants, this ability to produce seed without fertilization bypasses the requirement for a pollinator and produces many clones of an individual with characteristics adaptive to specific local conditions. Pollination, when it occurs, is likely to be by insects seeking pollen, including bumblebees, other bees, and syrphid flies, a bee-like mimic. The flowers do not produce nectar and are thus unattractive to butterflies and other nectar-seeking insects. Genetic studies on St. John's Wort from a range of latitudes in Europe and North America, including plantings in multiple common gardens, suggest that populations may be capable of rapid genetic adaptation to environmental conditions when the species spreads into new areas.

St. John's Wort, named for John the Baptist, is regarded as both harmful and beneficial, and it has both poisonous and medicinal properties. When eaten by livestock, St. John's Wort can cause blistering of the lips and eyes and severe skin irritation in areas with white hair that are exposed to bright light. Most, if not all, grazing animals are affected, though the intensity of response varies: sheep may require four times as much of the plant, relative to body weight, as cattle do before exhibiting harmful effects. The biologically active compounds retain their effects, with reduced potency, even when dry, and so may contaminate hay. Humans may also suffer skin irritation if exposed to bright sunlight after eating the plant. The plant is regarded in the horticultural trade as deer-resistant, and it repels rodents.

St. John's Wort has a long history of beneficial use in traditional medicine. Extracts have been used for the treatment of mild to moderate depression, though recent research on severely depressed patients showed no significant decrease in symptoms compared to a control group given a placebo. The antidepressant effects of *Hypericum* extracts are similar to standard antidepressants. Hyperforins, produced in the reproductive structures of the plant, and hypericins, concentrated in the leaves and flowers, are two compounds found in St. John's Wort with antimicrobial, antiviral, and anti-herbivore properties in vitro. Hyperforins are thought to be the primary ingredient contributing to antidepressive effects and have been associated with reduced alcohol intake by alcohol-preferring mice. Hypericins, the phototoxic components of St. John's Wort, may inhibit a variety of cancer cell lines, including retroviruses such as HIV. There is also interest in the antioxidant effects of extracts of St. John's Wort.

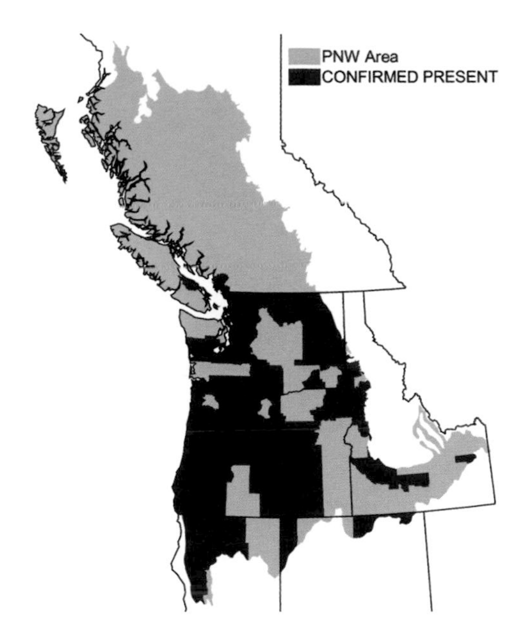

History of Invasiveness

St. John's Wort is native to western Europe, northern Africa, and parts of Asia. It was first introduced to the eastern US in 1696, where it was used for religious observances and planted in gardens outside of Philadelphia, PA. By the time of the Revolutionary War, it was a common roadside weed in New England. It was reported in OR around 1850 and in MT around 1880. It spread to CA around 1900, where it was given the name Klamathweed. It was widespread in eastern WA by the early 1900s and in ID before 1920. Genetic studies of St. John's Wort in western and central areas of the US provide evidence for multiple introductions of this species into North America.

Other Sources of Information: 37, 42, 89, 96, 126

References: 157, 242, 254, 304, 342

Author: Joseph Arnett

Tansy Ragwort *Senecio jacobaea*

Species Description and Current Range

The species name for Tansy Ragwort, *jacobaea*, refers to St. James (Jacobus). The word "ragwort" may apply to the deeply cut, ragged, and ruffled-looking leaves. Leaves are dark green on top and whitish-green underneath, and leaf stalks and stems have a reddish-purple tinge. Flowering plants can be 0.3–2 m tall. The daisylike heads of flowers form distinctive, flat-topped clusters. Each head consists of many tiny disk flowers surrounded by the bright yellow, petal-like ray flowers. Tansy Ragwort may form a distinct taproot.

In the PNW, Tansy Ragwort is more abundant in the western parts of northern CA, OR, WA, and southern BC. It often invades open areas in forests and woodlands, as well as grasslands, riparian areas, coastal sand dunes, and seminatural areas (particularly roadsides, pastures, and vacant lots). Tansy Ragwort is moving steadily eastward in the PNW, with scattered sites known in southeastern BC, eastern WA, eastern OR, and portions of ID.

Impact on Communities and Native Species

Tilling, excavating, logging, and digging by animals are a few of the disturbances to native plant communities that allow the opportunistic Tansy Ragwort to invade. Infestations can consist of a few scattered plants, long linear colonies along roads or under power lines, or dense patches that prevent native grasses, forbs, shrubs, and trees from establishing. Tansy Ragwort may prevent rare plant establishment: when a pasture in OR was cleared of Tansy Ragwort, the rare native Bristly-stem Checker-mallow, *Sidalcea hirtipes*, appeared and started spreading.

Tansy Ragwort was used historically in herbal medicines. However, because of the risk of liver damage, such use is not recommended. When broken down by the body, the pyrrolizidine alkaloids present in Tansy Ragwort cause liver damage in cattle, horses, and to a lesser extent, goats, llamas, and sheep. Younger animals are most vulnerable. Animals usually die after eating 3–8% of their body weight in Tansy Ragwort. Effects on deer and elk are unknown.

Native bees and butterflies collect Tansy Ragwort nectar and pollen, but the effects on the pollinators and the native plants they would otherwise visit are unknown. Honey made from Tansy Ragwort pollen can be off-colored, bad-tasting, and unmarketable. It can also contain pyrrolizidine alkaloids. Effects of eating this honey are unknown.

Control Methods and Management

Biological, mechanical, and chemical controls can be effective, depending on climate, infestation size, and eventual goals for the site. As with all noxious weeds, the best control method is to prevent the plants from establishing. To identify the best combination of controls for a particular infestation, contact provincial, state, or county weed control agents for advice. Monitor all sites after treatment to verify control success.

Control of Tansy Ragwort is one of the biggest success stories for biological controls. Biocontrol agents dramatically reduced Tansy Ragwort in OR by 90%, and by state estimates returned $15 for every dollar spent on control. Three insects attack different parts of the plant: the Tansy Ragwort Flea Beetle larva feeds on the roots, the Ragwort Seed Fly larva feeds on developing seeds, and the Cinnabar Moth larva feeds on leaves, buds, and flowers. However, these insects are only well-established in the moister portions of the PNW. Cinnabar moth larvae also reportedly feed on at least two native plants (*Senecio triangularis* and *S. pseudaureaus*).

Different herbicides can effectively control some infestations of Tansy Ragwort, depending on site characteristics and plant stage (rosettes or flowering plants). Always read herbicide labels before use to determine site suitability and learn about appropriate precautions. Improper herbicide use can be ineffective and may injure or kill desired plants, wildlife, pets, livestock, and humans. Large infestations can be expensive to treat with herbicide.

Small, easily accessible infestations can be effectively hand pulled if the site is periodically revisited to remove newly established plants. This method works best when the ground is soft, so that all roots, crowns, and seeds can be removed from the site. If grazing animals have access to the site, then the entire plant should be removed. Mowing is not effective, because the plants respond by flowering on stalks that are shorter than the mowing height.

Life History and Species Overview

Tansy Ragwort usually acts as a biennial. Seeds typically germinate in the fall and spring and develop into small rosette plants. The following year, the rosette bolts into a flowering stalk and then dies. In favorable conditions, a plant may complete its entire life cycle in one year. If damaged, the plant may set seed in a later year. Although a single plant may produce more than 150,000 seeds in one season, most seeds travel less than 10 m from their parent plant. The death of the parent plant provides open ground for the new seedlings to colonize. Seeds germinate best when exposed to light and moist soils and may be viable in the soil for at least 20 years.

Tansy Ragwort must cross-pollinate to set seed and relies on insects to carry pollen. In its native range of western Asia and Europe, bees, wasps, hoverflies, butterflies, and moths visit Tansy Ragwort. A similarly diverse mix of pollinators may visit Tansy Ragwort in its naturalized range.

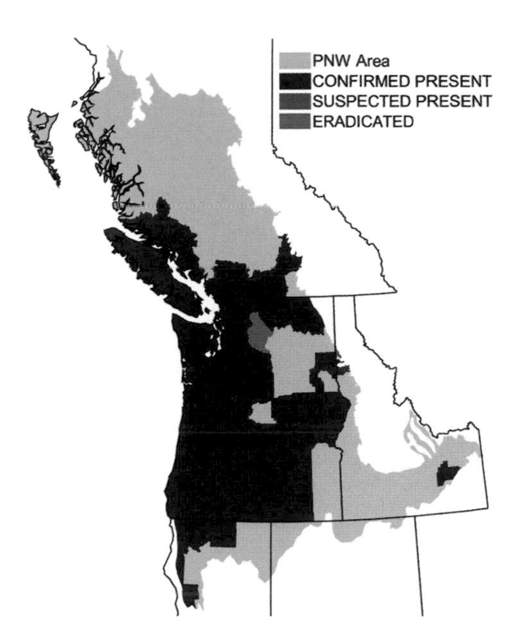

History of Invasiveness

Although its seeds do not travel far from the parent, Tansy Ragwort has spread from its native range in Europe and western Asia to the US, southern Canada, Argentina, New Zealand, Tasmania, Australia, and South Africa. Tansy Ragwort established in Australia and New Zealand by 1890 and in North America by 1910. Infected animal feed, fertilizer, vehicles, and machinery can all spread Tansy Ragwort.

Other Sources of Information: 12, 18, 43, 71, 125
References: 14, 42, 212, 213
Author: Tracy Fuentes

Cotton Thistle *Onopordum acanthium*
Musk Thistle *Carduus nutans*

Species Description and Current Range

Cotton and Musk (or Nodding) Thistle both have large purple to reddish flowers and thick spiny stems and grow singly or in very dense stands. Mature Cotton Thistle is 1–3 m tall and has leaves that are up to 60 cm long. Musk Thistle grows up to 2.5 m tall, with leaves up to 40 cm long. These thistles invade mainly seminatural, disturbed areas—pastures, rangeland, fence lines, roadsides, ditches, urban developments, waste areas—but are also found in riparian areas, woodlands, shrublands, grasslands, and other natural areas. Cotton and Musk Thistle are present from CA to BC and from the West Coast to ID, as well as just about everywhere else in North America.

Carduus nutans

Impact on Communities and Native Species

Both Cotton and Musk Thistle are highly competitive and spread rapidly. The large size and dense stands of thistles block sunlight from smaller native plants and take water and soil nutrients away from native species. When Cotton Thistle dies, the abundant litter it leaves behind smothers other plants. Musk Thistle secretes a chemical into the soil that inhibits the growth of some plants in an area 50–100% larger than that occupied by the thistle itself.

Cotton and Musk Thistles wreak havoc on agriculture and ranching. They can dramatically reduce pasture productivity by outcompeting desirable forage species and invade crops, particularly wheat. Spines make the thistles unpalatable and painful to livestock and wildlife, and the dense stands prevent animals from reaching food and water.

While invasive thistles can smother and outcompete native plants, thistles are known to support a broad array of species, from plant parasites to large predatory birds. These invasive thistles may have beneficial effects on native insects and birds. Larval Painted Lady and Mylitta Crescent butterflies feed on thistles, and many adult butterfly species, including the Western Tiger Swallowtail and the Painted Lady, feed on thistle nectar. Butterflies, wasps, flies, and many beetles use Cotton and Musk Thistles in addition to native thistles. The American Goldfinch eats thistle seeds and probably eats the seeds of invasive thistles. Studies have shown that predation by native insects helps control invasive thistles. However, if native insects and birds prefer invasive thistles, native plants may receive inadequate pollination.

Control Methods and Management

For both species, prevention of establishment is the best control. Thistles have some difficulty invading a healthy or annually cultivated ecosystem as they require adequate light and water for germination. Biological, mechanical, and chemical methods can control Musk Thistle once established. Two species of introduced weevils slow the spread of Musk Thistle, but use native thistles as hosts as well. There are currently no biological control agents for Cotton Thistle approved for use in the US. For small-scale spot treatments of both Cotton and Musk Thistle, cut the flowers as they begin to open or cut the root 2.5–5 cm below the soil. Do not till or plow, as new plants will regenerate from severed roots. Application of herbicides such as Finale (early in the season) or glyphosate (late in the season) after mechanical control helps kill plants. For large-scale agricultural invasions, more selective and complex combinations of herbicides are needed. These herbicides are not specific to thistles and should be used carefully around native species.

Life History and Species Overview

Cotton and Musk Thistles are usually biennials, forming a rosette in the first year and leaves, stalk, and flowers during the second year, but they can be annuals or short-lived perennials in warmer climates. For Musk Thistle specifically, under competition or drought during the seedling stage, growth is reduced and the time necessary to initiate flowering increases, leading to more biennials. This characteristic contributes to the persistence of these invasive thistles. Musk Thistle has 10,000–100,000 seeds/plant and Cotton Thistle 7,000–43,000 seeds/plant. Seeds can remain viable in the soil for 10–30 years. Cotton Thistle has a water-soluble germination inhibitor on the seed, requires a moist environment for growth, and has no specific light or temperature requirements. The growth of Musk Thistle is limited by light and moisture. It is therefore not successful in dense, healthy ecosystems. Both thistles are successful at low to mid elevations. Predators in their native habitats include native insects and birds.

Cotton Thistle was used to treat cancers and ulcers; the bulbs were eaten like an artichoke, the cottony hairs used to stuff pillows, and the oils burned or used in cooking. In medieval times, a concoction of thistles placed on a bald head was said to restore hair growth. Musk Thistle had medicinal purposes as well—leaves and seeds were used as a bitter tonic to stimulate liver function. The thistles are also edible, at their best early in the summer, and smell like almonds.

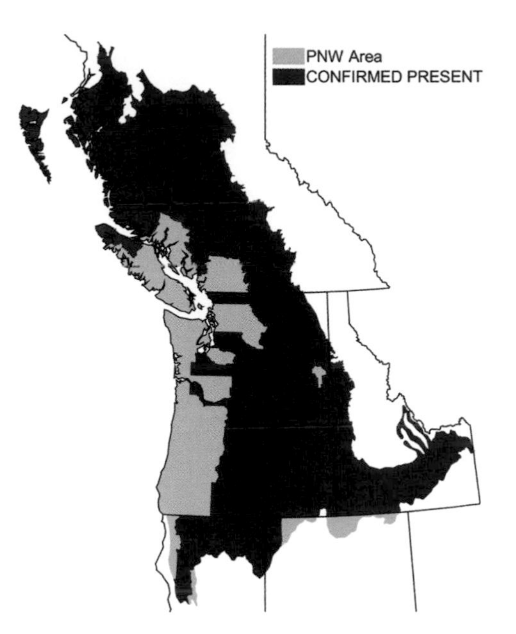

History of Invasiveness

Cotton and Musk Thistles are native to southern Europe, central and western Asia, and eastern Africa. Cotton Thistle was found in Roman deposits and likely made its way into northern Europe via the Roman Highway, a perfect disturbed area for weed colonization. From Europe, the thistles were brought to North America in the late 19th century as ornamentals. After introduction, seeds escaped and were rapidly dispersed by wind, taking hold and working their way westward as soil was disturbed. Today Cotton and Musk Thistles are prevalent almost everywhere in North America and are dispersed by wind, wildlife, livestock, and water. Humans also do their part to transport the species, by way of clothing, in hay or agricultural machinery, and possibly still as ornamentals or garden flowers. It is worth noting that in WA it is illegal to buy, sell, or distribute all plants, parts of plants, or seeds in packets of "wildflower mixes" of both Cotton and Musk Thistle. However, nurseries in the PNW still sell these thistles, as well as Canada and Bull Thistles (*Cirsium arvense* and *C. vulgare*, respectively).

Other Sources of Information: 12, 14, 48, 89
References: 75, 241, 276, 354
Author: Jackie L. Carter

Bull Thistle *Cirsium vulgare*
Canada Thistle *Cirsium arvense*

Species Description and Current Range

Cirsium thistles are exceptionally spiky even among thistles. The genus name comes either from *kirsion*, a kind of thistle with medicinal properties, or from the Greek *kirsos*, meaning "swollen vein," referring to an effect of being pricked by a thistle. Bull Thistle is depicted in Scotland's emblem with the motto "No one harms me without punishment."

Bull and Canada Thistles are often confused. Bull Thistles grow tall (1–2 m high), with a single stem bearing many large (4–7.5 cm), erect, dark purple blooms. Canada Thistles are smaller (usually 1 m high or less) and frequently have many stems with relatively small (1–2 cm) lavender, pink, or occasionally white blooms. Leaves of both species are alternate and deeply lobed. Bull Thistle stems, bracts, and leaves are covered with long spines (5–8 mm). Canada Thistles also bear long spines on their stems and bracts, but only prickles, not spines, on their leaves.

Throughout the PNW, both species are widely distributed and can reach high densities of several hundred individuals/m^2. Canada Thistles are considered a noxious weed throughout the PNW, and Bull Thistles only in OR and WA. Both species are common in grasslands and riparian and seminatural areas. Canada Thistle invaded the regenerating blast zone on Mount St. Helens.

C. arvense

Impact on Communities and Native Species

Both thistles crowd out and rapidly replace native vegetation, which consequently reduces animal diversity. Canada Thistle is considered one of the world's worst weeds due to its ability to invade nearly all crops, pasturelands, and disturbed grassland habitat, although it does not appear to invade intact natural habitats. Because of its remarkable ability to grow vegetatively, Canada Thistle is so pervasive that it is the defining species for a class C weed in WA. Bull Thistle impacts are similar but not as severe because this thistle spreads only by seed and thus not so aggressively. Both species are primarily a problem in rangelands and agricultural fields. They can, however, invade native grasslands and prairies if there has been human disturbance to create patches of bare ground. In rangeland habitats where grazing animals avoid the spiny *Cirsium* thistles, *Cirsium* easily becomes the dominant plant. Thistles are good competitors and reduce crop yields at a cost of millions of dollars annually. Both are widely used by native and non-native butterflies and bees, which produce a tasty honey.

Control Methods and Management

Cirsium thistles should be positively identified before control is attempted because of possible confusion with native thistles. Generally, *Cirsium* thistles are very difficult to control.

They will continue to make seed once they begin flowering. If there are few plants, cutting them before flowering each year will eliminate the infestation. Otherwise, the flowering heads must be cut off and burned rather than discarded which risks spreading the seeds. In agricultural areas, *Cirsium* thistles can be controlled through plowing before flowering, although Canada Thistles readily regenerate from root fragments, so plowing must be done repeatedly.

In prairies and meadows, control efforts aim at enhancing the growth of natives by direct seeding and by restricting grazing to protect natives. Controlled burns, if repeated annually for 3–5 years, can be effective in eliminating *Cirsium* thistles. Spot treatment with glyphosate also can be effective in areas where thistles are interspersed with desirable natives. Biological control with insect herbivores has been studied extensively and is not recommended. Unfortunately, some introduced insects attack and harm populations of native thistles, and given that Canada Thistle is not well controlled by insects in its native range, biocontrol is unlikely ever to be effective. Rust fungus infections, however, may be more promising.

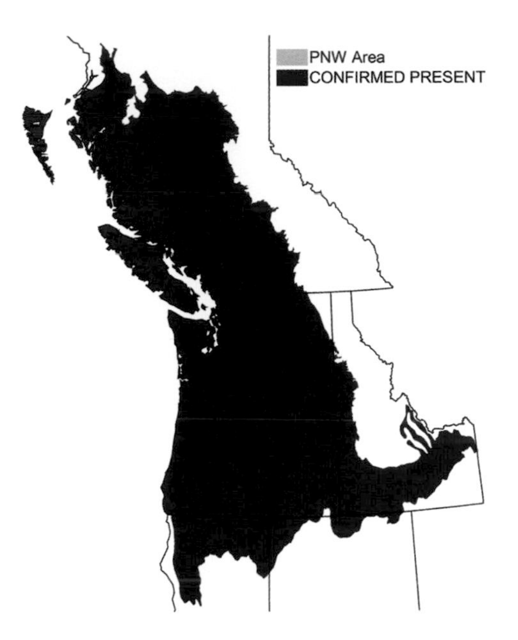

Life History and Species Overview

Cirsium thistles are native in nearly all of Europe as well as the British Isles and tend to be found in open, moist areas where temperatures rarely drop below freezing or hit highs over 32°C. Both are found in disturbed habitats in well-drained soils. They flower from July to September, require full or partial sun, and decline in full shade.

Canada Thistle is the only invasive thistle that is dioecious, with separate male and female plants. The ease of vegetative reproduction, with roots that can grow 4–5 m laterally per year and penetrate the soil up to 1 m deep, allows rapid spread from roots and root fragments. An established infestation can produce up to 250,000 seeds, which disperse readily by wind, live as long as 20 years in the soil, and germinate easily with soil disturbance. Most plants are short-lived, from a single year to a few years of age.

Bull Thistle is usually biennial, meaning that it lives 2 years, flowering only in its second year. Each plant may have up to 60 flowers, which can produce 15,000 seeds. These wind-dispersed seeds do not persist in the soil. Bull Thistle reproduces only from seed.

History of Invasiveness

In North America, both species were likely introduced as a crop contaminant by European colonists, Canada Thistle in the 17th century and Bull Thistle somewhat later. As Canada Thistle quickly became a pest in crops and pastures, control legislation was enacted in VT as early as 1795 and in NY in 1831. Since both spread readily by seed and are often mixed with crops or hay, it is not surprising that the thistles reached the West Coast by the mid-1800s. The two species are found throughout Canada and the US, except in the most northerly and southerly parts.

Other Sources of Information: 92, 93, 117, 125, 126

References: 94, 109, 182

Authors: Patricia A. Townsend and Martha J. Groom

Traveler's Joy *Clematis vitalba*

Species Description and Current Range

Traveler's Joy is a deciduous woody vine that commonly grows along fence lines, streams, and forest edges where soil is disturbed. Young plants can be expected to grow at least 2 m/yr, eventually reaching a length of 30 m. In the PNW Traveler's Joy occurs primarily on the west side of the Cascades, though it is currently spreading east along the Columbia River. The vine was suggested for listing as a class C noxious weed by San Juan County, WA, where it has become a major problem. It grows near the town of Eastsound on Orcas I., engulfing trees on the side of the road, and has spread to both the north and south edges of Moran State Park. It is also prevalent in King County, WA.

This species of clematis looks very similar to the native PNW variety, *Clematis ligusticifolia*, which is also called Traveler's Joy. Therefore, it is important to note a few key identification characteristics of *C. vitalba*. The leaves of the invasive Traveler's Joy (*C. vitalba*) usually comprise five (though sometimes three) leaflets. Generally the margins are smooth, though they may appear to be jagged due to the formation of one to three lobes. The leaves of the native *C. ligusticifolia* are more jagged and may have up to seven leaflets within one leaf. Both species of clematis have small white flowers that form in clusters and are approximately 2 cm in diameter. These flowers form a feathery seed head in the summertime, giving this vine its other common name, "Old Man's Beard." The key difference in these two species is that *C. vitalba* contains both male and female parts (stamens and pistil) in the same flower, while the native *C. ligusticifolia* has separate male and female plants. Female plants of *C. ligusticifolia* have flowers that contain a pistil and no stamens, while male plants have only stamens. Another key difference is in their distribution: *C. ligusticifolia* is only found east of the Cascades and *C. vitalba* primarily on the west. One exception to this is along the Columbia River where *C. vitalba* easily spreads upriver.

Impact on Communities and Native Species

Traveler's Joy is often mistaken for Kudzu (*Pueraria montana* var. *lobata*) because of its tenacious growing habit and tendency to smother trees and shrubs. It covers trees, blocks out available light, and puts tremendous weight on the tree, which eventually results in the death of the tree. After the tree has died and collapsed, the vine continues to grow, creating thickets of vegetation several m high. If there are no trees to climb, the vine crawls along the ground and inhibits regeneration. In New Zealand, detrimental effects on native forest include toppling trees over 20 m tall. Areas infested with the invasive Traveler's Joy usually have a higher amount of other weedy species, irregularly shaped native shrubs, and an overall loss of biodiversity due to the smothering habit of this plant. Native species that are most affected are those that are already endangered or are sensitive to changes in habitat.

Noxious-weed managers on Orcas I. consider the fast-growing Traveler's Joy to be as great a threat to native forests as ivies (*Hedera hibernica* and *H. helix*). It not only topples trees but it is also very difficult to control once it has invaded an area. It spreads by seed, root, and vegetation. In a sunny location the vine reaches reproductive maturity relatively early (after 1 year); a large specimen of Traveler's Joy can produce over 100,000 feathery seeds, dispersed by wind, water, and animals. Traveler's Joy also regenerates easily from damage or pieces of root or stem that touch the ground.

Control Methods and Management

A combination of cutting and chemical treatments is recommended to fully eradicate the vine. During the winter, cut climbing vines at waist height and leave the remaining foliage above to wither. The optimal time to spray the plant is in the spring once it has sufficient new growth to carry the herbicide down to the roots. Both glyphosate (Roundup) and triclopyr (Brush Be Gone) can suppress Traveler's Joy. Another option, for more mature plants, is to apply the herbicide to the cut stump of the vine. Chemical treatments can have negative effects, such as killing non-targeted plants, polluting the soil and water, and causing human health problems, so careful application is crucial.

Because the vine can resprout after being cut or damaged, any physical method of control should be used when the plant is young, and all parts must be burned or disposed of to prevent regrowth. Do not try to compost the debris. Because of the serious threat posed by Traveler's Joy, there is ongoing research to find biological controls of the plant, such as fungi (carried by flies) that are lethal to the vine. Any control method should be coupled with careful monitoring in subsequent years to make sure control is successful.

Life History and Species Overview

Traveler's Joy is native to Europe and southwestern Asia. It will tolerate moderate shade but prefers open, sunny areas such as forest edges, riparian zones, roadsides, and disturbed areas. Historically, the vine decorated summer huts and was used to bundle wood for transport. In western Europe, it is a common sight in hedgerows and along roadsides, which is how it got the name of "Traveler's Joy." This name was given not because of the plants' ability to travel quickly, but because its creamy, sweet-smelling flowers are thought to brighten the day of travelers who pass it on the roadside.

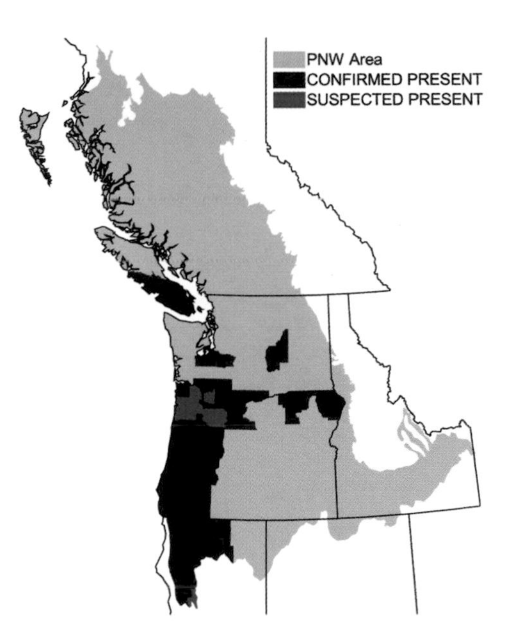

History of Invasiveness

Traveler's Joy was brought to Roche Harbor (San Juan I., WA) in 1903, though it is unclear if this was the very first introduction to the PNW. Both the transport of the plant and its subsequent seed dispersal have aided in the spread of the vine. Like many species of clematis, it was valued for its decorative appearance as well as its more benign behavior exhibited within its native range.

Other Sources of Information: 64
References: 79, 157
Author: Samantha Martin

Tree-of-heaven *Ailanthus altissima*

Species Description and Current Range

Tree-of-heaven, or ailanthus, is a small to medium-sized (10–25 m) deciduous tree, with relatively smooth gray bark and large (30–90 cm or more in length) pinnately compound leaves. The leaves bear a superficial resemblance to those of Black Walnut or Sumac but can be distinguished by their disagreeable odor, reminiscent of peanuts or cashews, when crushed. Tree-of-heaven is found in a variety of disturbed habitats, natural and artificial, in most of the US, particularly in the East and West Coast states. In the PNW it is discontinuously distributed, primarily along roadsides and in riparian areas, where it often forms dense thickets. Tree-of-heaven also occurs in southern BC.

Impact on Communities and Native Species

Tree-of-heaven grows quickly, sprouts extensively from stumps and roots, produces large numbers of viable seeds, tolerates extremely harsh conditions, and proliferates rapidly. It is also allelopathic, producing ailanthone and other compounds in its roots, leaves, and bark that inhibit establishment and growth of other species. This combination of traits results in rapid domination of invaded sites at very high densities, leading to the exclusion of other plant species through shading, root competition, and allelopathy. In riparian locations in the PNW, Tree-of-heaven may also degrade salmonid habitat through exclusion of larger tree species that provide more favorable shade and in-stream habitat features. The roots of Tree-of-heaven can also damage building foundations, pavement, and sewers.

Control Methods and Management

Control of Tree-of-heaven requires sustained efforts because of prolific sprouting from stumps and root fragments. Mechanical control is effective on small trees when the plant can be uprooted (e.g., with a weed wrench) without leaving extensive root material. Chemical control, or a combination of mechanical and chemical control, is necessary for trees that are too large to uproot. Summer-season foliar spraying of glyphosate, triclopyr, dicamba, or imazapyr herbicide is effective in situations where desirable species will not be damaged. Late winter or early spring application of triclopyr (ester form) in an oil solution to the bark at the base of the trunk and summer application of triclopyr (amine form), dicamba, or imazapyr to angled cuts in the sapwood are also effective. When trees are cut down, application of triclopyr, dicamba, or imazapyr to the stump immediately after cutting during the growing season controls stump sprouting, but is less effective than other methods in limiting root suckering. Cutting of trees without chemical control often results in proliferation of sprouts and should be avoided. Similarly, burning is not recommended because it stimulates sprouting. Regardless of the control technique used, follow-up removal or herbicide applications are usually required to suppress subsequent sprouting. Establishment of a native tree canopy above Tree-of-heaven seedlings may limit spread and assist in control.

Life History and Species Overview

Tree-of-heaven is native to China, where it has been used for firewood and in wood products and traditional medicines. It is able to reproduce both asexually, through root and stump sprouts, and sexually. The tree's small yellow-green flowers, pollinated by insects, form large clusters at the tips of branches. Most trees produce only male or female flowers. Female trees are prolific, producing 325,000 or more seeds annually. Following exposure to cool temperatures, the seeds have high germination rates in moist soil. Seedlings quickly develop a taproot and can grow more than 1 m/yr in height for the first several years. Root sprouts can grow even faster. In some locations rapid increases in height (1 m/yr) are sustained for the first 20 years, making Tree-of-heaven perhaps the fastest growing tree in North America. This ability to gain height quickly is probably the source of its common and scientific names (*Ailanthus*, "sky tree").

Tree-of-heaven can tolerate a wide range of adverse conditions, including nutrient-poor soils, highly acidic (pH <4.1) soils, low soil oxygen, compacted soils, seasonal drought, and high levels of air pollution. This tolerance is evidenced by the tree's frequent occurrence as a volunteer in sidewalk cracks and roof gutters. It grows quickly in full sun, but also tolerates partial shade. It cannot establish in deep shade, but sprouts can persist and can grow rapidly if light becomes available, for example, if disturbance opens the canopy. Tree-of-heaven is the only member of its genus that tolerates cold climates. Although very young trees are susceptible to freeze damage, older trees can withstand extremely cold temperatures (–33°C).

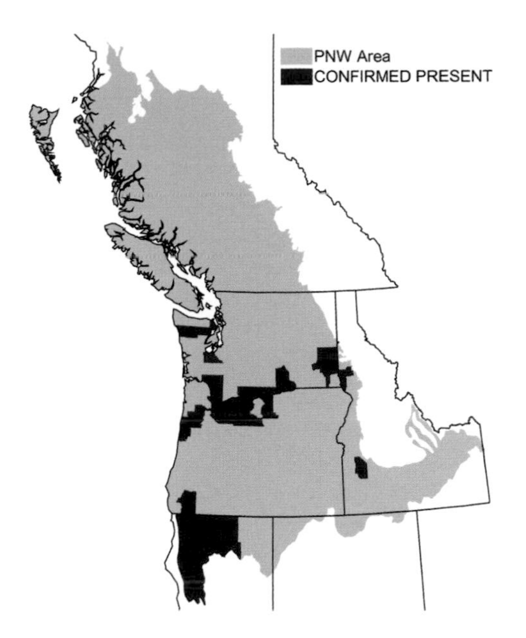

History of Invasiveness

Tree-of-heaven was brought to Europe in the mid-1700s because it was erroneously thought to be a source of lacquer. It arrived in the US in 1784, when it was planted by a gardener in Philadelphia, PA. By the mid-1800s, it was available in plant nurseries and was widely planted in the US and Europe as an ornamental tree because of its visually appealing foliage, rapid growth, and tolerance of harsh conditions. Its ability to thrive in places where many other trees could not survive, memorably described in Betty Smith's 1943 novel *A Tree Grows in Brooklyn*, led to its widespread planting as a street tree in cities in Europe and the eastern US. Tree-of-heaven was also introduced to many sites in CA by Chinese miners during the gold rush.

From the many locations where it is intentionally planted, Tree-of-heaven spreads locally by root sprouts, which can occur up to 15–25 m from the parent tree. It spreads farther by seeds, which are small (30,000/kg) and are attached to papery-winged structures allowing long-distance transport by wind. Birds and other animals, as well as human agents such as cars and farm machinery, may also spread the seeds. Intentional introductions also remain a threat, as Tree-of-heaven is still available in commercial plant nurseries. It increases in abundance with proximity to cities and often occurs along roads, which appear to facilitate its spread. Although individual trees are relatively short lived (ca. 30–50 yrs), the species can dominate a site for a much longer period through resprouting and allelopathy. An indication of this persistence is the existence of live sprouts from the original tree planted in Philadelphia in 1784.

Other Sources of Information: 18, 53, 61, 92
References: 166, 168, 240
Author: David Stokes

Hairy Whitetop *Lepidium draba* ssp. *draba* (*Cardaria draba*)

Species Description and Current Range

Hairy Whitetop is a perennial, rhizomatous, grayish-green plant standing 15–50 cm tall. Also called Hoary Cress, its leaves are oblong or arrowhead-shaped, 2–8 cm in length, and may have smooth or finely toothed edges. The upper leaves are typically smaller and clasp the stem, while the lower leaves are larger and taper to a short stem. Flower clusters are dense and flat-topped, consisting of numerous four-petaled white flowers, each 4–8 mm across. When in bloom, dense infestations of Whitetop often look like patches of snow from a distance. Mature seedpods resemble upside-down hearts, distinguishing it from other closely related varieties of whitetop. Hairy Whitetop occurs throughout most of the PNW but is particularly concentrated in the more arid portions of the region. It is most aggressive in open habitats that are seasonally moist, such as wet meadows, streamsides, or irrigated fields, but can also grow in drier upland locations. Whitetops are salt-tolerant and are frequently found in habitats with salty, alkaline soils.

Impact on Communities and Native Species

Hairy Whitetop impacts our native ecosystems by reducing biodiversity, inhibiting natural recovery of disturbed areas, and competing with rare species. It has invaded large acreages in much of the PNW, including over 15,000 ha in ID and more than 4,000 ha in WA. In OR, it is estimated that over 500,000 ha are affected by this plant. Hairy Whitetop grows very rapidly and often forms dense monocultures, outcompeting and displacing native plant species. One plant can spread by rhizomes into a 4-m-diameter patch with hundreds of stems in a single growing season. Its extensive root system penetrates up to 10 m deep into the soil and rapidly produces numerous lateral roots. Because it begins its rapid growth earlier in spring than many native plants, Hairy Whitetop has the ability to deplete water and nutrients before native species have a chance to use them. In addition, its roots secrete chemicals, known as glucosinolates, which reduce the germination of wheat, alfalfa, and the native Bluebunch Wheatgrass. In WA, Hairy White-top is known to threaten at least two rare and endangered plant species: Wenatchee Mountains Checker-mallow and Wenatchee Larkspur.

Hairy Whitetop also has considerable impacts on agriculture, reducing yields of crops, such as wheat and alfalfa, and forage availability on rangelands. The same glucosinolate chemicals that inhibit germination of plants can also be toxic to livestock if eaten in large amounts. They contain a high concentration of sulfur and form compounds in the digestive tract that can inhibit the absorption of iodine. These chemicals also are believed to taint the flavor of milk from livestock that feed on Whitetop. Economic losses due to Hairy Whitetop are estimated at over $9 million/yr in OR's economy alone.

Control Methods and Management

As with all weeds, prevention is the most effective management strategy. When leaving infested areas, be sure to clean clothing, shoes, vehicles, and other equipment. Limit soil disturbances, particularly in areas with nearby Hairy Whitetop infestations. Manage grazing animals appropriately to maintain healthy, competitive vegetation, and avoid grazing Whitetop when it is in seed. Once it is established, the most effective methods for controlling Whitetop are herbicide application and mechanical removal. Consult your local extension office for a list of effective herbicides. Herbicides are most effective if applied when the flowers are in full bloom or when rosettes form in the fall. For small infestations of seedling plants or locations with only a few mature plants, extensive digging that ensures removal of all the roots may be successful. This type of mechanical method must be repeated frequently throughout the growing season to ensure control. Mechanical methods are rarely effective on established stands. Several insects and soil-dwelling organisms are being actively researched as biological controls, but none has been released.

Life History and Species Overview

Hairy Whitetop is native to the southern portions of central Europe and western Asia. It is considered a problem weed in portions of its native range, as well as throughout much of Europe and North America. It is found as far south as Guatemala, and as far north as northern Finland. While it is a troublesome weed, seeds are useful as a pepperlike seasoning, and honeybees make high-quality honey from the flowers.

Whitetop seeds typically germinate in fall, although some spring germination also occurs. Seedlings develop into rosettes, which then develop lateral roots, stems, and rhizomes. Lateral roots and rhizomes can grow long distances from the original plant, producing new rosettes and above-ground stems. In this manner, a single plant can develop into a large mat with hundreds or even thousands of flowering stems. Plants flower from late May through July, depending on habitat and elevation, and produce up to 5,000 seeds/plant. Seeds have high germination rates initially but are viable for only 2 years once they are in the soil. They likely remain viable for a longer time when not in contact with the soil, for example in hay or seed shipments. New plants can also establish from root fragments as small as 1 cm long.

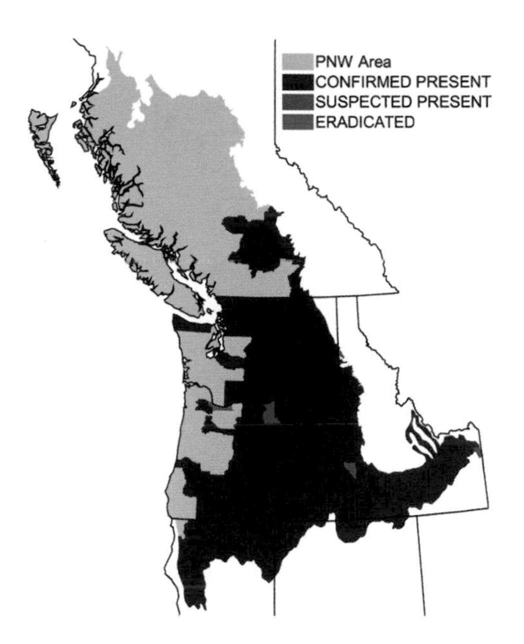

History of Invasiveness

Hairy Whitetop was introduced to North America through contaminated ship ballast and/or alfalfa hay. It was first documented in North America in Yreka, CA, in 1876 and was found in the eastern and southwestern US by 1900. The first introduction in eastern OR, probably through contaminated hay, was in 1909, and it was found in the Portland, OR, area in 1914, most likely introduced from ship ballast. By the 1950s it was found throughout most of the arid portions of the PNW, and it subsequently spread into portions of the coastal and mountainous areas. Most of its spread has been due to contaminated alfalfa hay and seed; seeds are also spread via farm equipment, vehicles, livestock, and moving water. Hairy Whitetop continues to expand its range throughout much of the PNW.

Other Sources of Information: 18, 43, 67, 91, 125
References: 267, 342
Author: David L. Wilderman

Asian Freshwater Clam *Corbicula fluminea*

Species Description and Current Range

The Asian Freshwater Clam (AFC) has two thick shells with coarse, concentric ridges and ranges from yellow-gold to dark brown in color. Young, smaller shells are light yellow-gold, and most large shells are worn near the hinge, exposing the white shell layers underneath. Adults rarely grow larger than 40 mm and are usually about the size of a quarter. They have very short siphons, translucent white foot and mantle tissue, and strong shells. No native freshwater clams in the PNW closely resemble the AFC. Our native clams are thin, smooth, smaller than 25 mm, and their shells easily break when squeezed.

AFCs are found in the shallow portions of lakes, canals, reservoirs, rivers, and streams. They live at the sediment surface in gravel, sand, and mud, usually at very high densities. Aggregations of these clams are easily mistaken for gravel when viewed from above the water surface. The clams do not survive in water that approaches freezing, and this trait may slow their spread to BC and higher elevations. The AFC is very abundant in the Columbia River, Snake River, Willamette River, Lake Shasta, Lake Washington, and tributaries. These clams also live in slightly brackish waters in the tidal portion of rivers entering Willapa Bay and Grays Harbor.

Known as "Prosperity Clams" in their native Asia, AFCs are eaten fresh or as a dry powder to aid liver function and have been deliberately introduced throughout the world. Found in 43 states (including HI), eastern Canada, and northern Mexico, this is the most widespread introduced freshwater mollusk in North America.

Impact on Communities and Native Species

AFC populations reach densities of 130,000/m^2 in parts of CA. These clams are strong competitors because they eat suspended food particles by filtering water and organic material in the sediments by digging with their foot and passing this material to their mouth. Although high densities of AFCs displace native species, at low densities impacts to native communities are not known.

The AFC has been in North America at least 70 years. It is used as an index organism in pollution studies, as a food source for native fish, crayfish, birds, and mammals, as commercial fish bait, and as locally produced live and powdered products for human consumption. This clam is an important food for juvenile White Sturgeon in the Columbia River and for diving ducks worldwide. AFCs cause over $1 billion damage/yr in the US by blocking irrigation systems and water intakes to power generating facilities (including 11 of the 38 US nuclear facilities) and by clogging other industrial water systems. Prior to the introduction of Zebra Mussels, AFCs were the most damaging molluscan invader in North America.

Control Methods and Management

The spread of this species is not carefully managed. BC, WA, CA, ID, and OR list it as exotic, but only OR enforces regulations governing intrastate transport and possession. Industrial facilities control infestations through combinations of thermal, chemical, and mechanical techniques. Managers use heated water, chlorine, bromine, and screening/particle traps to prevent these clams from entering industrial water systems. However, these techniques are not generally feasible (or legal in the case of chlorinating public bodies of water). The only documented successful eradication of AFC was from a small, private Koi pond in HI. People encountering this species should report the location to their state's Fish and Wildlife Dept.

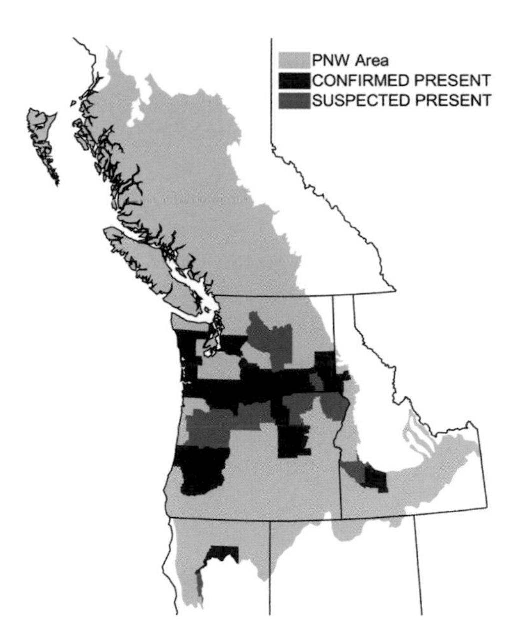

Life History and Species Overview

AFC is hermaphroditic (both sexes in one individual) and is capable of self-fertilization. Larvae are brooded internally for up to 5 days, and over 350 crawling juveniles are released per day. Crawling juveniles enable this species to spread upstream or downstream. A single clam can produce up to 70,000 juveniles/yr. Spawning can occur continuously in warm waters but is most common from July through September. Juveniles reach sexual maturity at 7-mm shell length, which is possible in 3 months in typical conditions, and adults live up to 4 years. Populations reach highest densities in warm areas of the South and Southwest, where spawning can occur year-round.

History of Invasiveness

Native to Southeast Asia, the AFC has spread to Europe, Africa, Australia, the Pacific islands, South America, and North America. Empty shells were first collected in Nanaimo, BC, in 1924. Live adults were discovered in 1938 in the Columbia River, WA. Unknown pathways spread the invasion from the PNW to the Ohio River in the late 1950s. This second point of invasion and subsequent natural expansion has led to naturalized populations in all states except AK, NH, VT, ME, the Dakotas, and MT. This amazing rate of spread corresponds to crossing a new state every 2 years, with the wave covering the length of the Mississippi River in less than a decade.

Humans are very effective at spreading AFCs, both intentionally and accidentally. The original introductions were coincident with significant immigration of Chinese laborers, in both WA and CA. In addition, clams have been illegally spread to streams and reservoirs throughout HI for food stocks. Adult clams are sold in CA and some other states as fishing bait, accidentally transported in sand and gravel for construction, and sold by pet stores for aquaria. The invasion continues to grow each year and is expected to be present in all US states and BC during this decade.

Other Sources of Information: 111
References: 34, 72, 178, 250
Author: Alan C. Trimble

Red Swamp Crayfish *Procambarus clarkii*
Rusty Crayfish *Orconectes rusticus*
Virile Crayfish *Orconectes virilis*

Species Description and Current Range

Crayfish, also called crawfish, crawdads, mudbugs, yabbies, or creekcrabs, are small, lobsterlike crustaceans that live in streams, lakes, and estuaries throughout the world. Crayfish often brawl over food, shelter, and mates, using pinchers to wrestle with their opponents and also urinating at their opponents as a form of intimidation. Like lobsters, crayfish are a delicacy and are farmed and harvested. Crayfish are eaten in huge quantities in some regions of the world—one crayfish-eating contest in the southeastern US was won by a man who ate 25 kg of crayfish in an hour! Although there are several native species in the PNW, three crayfish species threaten to invade or are in the process of invading: the Red Swamp Crayfish, the Rusty Crayfish, and the Virile Crayfish. The Red Swamp Crayfish is already in a few places in ID, CA, OR, and most recently WA, but the Rusty and Virile Crayfish have not yet invaded the PNW.

The Red Swamp Crayfish, often cooked in Cajun dishes, grows up to 12 cm long, not including its long, narrow pinchers. This crayfish was a war totem of the Houma and Chakchiuma tribes of the southeastern US. It is dark red, nearly black on its carapace, and has a wedge-shaped black stripe on its abdomen. Found in lowland ponds, wetlands, and lakes, and occasionally in slow-moving streams, the Red Swamp Crayfish survives in seasonal ponds by burrowing into the mud to avoid droughts. Rusty and Virile Crayfish need permanent fresh water because they don't burrow to avoid drying. Adult Rusty Crayfish are about 2 cm shorter than the Red Swamp Crayfish, with a dark spot on each side of the carapace and stout, black-tipped claws. Virile Crayfish are even smaller and can be identified by their bluish claws that are covered with white warty bumps.

Procambarus clarkii

Impact on Communities and Native Species

Invasive crayfish wreak havoc on native species because they feed on everything, including aquatic plants, snails, fish eggs, amphibians, and carcasses. Consequently, crayfish drastically change the community composition and characteristics of lakes that they invade. Invasive Rusty Crayfish in midwestern lakes ate 80% of all the aquatic plants, ate many of the aquatic invertebrates, and eliminated the native species of crayfish. Rusty Crayfish outcompete native crayfish for food and evict the native crayfish from safe hiding locations, exposing them to predacious fish. Rusty Crayfish are aggressive and have been known to pinch human swimmers and to attack the nests of bluegill fish to eat their eggs. Although less aggressive than Rusty Crayfish, Virile Crayfish also can have large impacts on communities that they invade. For example, invasive Virile Crayfish in AZ have caused concern because of their propensity for attacking rare native fishes.

Red Swamp Crayfish also can have large impacts on native communities. In CA, these crayfish fed on so many newt eggs and larvae that they probably caused newt populations to decline. The Red Swamp Crayfish also acts as a vector of diseases that attack native crayfish. It is an aggressive burrower, breaching earthen dams needed for rice farming, which makes it a severe agricultural pest in some regions. Introduced crayfish don't always have negative impacts—the Red Swamp Crayfish was used as a biological control, with unknown effectiveness, in Africa to feed upon the snails that harbor the parasite that infects humans with schistosomiasis.

Control Methods and Management

The best method to control invasive crayfish is to prevent introduction by anglers, aquarium owners, and aquaculturists. After an invasive crayfish species is established, there is little that can be done that wouldn't also harm rare and native crayfish. Managers have experimented with Largemouth Bass to control invasive crayfish; however, the larger and more aggressive invasive species seem more resistant to fish predation than their native counterparts. Trapping crayfish removes some individuals but is time-intensive and not that effective. Detection of these invasive species is often difficult as well, and the ranges of both invasive and native crayfish are not well described.

Life History and Species Overview

The female of these crayfish lays 100–500 eggs, which she fertilizes with sperm stored from a previous mating. She attaches the fertilized eggs to the underside of her tail on her swimmerets for 3–6 weeks until they hatch. The baby crayfish cling to the underside of their mother for several more weeks until they are ready to live on their own. These three species of crayfish live 3–5 years. Red Swamp Crayfish are native to the swamps of the southeastern US; the Rusty Crayfish is native to the Ohio River basin; the Virile Crayfish is native to central Canada and WI, MN, and MI.

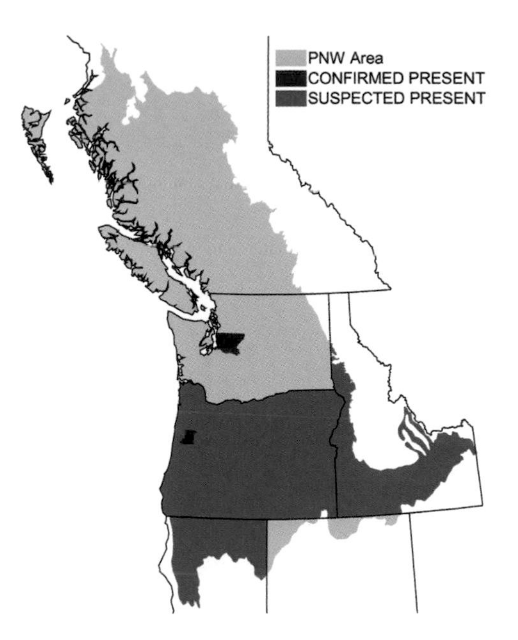

History of Invasiveness

Crayfish introductions are both accidental and intentional. Aquarium owners, anglers using them as live bait, and people interested in harvesting or using them in aquaculture have all facilitated the spread of these invasives. Because female crayfish can store sperm over the winter, one female with sperm can potentially start a new population.

The Rusty Crayfish has escaped its native range in the Ohio River basin to invade MI, MO, NM, MN, and all of New England (except RI). Although not yet in the PNW, it is likely to arrive. Similarly, the Virile Crayfish also threatens to invade the PNW. It has already invaded AZ, NV, UT, HI, and parts of CA. The Red Swamp Crayfish has reached the PNW; it came to CA in the 1920s, ID in 1975, OR in 1980, and was found in WA in 2000. The Red Swamp Crayfish has an impressive history of invasiveness—in the 1970s, it arrived in Spain and has since spread throughout Europe, decimating native European crayfish and the associated aquaculture industry. Among other countries, this crayfish has also invaded Costa Rica, Uganda, Kenya, China, Taiwan, and Japan.

References: 193, 194
Author: Jonathan W. Moore

New Zealand Mudsnail
Potomopyrgus antipodarum

Species Description and Current Range

Shells of adult New Zealand Mudsnails in the PNW are 3–6 mm long and about twice as long as wide, with 5–6 spiral turns or whorls. Some populations have a faint ridge that spirals around the shell, mid-whorl. The shell opening is on the animal's right side and is sealed by a horny disc attached to the top of the rearward-facing portion of the foot. In the PNW, a female can give birth 2–3 times a year to about 90 live juvenile snails that are about the size of small grains of sand.

The Mudsnail inhabits lakes, ponds, streams, rivers, reservoirs, and estuaries with low salinity. Mudsnails colonized WA, OR, and ID in the late 1980s. They can live in extremely high population densities: the highest recorded is a staggering 800,000/m^2 in Lake Zurich, Switzerland. At Banbury Springs, near Hagerman, ID, densities reach 500,000/m^2 and comprise 90% of the visible invertebrate biomass. Under laboratory conditions, the Mudsnail's growth rate averages 0.1 mm/day at 21°C, and they reach reproductive maturity at 3 mm. In the PNW the Mudsnail's growth rate is slowed in the winter unless it is in a geothermal spring habitat. Populations in the PNW are composed of female clones that do not require a male for reproduction. Relocating a single snail may start a new population.

Impact on Communities and Native Species

In New Zealand, the Mudsnail increases in density in degraded habitats and can be an indicator species for poor water quality and higher water temperature. Mudsnails can change the makeup of the community they invade by consuming most of the primary production and by becoming the dominant macro-invertebrate. In one system, they consumed roughly 75% of the available plant matter and accounted for 66% of the ammonia excreted. The snail's consumption can reduce diatom and algae biomass, and at high densities it may make up most of the macro-invertebrate biomass. In the Madison River, MT, it reduced native macro-invertebrate densities and biomass, including mayflies, caddisflies, and chironomids. In ID, it competes directly with two endangered species in the Snake River, the Snake River Physa Snail (*Physa natricina*) and the Idaho Springsnail (*Pyrulogopsis idahoensis*). The Mudsnail was also one of the reasons for a threatened listing of the Bliss Rapids Snail (*Taylorconcha serpenticola*). In New Zealand, fish eat the Mudsnails. In the PNW, one Mudsnail passed through a trout's digestive tract undigested, unharmed, and able to immediately give birth to live young. Whether trout in the PNW will feed in the wild on Mudsnails is unknown but with Mudsnails at densities that are likely to displace native prey species, fish could face food scarcity while surrounded by the inedible abundance of what has been called a trophic "dead end." Changes in the connections between primary producers and secondary consumers can have dramatic, and likely negative, impacts on native species. Research is needed to determine the extent of displacement of native species by the Mudsnail's introduction and its impact in the PNW.

Control Methods and Management

No effective method exists to eradicate or control the Mudsnail in the PNW. All New Zealand Mudsnails in the western US appear to have originated from one clone. Such genetic similarity may make this snail vulnerable to an as yet undiscovered biological control. To slow its invasion in the PNW: (1) remove all vegetation that may have fouled equipment and leave it on site; (2) clean all potentially contaminated equipment, waders, and boots thoroughly and allow to dry completely before using in another body of water (cleaning removes larger adults, while smaller juveniles that may remain succumb more quickly to dehydration). When there isn't sufficient drying time, equipment can be frozen, immersed in hot water or a bleach solution, or heated to accelerate drying. Clean fish from infested waters and leave the internal organs on site or dispose of them where they will not have access to water because Mudsnails in the fish's digestive tract may still be viable.

Life History and Species Overview

In their native New Zealand, Mudsnails occur in two types of populations. One is made up almost entirely of females that readily reproduce without males but includes a small number of males (<10%), and another population is made up largely of males and females that require males to reproduce. The Mudsnail follows a common pattern of multiple sets of chromosomes in parthenogenic populations but the number of chromosomes in males varies. Males in the low-male populations may be from rare matings of the parthenogenic females. In Europe the invasive populations are all clonal, have few males, and are not diploid. All known populations in the PNW are parthenogenic and are composed almost entirely of cloned females; only rarely are single males found in these populations. Mudsnails do not have an egg or larval stage, instead bearing offspring that develop directly into adults. Young are born every 3 months in New Zealand but are most commonly produced during summer and autumn months in the PNW. Adult snails collected in the PNW contain 10–90 young in their brood pouches and can release young 2–3 times/yr. The

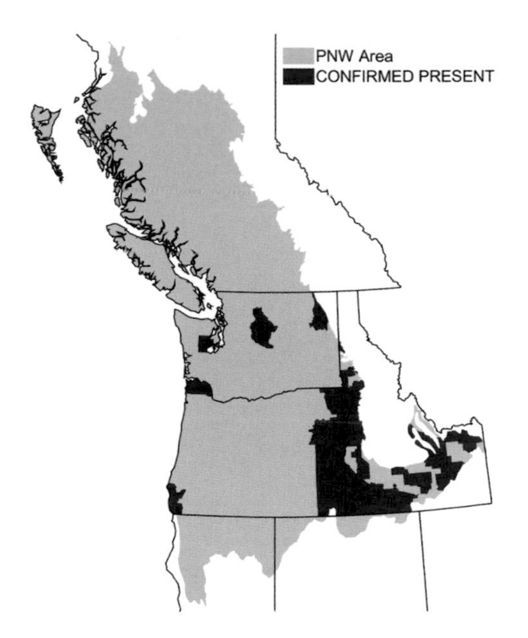

Mudsnail grazes on diatoms, detritus, and attached algal mats by scraping rocks and vegetation with its radular teeth.

History of Invasiveness

The Mudsnail, native to New Zealand, now occurs in Europe, Asia, Australia, and North America. Introduced to Britain in the mid 1800s, it spread rapidly and within 100 years covered Britain and reached Europe. The Mudsnail is now the dominant snail in many of the freshwater and low-salinity locations where it is found in Europe. It reached the western US by 1987, when it was found in the Snake River near Hagerman, ID. By 1989 it was the most common snail in that section of the river and is now found in MT, WY, WA, OR, CA, AZ, and UT. The only known population in the eastern US is in Lake Ontario. Small, abundant, and able to survive short periods out of water, the snail is particularly well suited to hitching rides to new locations with humans and possibly on the feet of birds. Active expansion after introduction can be rapid.

Other Sources of Information: 5, 116, 127
References: 96, 135, 263, 347, 360
Author: Trevor R. Anderson

Siberian Prawn *Exopalaemon modestus*

Species Description and Current Range

The Siberian Prawn has truly gone where no prawn had gone before. Prior to its introduction to the lower Columbia River on the border of OR and WA, freshwater prawns (or shrimps) were entirely absent from the PNW. The paucity of similar species makes identification of the Siberian Prawn unequivocal in freshwater areas of that river, but the native Bay Shrimp, *Crangon franciscorum*, occurs in the estuary, where fresh and marine waters mix. The Bay Shrimp is speckled with light and dark spots and has a broad, flattened body. By contrast, the translucent Siberian Prawn has a narrow, deep body and a long, toothed rostrum (or head spine) that protrudes between the eyes. The Siberian Prawn also has pincers on the second pair of walking legs, and these are absent in the native shrimp.

In Asian populations, male and female prawns reach a maximum size of 60 mm total length, yet individuals as large as 76 mm total length have been recorded from the Columbia River. Their larger size in the PNW may reflect better health, since introduced populations have escaped most of the pathogens that afflict them in other parts of the world. Siberian Prawns are currently restricted to the lower Columbia River and its tributaries, from the estuary to Bonneville Dam at river kilometer (Rkm) 234. The dam, which also hinders the movement of native species upriver, is an effective temporary barrier to their spread, but the prawns will probably bypass this obstacle and reach suitable habitats in the upper Columbia River basin unless a containment policy is implemented. Throughout the lower river, the prawns are distributed widely in a variety of habitats, ranging from deep (>24.4 m) sandy pools to shallow (<0.05 m) marsh channels laden with woody debris. They have even managed to invade a turbid, vegetated lake on Sauvies I., OR, by exploiting a very small outlet into the adjacent river; adult shrimp are proficient swimmers and are apparently capable of migrating between the areas.

Impact on Communities and Native Species

The Siberian Prawn's frequently high density (>72 prawns/m^3) in invaded habitats and its mixed diet of plant and animal material put it at odds with native species in the PNW. The lower Columbia River, for example, figures prominently in the early life history of several endangered or threatened salmon species. Juvenile salmon use this portion of the river as a nursery, gorging themselves on rich food resources before migrating out into the open ocean. Scientists fear that expanding Siberian Prawn populations may decrease the availability of critical salmon prey like small crustaceans, thereby reducing the growth and survival of these ecologically and economically important fish. Siberian Prawns also feed on fish eggs, which may further impact some species already reeling from pollution, overfishing, and habitat loss or degradation.

Control Methods and Management

In the PNW, growing prawn populations have gone relatively unnoticed by state management agencies, and neither OR nor WA has a specific program in place to control them. Both states do have general regulations prohibiting the transport and release of live nonindigenous aquatic animals to avoid further spread of the species. Members of the public who encounter Siberian Prawns should record the time, date, and location of the occurrence, retain frozen specimens if possible, and contact local Department of Fish and Wildlife personnel. Although Siberian Prawns are edible and support subsistence fisheries in China and other parts of Asia, they are not currently managed for recreational or commercial harvest in the PNW.

Life History and Species Overview

Siberian Prawns are native to fresh waters of the western Pacific, including the Amur and Ussur basins of Siberia, Manchuria, Korea, China, and Taiwan. Throughout this region, they occupy slow-moving rivers and streams, ponds, lakes, and reservoirs. While large populations are common in undisturbed waterways, the prawns may become abundant in highly modified drainages and canals as well. Like most shrimps and crabs, they have a complex life cycle characterized by a planktonic (drifting) larval stage and a benthic (bottom-dwelling) adult stage. Their life span is relatively short at 1–1.5 years, and so they waste no time. Prawns become sexually mature before they're 6 months old. Spawning begins as water temperatures increase in spring and is more or less continuous throughout summer. After mating, a female prawn deposits a cluster of 60–350 eggs onto the paddle-like swimmerets of her abdomen, where they remain for several weeks. Although the brood is small, it may represent up to 20% of the female's body weight, and she is heavily invested in the care of her eggs. In fact, an attentive female will spend much of the incubation period grooming and cleaning the developing embryos using special brush-tipped legs. Once the eggs hatch, free-swimming larvae disperse and circulate through the water, feeding ravenously on microscopic plankton. They grow rapidly and eventually metamorphose into a benthic juvenile form, which looks like a miniature adult and spends more time crawling on the bottom than swimming.

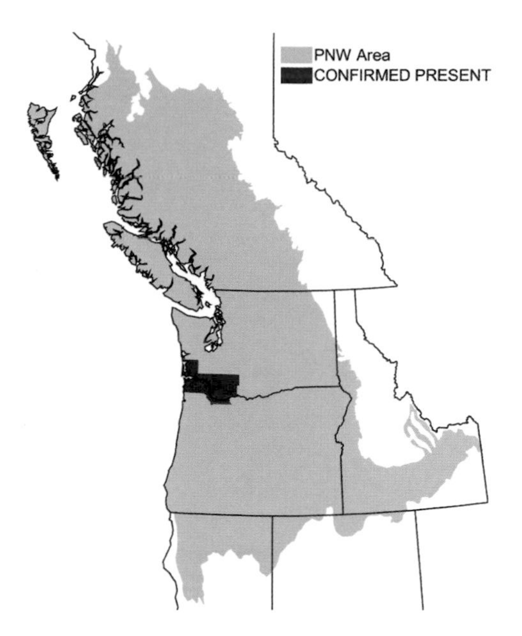

History of Invasiveness

Siberian Prawns were first discovered in the lower Columbia River in 1995, but little is known about the specific origin of the population. It is widely accepted that the initial colonists were transported as larvae in ships' ballast water from an Asian port. Based on their broad distribution in the river, the size range of individuals present, and the presence of brooding females over multiple years, scientists believe the prawns were well established prior to their detection. A large population of Siberian Prawns was discovered in the San Francisco estuary in the late 1990s, and it is unclear whether they originated from Asia or the Columbia River. Stricter ballast water regulations for international and coastal shipping will decrease introductions from foreign ports and may reduce the spread of some species between ports along the Pacific coast. However, the movement of Siberian Prawns into the upper Columbia River is little affected by these rules since ballast water exchange is not regulated within the waterway. The species' spread might be halted if vessels were required to empty or treat ballast water before passing Bonneville Dam.

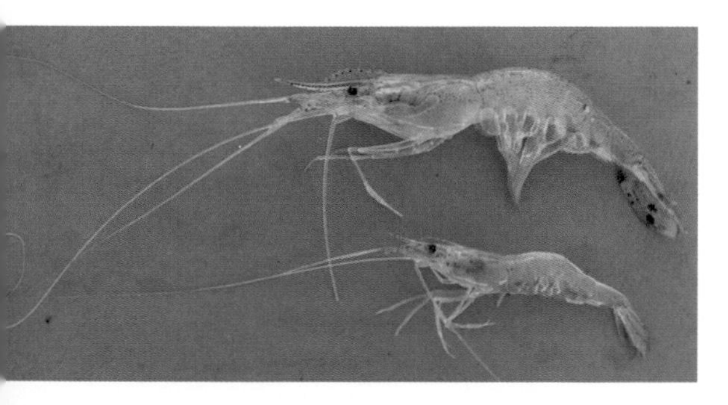

Other Sources of Information: 75, 118
References: 105, 160, 237, 364
Author: P. Sean McDonald

Botrylloides violaceus
Botryllus schlosseri
Molgula manhattensis

Species Description and Current Range

At least five ascidian species have invaded local marine waters, and three are very abundant. *Botrylloides violaceus* forms violet or orange sheets, 3–4 mm thick, of clonal individuals arranged in linear patterns. The Star Ascidian, *Botryllus schlosseri*, grows in gray, black, orange, and maroon sheets of individuals arranged in flowerlike clusters. Both species are present from CA to AK. *Molgula manhattensis* is common in Coos Bay, OR, and the Puget Sound and Willapa Bay, WA. This ascidian looks like translucent beige balls about 2 cm in diameter with two small siphons protruding from the surface.

Botryllus schlosseri

Ascidians are common in pleasure-craft marinas, fouling boat bottoms, docks, and pilings. They also occur in protected natural areas, near and below the low tide line, on rocks, eelgrass, and shells. They are easily observed as the most brightly colored, slimy, motionless creatures in these areas. This simple appearance does not hint at their widespread use in scientific research on immune systems, animal development, metal sequestration, toxin accumulation, and AIDS. Ascidians share many genetic, molecular, and developmental characteristics with humans; they are our distant relatives. Ascidians, however, have some amazing characteristics that make them ideally suited for many types of research. They can clone themselves. These clones can recognize and distinguish between self and siblings.

Impact on Communities and Native Species

Ascidians are fouling organisms, so they require a surface to settle on. They can compete with native organisms for space and grow on the native organisms themselves. While this competition hasn't attracted much research attention, fouling of ship hulls and aquaculture facilities has a strong research history, including a dedicated journal, *Biofouling*. Locally, invasive ascidians are partly responsible for the continued rarity of the native Olympia Oyster, *Ostreola conchaphila*, through space competition and smothering of juveniles. In addition, where these invaders are common, native ascidians have become less abundant.

Control Methods and Management

Shipping traffic is the major means of transport and introduction of ascidians worldwide. While local and international regulation of ballast water is developing, hull-surface fouling has yet to be addressed. Centuries of global travel by large vessels have spread ascidians to major ports worldwide. Recently, growing numbers of pleasure craft, floating dry docks, drilling platforms, barges, and translocation of retired navy vessels have dramatically increased the extent of invasions. In addition, many of the previously effective coatings (abalative copper and TBT bottom paints) that historically prevented fouling of submerged surfaces are no longer used due to their widespread toxic effects on marine ecosystems. However, private vessel owners can control the spread of ascidians by three methods: weekly, intensive manual cleaning of submerged surfaces, moving vessels to freshwater moorage for several days prior to transit, or moving vessels to dry storage several days before transit. Ascidians are intolerant of fresh water and exposure to air.

Life History and Species Overview

Ascidians are named following the Greek *askidion*, which refers to a wineskin or bladder. They are also called "tunicates," implying a covering or tunic surrounding the main body parts, and "sea squirts" because larger animals squirt liquid when poked. They are primitive chordates, with early developmental stages similar to fish and humans. Over 2,000 species are found worldwide as solitary individuals and as clonal colonies.

Ascidians can reach sexual maturity and produce new offspring within weeks after initial settlement. Brooding adults (*Botryllus*, *Botrylloides*) incubate fertilized eggs and release nonfeeding, tadpolelike larvae that settle quickly, usually less than 1 m from their parent. Settlers clone themselves into sheets of genetically identical individuals. Fast reproduction, short dispersal, and cloning enable these invaders to dominate small areas very quickly. Invasions can spread through areas as large as pleasure-craft marinas in months.

Nonbrooding adults (*Molgula*) release sperm and eggs into the surrounding water, where fertilization and development of nonfeeding swimming larvae occur. This reproductive trait allows for wider dispersal and transport in ballast water.

Adult ascidians are filter feeders, capable of capturing particles as small as 0.3 microns (0.00001 inch) from the water they live in. Ascidians are also well known for bioaccumulating vanadium at 100 times the concentration found in seawater. Vanadium compounds in ascidian tissue produce bright colors of yellow, green, blue, orange, red, and purple.

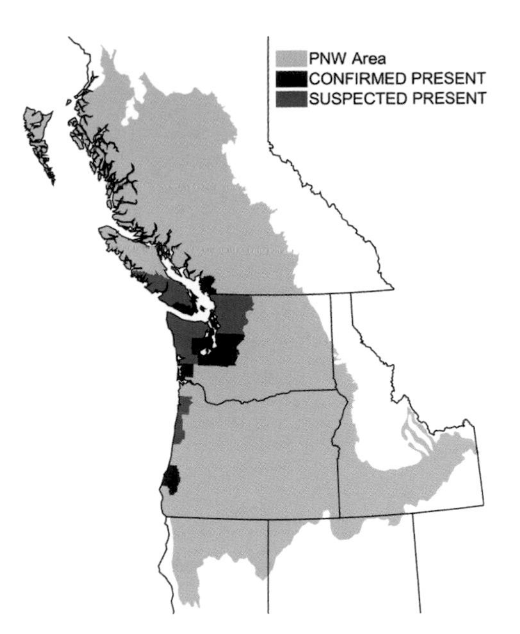

History of Invasiveness

Botrylloides violaceus, native to northern Japan, was first recorded in Puget Sound in 1977. It is found from AK to CA, ME to VA, and throughout northern Europe. *Botryllus schlosseri*, native to Europe, was originally found in an aquaculture facility in the San Juan Is. in the 1980s. It is a common port resident in temperate sea areas worldwide. *Molgula manhattensis*, native to the East Coast of the US, was found in Coos Bay, OR, in 1995 and in Puget Sound in 1998. It is also common in Willapa Bay, WA. Although ascidians are being spread worldwide, the majority of these invasions have not proliferated great distances beyond the artificial surfaces in ports onto natural substrates. The mechanisms preventing this spread are unknown but give hope that controlling ship-borne introductions will be an effective management tool.

Other Sources of Information: 11, 88a, 88b
References: 170, 280, 359
Author: Alan C. Trimble

Atlantic Slipper Limpet *Crepidula fornicata*

Species Description and Current Range

The Atlantic Slipper Limpet, a mollusk and gastropod, is a relative of the common garden snail. Its shell is very different from the classic garden-snail shell, as it is convex and has a wide opening, making it look like a flattened suction cup. The Atlantic Slipper Limpet shell has exterior growth lines and white or pale tan interior (maximum length 59 mm), and its mantle has a white border. The three PNW native limpets (*Crepidula nummaria*, *C. perforans*, and *C. aculeata*) live mainly on the outer coast; the Atlantic Slipper Limpet, in contrast, is most abundant in muddy protected sites such as estuaries and bays because it can tolerate fresh water and variable salinity. Atlantic Slipper Limpets form stacks of up to 15 individuals, with the older females at the bottom and the younger males at the top of the pile.

The Atlantic Slipper Limpet, introduced probably in the late 1890s with oysters, is abundant in Puget Sound and present in Grays Harbor and Willapa Bay, WA, and in Boundary Bay and Victoria, BC. These limpets can reach densities of $141/m^2$. The native range of the Atlantic Slipper Limpet is the Atlantic coast of North America, from Prince Edward I., Canada, down to TX and around Mexico. Atlantic Slipper Limpets are most abundant in shallow, muddy, protected, sublittoral sites such as estuaries and bays. They prefer convex surfaces where they are exposed to less sediment, so typically they attach to shells of mussels and oysters, and to stones.

Impact on Communities and Native Species

Atlantic Slipper Limpets breed very quickly around oyster farms, fowling nets, and oyster shells. Where fishing nets scrape the bottom and create open space, limpets can move in and prevent oysters and other species from settling, all behaviors that reduce the economic value of these resources. Limpets are a serious pest because they increase sedimentation and slow the currents at the bottom, burying oysters, reducing oyster feeding success, hindering oyster recruitment, and changing the structure of benthic communities. When Atlantic Slipper Limpet shells litter the bottom, species that burrow in the sediment cannot burrow easily, spend more time exposed, and become more vulnerable to predators. Atlantic Slipper Limpets can alter ecosystem function through their feeding activity and defecation, increasing the temporary retention of silica. This may have important cascading effects on community structure by decreasing the abundance of basal organisms in the food chain that use silica, such as diatoms.

Control Methods and Management

In Europe, dredging efforts to clear Atlantic Slipper Limpets from oyster beds were unsuccessful at preventing their spread. Manual removal was also slow, costly, and ineffective in Ireland. Dipping oysters covered with limpets in a saturated salt solution for about 15 minutes kills the limpets without harming the oysters. However, dipping oysters in salt is time consuming and does little to remove the limpet from the wild.

Life History and Species Overview

Atlantic Slipper Limpets filtrate water through their gills at a rate of 250 ml/hr, extracting small algae and animals (phytoplankton and zooplankton). They aggregate in a sort of pyramid, where large females are located at the bottom and the smaller males at the top of the pile in a proportion that varies around 1:3. The males at the top can copulate with any of the females below even with six or more individuals in between. They reproduce every year between February and October. Each female usually spawns twice. Females can store sperm in a specialized compartment for several months, waiting until the conditions, such as temperature and salinity, are appropriate for reproduction. Interestingly, Atlantic Slipper Limpets not only form piles to reproduce but also to change sex. As the small young males grow, they turn into females, according to size, position in the pile, and other biotic variables. As a result, females at the bottom of the pile attached to the substrate are the oldest and largest individuals in a population.

Females produce millions of small eggs that are deposited in balloon-shaped capsules. Capsules are attached to the substrate and covered by the parent, who cares for the eggs by aerating and cleaning the embryos during the month they take to develop. At hatching the young larvae swim and drift with the current for 2–3 weeks, feeding in the plankton, until they go back to the coast and settle. Once they settle, they metamorphize into their adult form, staying in that habitat for the rest of their 10 year lifespan. Exactly why larvae stay in the plankton or return to the coast is unknown; however, we know that the planktonic stage is the main method of dispersion in natural conditions and that larvae may travel distances of at least 10 km.

History of Invasiveness

The Atlantic Slipper Limpet was introduced with Atlantic oysters in Puget Sound probably before 1890 and showed up in Grays Harbor and Willapa Bay, WA, and BC during the last century. The first specimen was collected and cataloged in 1905 by Trevor Kincaid, a zoologist at the University of Washington, proving that the Atlantic

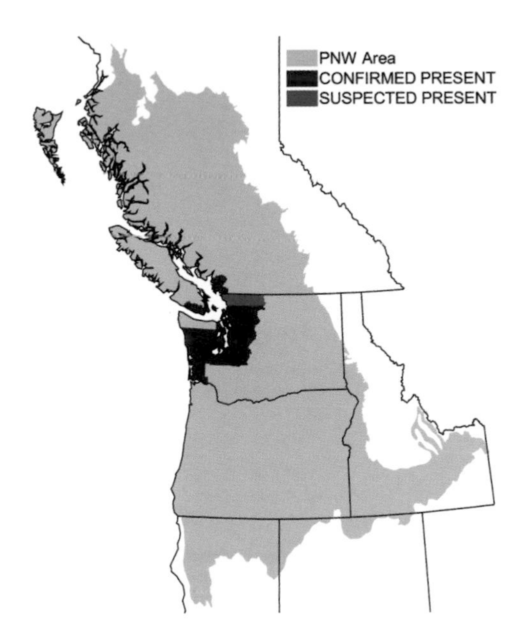

Slipper Limpet had established itself in Puget Sound. It was introduced, with oysters, to northern Europe in the 1870s, established throughout the northeastern Atlantic and Mediterranean, and has become extremely abundant in France, southern England, and the southern Netherlands. It has also been introduced to Japan. There is extensive American and European literature about the biology and ecology of this species, yet the impact of its presence along PNW coasts is unstudied. The mode of reproduction and the dispersal ability of the larvae are the major reasons why the Atlantic Slipper Limpet has been so successful in its invasion. However, humans play an important role facilitating invasion as we move the limpet with oysters, contributing to its rapid spread.

Other Sources of Information: 81, 85, 116, 121
References: 50, 66, 68, 306, 323
Author: Fernanda X. Oyarzun

European Green Crab *Carcinus maenas*

Species Description and Current Range

Despite their name, European Green Crabs are seldom simply "green." Adults tend to be multicolored, often mottled olive-brown with black and yellowish spots. The crabs' underside may become pale orange or red with age. Yet adults are drab compared to juveniles; the youngest crabs are speckled with green, white, black, and rust, which acts as camouflage in their rock and seaweed habitat. When coloration is not a reliable character for identification, shell features can be used. Green Crabs are easily distinguished from native species by five evenly spaced curved spines (or teeth) that extend beyond the eye along the front margin on both sides of the upper shell (or carapace).

Green Crabs are now found on four of the five continents, having spread to more places than any other crab species. They grow largest in the PNW, probably because of the extended growing season and lack of parasites and diseases that affect them in other regions. Male Green Crabs typically grow to 89 mm measured across the width of the carapace, while females are smaller (76 mm). In the PNW, they range north to Clayoquot Sound, Vancouver I., and occupy high intertidal areas and marshes in coastal estuaries and wave-protected embayments. Green Crabs seek shelter at low tide and hide under boulders or debris. Disturbing a resting crab may provide insight into the origin of the species' scientific name, *Carcinus maenas*, which literally means "raving mad crab." Males often strike ridiculously pugnacious poses when agitated; standing with legs extended and claws spread wide, a threatened crab attempts to make itself look as large and imposing as possible.

Impact on Communities and Native Species

Green Crabs are opportunistic feeders, and more than 150 types of animals and plants have been recorded in their diet. They are also frequently cannibalistic. The species' appetite, excavating habits, and abundance can be a disastrous combination for ecosystems in which they become established. In New England, for example, Green Crabs are blamed for the collapse of the soft shell clam fishery and their digging activities have impeded efforts to restore eelgrass habitats. Reports from central CA likewise depict Green Crabs as voracious predators in some locales. However, their impact along the west coast of North America is largely moderated by the aggression of native crabs. Green Crabs are notably absent from many preferred habitats and are apparently relegated to areas neglected by the natives. These interactions limit Green Crab impacts in some places but intensify them in others.

Control Methods and Management

In the PNW, little is being done to reduce burgeoning Green Crab populations since government budget cuts eliminated all formal management programs. Even WA, which enacted rules making it illegal to import, possess, or transport live Green Crabs (the public is instructed to retain killed crabs and contact local Fish and Wildlife personnel), now relies solely on volunteers and shellfish growers to control the species. Baited traps and Crab Aggregating Devices (CADs) are used to collect and remove Green Crabs from coastal areas and monitor for spread into the inland waters of Puget Sound; these methods are inexpensive and far safer for non-target organisms than poison bait or biological control agents. Although Green Crabs are fished commercially in Portuguese waters, they are small relative to local crabs and unlikely to gain culinary acceptance here.

Life History and Species Overview

Green Crabs have a complex life cycle characterized by a planktonic (ocean-drifting) larval stage and a benthic (bottom-dwelling) adult stage. They begin life as bright orange eggs, each smaller than the head of a pin. Female crabs deposit these eggs onto their abdomen as a single, golf ball–sized cluster after mating. While not generally known for maternal care, females do clean and protect the incubating young, which may number 200,000 or more, for up to 3 mo. Eggs hatch with the ebbing tide and larvae are swept out into coastal waters. Despite their minute size and weak swimming ability, larvae may travel more than 1,500 km, carried on strong currents, before moving back to nearshore habitats. After up to 2 months at sea, they settle to the bottom and metamorphose into juvenile crabs. They usually reach maturity within their first summer and may live as long as 5 years.

Native to the eastern Atlantic from Norway to northern Africa, Green Crabs now inhabit both coasts of North America, southeastern Australia, Tasmania, South Africa, and Patagonia. The species' success is due, in part, to its resilience and adaptability to novel environments. Throughout the world, Green Crabs thrive in habitats variously characterized by sand and mudflats, shell, cobble, algae, and rock, and they do equally well on pristine shores or in highly modified harbors and marinas. Green Crabs also tolerate a broader range of physiological conditions than many of their relatives. Adults and juveniles can survive warm summer days (86°F/30°C) and yet frequently overwinter at temperatures below freezing (32°F/0°C). They can even survive more than 2 months out of water beneath damp algae.

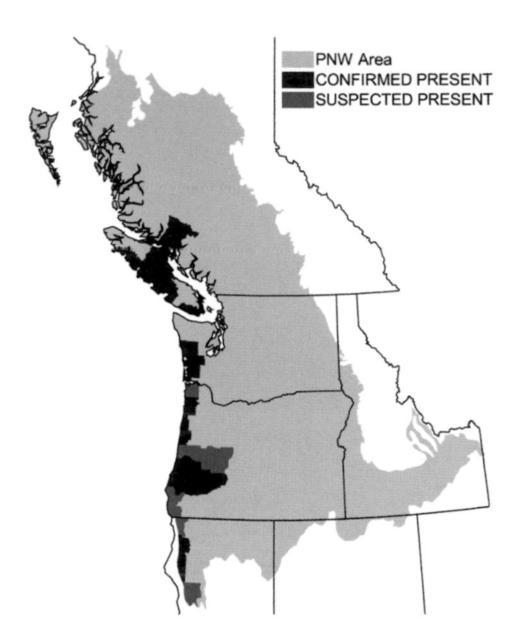

History of Invasiveness

Green Crabs arrived in the PNW "naturally" in the late 1990s as larvae carried on strong El Niño currents. However, they did not come all the way from the Atlantic. These larvae were progeny of populations in central CA that had grown relatively unimpeded for about a decade. In fact, a large breeding population was already well established in San Francisco Bay by the time it was discovered in 1989. Genetic tests showed that CA Green Crabs descended from populations on the East Coast of North America, but scientists have debated exactly how they were introduced. Although Green Crabs have been transported to many regions in ballast water, it is likely that the initial colonists of San Francisco Bay were accidentally introduced with discarded seaweed used to pack Atlantic bait-worms or seafood products such as lobsters. Their ability to survive long periods out of water undoubtedly aided them in making the transcontinental journey.

Other Sources of Information: 36, 75, 114, 115, 118
References: 67, 126, 172, 211, 351
Author: P. Sean McDonald

Japanese Mussel *Musculista senhousia*

Species Description and Current Range

This marine invader is a mussel in the family Mytilidae, which includes other familiar PNW species such as the Bay Mussel (*Mytilus trossulus*) and California Mussel (*Mytilus californianus*). *Musculista* is relatively small, with a maximum length of about 3.5 cm. It has a thin, greenish shell, often with brown concentric markings. Like other mytilid mussels, *Musculista* produces sticky threads (called byssus) from a special gland on the foot. Unlike more familiar mussels, however, *Musculista* typically lives in soft sediment and uses its byssus to create a bag or "cocoon." It has a very broad native range within the Pacific and Indian oceans and can be found from southern Siberia, through Japan, China, the Malay Peninsula, India, and all the way into the Red Sea. It has successfully invaded the Mediterranean basin, Australia, New Zealand, and North America. On the Pacific Coast, it is currently found from southern BC to Estero de Punta Banda, just south of Ensenada, Baja California, Mexico. *Musculista* has no well-established English common name and is variously called Japanese Mussel, Asian Mussel, Asian Date Mussel, and Bag Mussel. Its Japanese name, *hototogisu-gai*, means cuckoo shell, and it is so named because its shell coloration resembles the plumage of the cuckoo bird.

Impact on Communities and Native Species

The Japanese Mussel is one of the better studied marine invaders, although most of what we know is from work done in southern CA and New Zealand (as well as studies within its native range of Japan and China). The most dramatic effect of the mussel results from its ability to achieve high densities, whereby the individual byssal cocoons become woven together to form a mat on the surface of the sediment. Typical densities are around 5,000–10,000 mussels/m^2, although densities in excess of 150,000/m^2 have been found in San Diego, CA (one of the highest densities ever reported for a marine bivalve). The mussel is a densely living suspension feeder and may be able to filter large amounts of water and increase the flow of material to the seafloor. There has been little investigation of its potential to affect water column properties; however, it likely has an effect, particularly at small scales.

The mussel is abundant and an important food for a variety of fish (e.g., croakers and sargo), shorebirds (e.g., willets and godwits), and snails (e.g., the Festive Murex in southern CA). The feeding activities of these native predators may be able to control mussel populations locally. When mussels achieve densities of more than a couple of thousand per m^2, the mats that form can affect other seafloor organisms. Within mats, the mussels benefit small fauna, including worms, snails, and crustaceans. This likely results from a process called ecosystem engineering, whereby an organism substantially modifies the physical nature of a habitat. For small organisms, the mats provide a complex physical setting, offer refuge from predators, reduce environmental stress, and increase food supplies by slowing water flow and biodepositing organic material. Different patterns emerge at larger species scales, however. Larger animals, such as surface-dwelling clams, are harmed because they can't live within the dense mat matrix and may be outcompeted for suspended food supplies. This negative interaction with clams may have led to the long-term decline in clam populations seen in some CA tidal wetlands. The mussel also has a complex relationship with eelgrass. At low densities, the mussel can facilitate eelgrass-shoot growth through localized provision of nutrients. At higher mussel densities, however, effects are similar to those for surface-dwelling clams, and the thick mats can impede vegetative propagation of eelgrass.

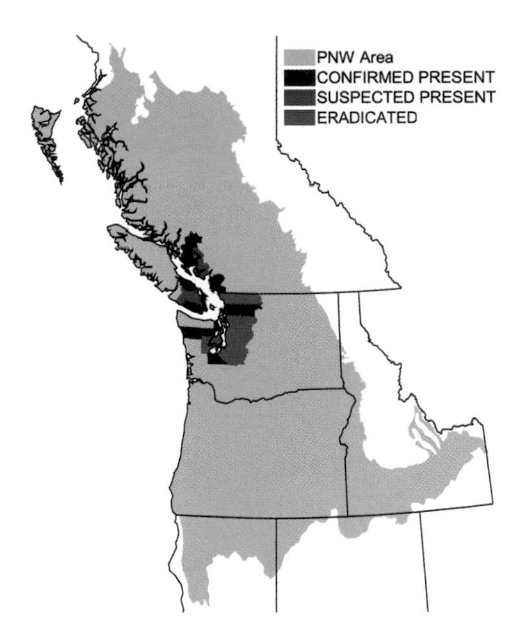

Control Methods and Management

Large-scale control of *Musculista* is unlikely, given its densities and the habitats where it is found (soft intertidal and subtidal muds). In its native range, there has been some harvesting of *Musculista* for animal feed, and there are historical reports of its use as a human food source. Where introduced, the mussel has not been harvested. Potential control methods, such as harvesting, need to be balanced against the inevitable impacts to marine habitats. Given that the mussel is sometimes associated with degraded environmental conditions, improving habitat quality in coastal bays and estuaries may limit the success of this invader. As a general rule, the best management option is to prevent invasion in the first place by instituting vector control.

Life History and Species Overview

The Japanese Mussel is a typical opportunistic, weedy species. It is very short-lived for a bivalve, living no more than 2 years, although most adults survive no more than a year. It is also small (3.5 cm max) and thin-shelled, putting much of its energy into reproduction. The mussel has separate sexes and is a broadcast spawner with a planktonic larval phase that remains in the water column up to a month. The species is most often associated with intertidal and subtidal soft sediments (mud and sand) within bays and estuaries, and can be found in unvegetated and vegetated habitats. The mussel is tolerant of degraded environmental conditions such as those found in the poorly flushed reaches of estuaries. It also can be found occasionally in fouling communities, such as on boat bottoms and pier pilings. It is typically found in full-strength seawater, although it can tolerate periodic brackish-water conditions. Its natural distribution suggests broad temperature tolerances, as it has been found from southern Siberia to the Red Sea.

History of Invasiveness

Musculista takes advantage of many of the primary vectors of invasion in its spread around the world. It was probably introduced to the Pacific coast with Japanese Oysters brought to WA for outplanting, although it is possible that other vectors were involved in its movement from Asia to North America. Subsequent spread on the West Coast has likely occurred both naturally, through long-distance transport of larvae, and in association with ship ballast and/or fouling. This mussel also has been introduced through oyster transport into the western Mediterranean, and has spread via ballast water to Australia and New Zealand. *Musculista* was able to traverse the Suez Canal and invade the eastern Mediterranean from the Red Sea. Given its demonstrated success as an invader, it may continue to spread around the world.

Other Sources of Information: 84, 94
References: 1, 74, 80, 140, 275
Author: Jeffrey Crooks

Eastern Softshell Clam *Mya arenaria*
Manila Clam *Venerupis philippinarum*
Purple Varnish Clam *Nuttallia obscurata*

H

Species Description and Current Range

Eastern Softshell Clams (bottom of image) can reach 15 cm in shell length. Their chalky-white shells are rounded at the foot end, pointed at the siphon end, and the leathery siphon is not able to be fully withdrawn into the shell. They bury themselves 20–35 cm in mud or sand bottoms and are common in bays and estuaries from AK to Elkhorn Slough, CA, in the high to middle intertidal zone.

Manila Clams (center of image) are similar in appearance to native "littleneck" clams and grow to 8 cm in length. The shells of both species have concentric rings and radiating ridges and allow complete siphon retraction when closed. However, Manila Clams have oblong shells, often colored and patterned, with purple staining inside. Native littlenecks generally have round, chalky-white shells. Manila Clams are common from Laredo Inlet (52.6°N) in BC to southern CA in the mid intertidal zone.

Purple Varnish Clams (top of image) grow to less than 7.5 cm in shell length. The outer surface of the shell is covered with a thin, shiny brown layer, like varnish, and the inside of the shell is purple. The two long, thin siphons and large foot are translucent white and can be completely withdrawn into the closed shell. The current range extends from the Discovery Is. (51°N) in BC to Alsea Bay (44.4°N) in OR, including Barkley Sound, Puget Sound, and the Strait of Georgia, Willapa Bay, Grays Harbor, and Nehalem Bay, WA. Purple Varnish Clams live buried 20–25 cm deep in the high to mid intertidal, in sediments ranging from cobble to muddy sand, generally at higher elevations than Manila Clams.

Impact on Communities and Native Species

All three of these invasive clams have become locally abundant and have displaced native clams from their former ranges. The Eastern Softshell Clam was once so abundant in Willapa Bay and Grays Harbor, WA (>100 adults/m^2) that large fisheries were forecast in the early 1900s, just 15 years after initial plantings. Instead, disease and predation reduced population densities from dominant to common and the displaced native clam populations have not recovered.

Manila Clams have become the primary aquaculture species locally, with WA now producing more clams each year than any other state. The extent of this industry includes massive hatchery seeding of over 1 billion juvenile clams each year on local tidelands in addition to naturally occurring spawning. Native littleneck clams are rare on beaches with large numbers of Manila Clams.

The Purple Varnish Clam has become sufficiently abundant to allow commercial fisheries in BC and legal recreational harvest in OR. This invasion is spreading rapidly and primarily impacts the upper intertidal zone, which already suffers depleted stocks of native clams. The long-term impacts of this invasion are unknown.

Control Methods and Management

There are currently no local control efforts for invasive marine clams. Governmental fisheries management agencies, however, periodically perform stock and range assessments to forecast commercial and recreational exploitation opportunities. While Eastern Softshell and Manila Clams are common, the Purple Varnish Clam is spreading, and its invasion is being monitored by government and university scientists. All sightings should be reported.

Life History and Species Overview

These three invasive clams typically reproduce once each year, with peak spawning from June through August. Adult females are highly fecund, with over 100,000 eggs released per spawn. Both sexes free-spawn gametes directly into the water column where fertilization occurs and the embryos develop into planktonic larvae that swim and feed for a few weeks before settling to the bottom. Once settled, they slowly develop (over several days to weeks) the ability to burrow in the sediment and filter water to feed. Unburied larvae suffer high mortality, often greater than 99% of individuals, from predation by crabs and sea stars while still exposed on the substrate. Buried juveniles grow quickly and mature to reproductive size within a year. Adult clams of all three species can live more than 4 years, but their maximum life span is unknown.

History of Invasiveness

Manila Clams were accidentally introduced from Japan in the 1930s, probably with shipments of Pacific Oyster seed planted in the northern regions of Puget Sound. They rapidly naturalized and spread throughout the coastal waters.

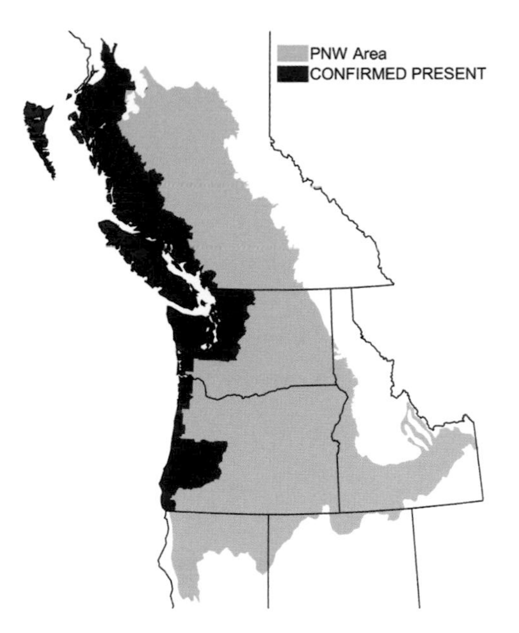

Eastern Softshell Clams were first shipped by rail from Chesapeake Bay to San Francisco in 1869 and placed in San Francisco Bay to keep them alive until sold. These clams rapidly spread and by 1888 over 2 million pounds were harvested each year from the Bay for local markets, completely replacing local species. Captain Stearns transported 12 barrels of adult clams from San Francisco to Willapa Bay by schooner in 1881, and these clams had spread throughout Willapa Bay and Grays Harbor, WA, by 1884. In 1889, a number of adult Eastern Softshell Clams were transported by steamer from Willapa Bay to Tacoma and planted there. Eastern Softshell Clams have subsequently spread along the entire Pacific coast.

Purple Varnish Clams were introduced via ballast water near Vancouver, BC, in the late 1980s. They have subsequently spread extremely rapidly throughout the Strait of Georgia and Puget Sound, with distant colonization along the coasts of OR and WA, via widespread transport of planktonic larvae.

References: 45, 65
Author: Alan C. Trimble

Mediterranean Mussel *Mytilus galloprovincialis*

NOT LISTED

Species Description and Current Range

Although the Mediterranean Mussel is native to the warmer, saltier waters of its namesake—*galloprovincialis,* meaning "from the land of the Gauls"—it now lives in the PNW, Japan, southern Africa, and South America. Its shells are a smooth, dark blue-black on the outside and mother-of-pearl shiny on the inside. In the wild, this species attaches to boat hulls, marker buoys, marina floats, and wharf pilings, as well as to its more natural home between the tidelines of a rocky shore. Not all the mussels in the PNW are this invader. The Mediterranean species can be visually distinguished from the larger, heavier native California Mussel (*Mytilus californianus*), but it very closely resembles its native sibling species, the Bay Mussel (*M. trossulus*). Not until the 1980s were scientists using genetic analyses capable of distinguishing these two species reliably.

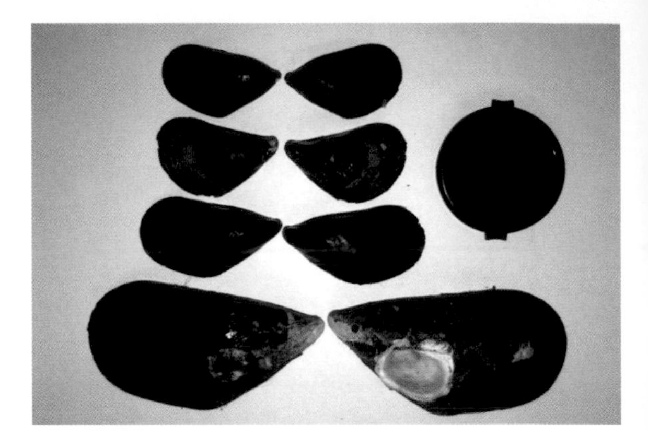

Impact on Communities and Native Species

Wild non-native populations of the Mediterranean Mussel live in intertidal and subtidal ocean habitats in both hemispheres of the Atlantic and Pacific. In Africa, the invader was first collected in the 1970s and now outcompetes native mussels and grazing limpets along rocky shores from Namibia to Cape Horn. Recently, it has begun spreading into the sandy wetlands of South Africa's West Coast National Park. In Japan, it had arrived by the 1930s and in California by the 1940s; on both sides of the Pacific it established an extensive population just to the south of the native Bay Mussel. The invader's impact in the PNW has yet to be studied. Historical museum specimens show that the native mussel used to live as far south as Monterey Bay, CA, where the Mediterranean Mussel now dominates. Whether the invader displaced the native species is unknown; where the two overlap today, they form hybrids. An additional aspect of this introduction are the natural drifting mussel-culture rafts, with their hanging population of thousands of feeding animals. These rafts are known elsewhere to alter food-web structure, water flow, and nutrient, oxygen, and microbe levels, but have not yet been studied in detail in the PNW.

Control Methods and Management

Like Pacific Oysters and Manila Clams, Mediterranean Mussels are a "regulated aquatic animal species" that may not be released into WA waters. Nonetheless, all three species are legally farmed from local hatchery broodstock. In *Association to Protect Hammersley, Eld, and Totten Inlets v. Taylor Resources,* a nonprofit residential group sued a Puget Sound mussel farm on the grounds that the biological materials emitted from the farm—mussel shells, feces, pseudofeces, and dissolved ammonium and inorganic phosphate—constituted pollutants under the federal Clean Water Act. The resulting 9th Circuit Court opinion found that these biological materials were not pollutants, noted that the protection and propagation of shell-fish is a goal of the Clean Water Act, and added that mussel filtration can help reduce excess nutrients in marine waters. Indeed, an adult *Mytilus* mussel can filter a liter of water in 20 minutes, which at the scale of thousands of kilograms of farmed mussels per year becomes a considerable process. However, the downside to this profligate activity is that mussels may accumulate large amounts of biological toxins and industrial pollutants, concentrating these poisons and moving them up the food chain. Whether Mediterranean Mussels are established independently of the reproduction supplied by farm populations is not known. A major obstacle to any proposed control effort would be identifying individuals in the field, because they are visually indistinguishable from the native mussel.

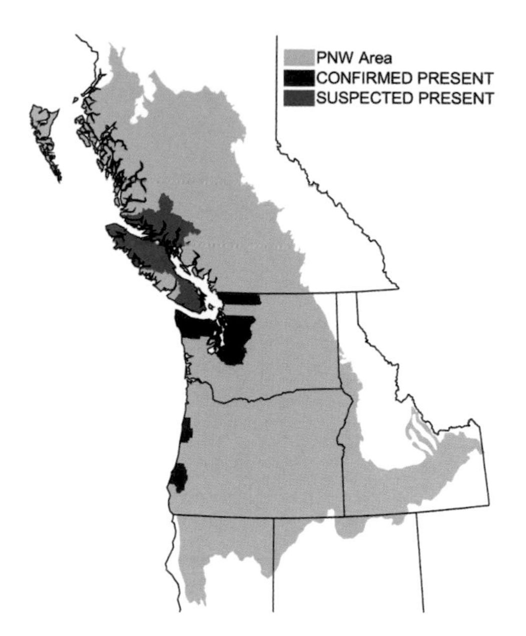

Life History and Species Overview

In general, *Mytilus* mussels are reproductively mature at 1 year, and may live anywhere from 1–2 years to over 20 years. Females and males release eggs and sperm directly into the water, where the eggs are fertilized. The microscopic developing larvae drift with water currents for several weeks before they settle and metamorphose into the familiar shelled animals we recognize. As they metamorphose, the mussels begin to secrete short, tough threads that stick onto the substrate—a rock, a boat hull, another mussel—and hold them in place for the rest of their lives. They make a living by filter feeding, pumping water through their sieve-like gills and trapping tiny particles of food that are pushed by little hairs up to their mouths. The mussels are in turn eaten primarily by snails, starfish, fish, and birds, as well as raccoons, bears, and other mammals that venture down to the shore.

History of Invasiveness

Mussels are harvested from the wild and grown on farms all over the world. In the PNW, native mussels proved difficult to grow in culture because they succumbed to a mysterious disease, known as disseminated hemic neoplasia, which caused their tissue to soften and liquefy before they reached market size. In the mid-1980s, a sample of live mussels in CA were inoculated with the disease but didn't become infected. A thriving industry was launched on the importation of these super-mussels. Although it wasn't known at the time, the disease-resistant mussels were in fact harvested from a population of the Mediterranean Mussel that had established in southern CA. How did this species cross an ocean and a continent to arrive on the Pacific coast? No one knows for certain, but there are two clues. First, the floating larvae of Mediterranean Mussels have been collected and identified from the ballast water of commercial cargo ships. Ballast water is harbor water that a ship loads in one port and discharges in another in order to adjust its trim and stability. In the process, it picks up and drops off a teeming soup of live marine organisms in new locations day after day, some of which survive to begin a biological invasion. The second clue comes from the voyage of the *Golden Hinde II*, a replica tall ship that sailed down the Pacific coast in the early 1990s in a reconstruction of Sir Francis Drake's famous voyage. After a few days of sailing, mussels began to settle out and grow on the ship's wooden hull, and they hitched a ride all the way from OR to CA. Mussels are also regularly scraped from the hulls of ferries, fishing boats, and private yachts. These clues tell us that mussels are very effective stowaways, and suggest that they have probably been sailing the world since the time of Columbus.

Other Sources of Information: 116
References: 6, 51, 52, 313
Author: Marjorie J. Wonham

Asian Mudsnail *Batillaria attramentaria*
Eastern Mudsnail *Ilyanassa obsoleta*

Species Description and Current Range

The Asian Mudsnail is an elegant, cone-shaped snail reaching 3 cm in length. The shell is dark gray to tan, sometimes with white stripes. It has a prominent "spout" or siphonal canal and 8–9 whorls with axial ribs that give it a beaded appearance. The Asian Mudsnail occurs in the Strait of Georgia, Puget Sound, Hood Canal, and Willapa Bay, WA. The Eastern Mudsnail is oblong, up to 3 cm in length, with its dark brown or blackish periostracum (the tough outer layer of the shell) often cracked and peeling. It is found in Boundary Bay, BC, and Willapa Bay, WA. Both species inhabit intertidal mudflats in marine and estuarine areas.

Impact on Communities and Native Species

Although little is known about the impacts of these two species in the PNW, both have ecological effects in other regions where they are introduced. In northern CA, both species displace the native Horn Snail (*Cerithidea californica*) by competition for food (Asian Mudsnail) or by physical disturbance and predation (Eastern Mudsnail). Our region lacks ecologically similar native gastropods, so any impacts are likely diffuse and community wide rather than species specific. Both species are locally extremely abundant; for example, the Asian Mudsnail reaches densities of over 1,000/m^2 in Padilla Bay, WA, making it one of the most conspicuous mudflat organisms. At such high densities, the feeding activities of these snails likely alter sediment characteristics and benthic algal and invertebrate communities, and could potentially facilitate the establishment of other mudflat invaders. For example, a study in Padilla Bay found that Japanese Eelgrass (*Zostera japonica*) abundance declined when Asian Mudsnails were experimentally removed from a small area.

Control Methods and Management

No coordinated efforts are being made to control Asian Mudsnails or Eastern Mudsnails in the PNW. Removing established populations is generally impractical, but efforts should be made to avoid spreading these species to new locations. Carefully clean and dry clamming, crabbing, and fishing gear, and do not collect or release live individuals.

Life History and Species Overview

The Asian Mudsnail occurs on muddy and sandy substrates in the mid to upper intertidal zone of sheltered bays and estuaries. This species is astronomically abundant in some locations, with densities up to 10,000/m^2 reported from Elkhorn Slough, CA. This snail grazes primarily on diatom films growing on the sediment surface. In CA populations, reproduction occurs in summer, with each female laying several hundred egg capsules. Juvenile snails emerge the following spring without undergoing a planktonic larval phase.

The Eastern Mudsnail occupies habitats similar to those of Asian Mudsnails, but has a broader tidal distribution, extending from high intertidal lagoons and marshes into the lower intertidal and shallow subtidal. Like the latter species, Eastern Mudsnails reach local densities up to several thousand individuals per m^2. The Eastern Mudsnail is catholic in its dietary preferences, consuming a variety of living and dead materials. It breeds in spring and summer, with egg capsules releasing planktonic larvae that develop for 20–30 days before metamorphosis.

Asian and Eastern Mudsnails lack specialized predators in the PNW, but are eaten by shorebirds, crabs, and moon snails. Some adults of both species are parasitized by "castrating" flatworms that destroy the gonads, cutting off reproduction. In CA, Asian Mudsnails experience lower infection rates than the native California Horn Snail, but studies indicate that this differential parasitism plays a relatively minor role in the invader's success.

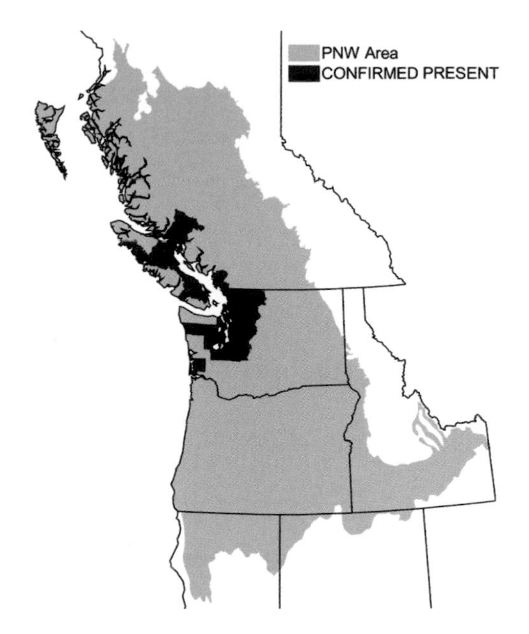

History of Invasiveness

Both of these snails were accidentally introduced to our region with commercial oyster shipments, but their invasion histories are quite distinct. The Asian Mudsnail is native to Japan and arrived in the PNW with shipments of Pacific Oysters (*Crassostrea gigas*) during the early 1900s. It was first recorded from Samish Bay, WA, in 1924 and now occurs in the Strait of Georgia and Puget Sound, as well as several bays and estuaries in northern CA. Given the absence of a planktonic larval phase in this species, its range extension probably reflects multiple introductions from the native range or secondary transfers within the Northwest, rather than natural dispersal. There are no region-wide estimates of the rate of spread of Asian Mudsnails, but their distribution in the PNW appears to be stable or increasing.

The Eastern Mudsnail is native to eastern North America from the Gulf of St. Lawrence to FL. It was first reported from San Francisco Bay in 1907, having arrived with shipments of the Eastern Oyster (*Crassostrea virginica*). One of the many hitchhikers that have persisted long after *C. virginica* failed to become established on the West Coast, Eastern Mudsnails were discovered in Boundary Bay, BC, and Willapa Bay, WA, in the 1940s-1950s. These introductions likely resulted from independent imports of eastern oysters or possibly oyster shipments from San Francisco Bay, rather than natural dispersal. Despite its planktonic larval development and locally high abundance, the Eastern Mudsnail has not expanded its range in the PNW. There is evidence that the species has disappeared from some CA localities where it previously occurred. Interestingly, native populations of Eastern Mudsnails in New England have suffered competitive displacement by the introduced European Periwinkle, *Littorina littorea*.

A third nonindigenous mudsnail, Japanese Nassa (*Nassarius fraterculus*), was also introduced to our region from Japan with shipments of Pacific oysters. It has been reported from Boundary Bay, BC, Padilla and Samish bays in Puget Sound, and Willapa Bay, WA. Virtually nothing is known of its ecology or impacts.

Other Sources of Information: 66
References: 31, 45, 50, 256
Author: Eric R. Buhle

New Zealand Isopod *Sphaeroma quoyanum*

Species Description and Current Range

New Zealand Isopods (NZ Isopods) are small crustaceans that look like a marine version of the common pillbug, which is also an introduced species. They live in tiny, densely populated interconnected burrows along the muddy banks of salty bays and wetlands. As many as 10,000 isopods can inhabit 1 m^2 of soil. Considered "bioeroders," the isopodes can more than double sediment erosion rates in coastal wetland habitat.

The NZ Isopods (also sometimes called Australian Isopods) live primarily in the intertidal zone and tolerate changing salinity levels well. They can bore into wood and soft rock but prefer firm peat and the muddy banks of creeks and marsh edges. The isopod colonizes polystyrene in docks and buoys, thereby earning its nickname "Styropod." Isopods are widespread in both San Francisco Bay and San Diego Bay, CA, and were identified in the PNW for the first time at Coos Bay, OR, in 1995.

NZ Isopods seem to congregate most heavily where the wetland vegetation is dominated by native Pickleweed (*Salicornia virginica*), and are less dense where *Spartina*, an invasive cordgrass, has taken over. Populations of the isopod decline in winter months, but the burrows, which average about 2 cm long and 5 mm wide, remain even during seasonal population fluctuations. NZ Isopods are also more likely to be found along steep vertical or undercut banks and usually reside at higher tide levels.

Impact on Communities and Native Species

When the isopods burrow into the soft banks along the edges of a bay, the soil becomes more susceptible to erosion from lapping waves or stream currents. As soil washes away, shoreline plant roots are exposed, and soon the plants die and wash away. In parts of San Francisco Bay, 90 cm of riverbank washed away in 9 months, a 250% increase over normal erosion rates.

Wetlands serve many important environmental functions: as resting and feeding spots for migratory birds, as nurseries for young fish before they head out to sea, as shoreline stabilizers, and as a filter for contaminated waters. Preserving existing wetlands is a high priority for conservation. This is especially true on the West Coast of North America; in San Francisco Bay, over 80% of the original wetland habitat was developed by the turn of the century. The NZ Isopod is a major threat in the battle to protect wetlands because of the erosion it causes. Perhaps the greatest damage caused by these invasive isopods entails the destruction of underground plant matter. Wetlands are comparable to forests in terms of carbon storage, and any loss of this storage exacerbates global warming.

Though unstudied, the NZ Isopod also likely increases the degree of turbidity in surrounding waters. It may even insidiously cause pollution when it bores into docks and releases millions of tiny particles of polystyrene into surrounding waters. Styrene is a known neurotoxin and a potential carcinogen.

Control Methods and Management

Sloped banks are less hospitable than vertical banks to the NZ Isopod. In the late 1990s, when the navy planned new channeling in

CA marshlands, it agreed to make the banks sloped instead of vertical. No other control or management approaches have been proven effective.

Life History and Species Overview

Isopods make up part of the superorder Peracarida, also known as the marsupial crustaceans because they, like kangaroos and koala bears, carry their young in a frontal pouch until the brood emerges. There are about 10,000 species of isopods, and though their bodies and habitat vary widely, they are all at least slightly flattened and segmented, and they each have seven sets of legs and two pairs of antennae. The NZ Isopod lives head-first in its burrow, using its strong mandible (jaw) to break off bits of substrate and bore into marshy banks. Two appendages called "pleopods" act as gills, and as they flap back and forth they create a mini-current in the burrow. This current brings in food particles, which catch on feeding brushes on the isopod's first and third pair of legs.

When threatened by passing fish, polychaetes, or other predators, the isopod rolls into a ball (just as a pill bug does). It will swim if heavy rainfall floods its burrow, exposing it to predator fish. Frequently the NZ Isopod is accompanied by a much smaller, rounder isopod, *Iais californicus*, which attaches itself to the larger isopod's underside. *Iais* was probably introduced along with the NZ Isopod and is found in most of the same bays. When it is found with the native isopod *Gnorimosphaeroma* it is always in an old NZ Isopod burrow. *Iais* feeds off food particles that are drawn into the burrow, and it is not known to cause any harm to the environment or to its host.

History of Invasiveness

The NZ Isopod was first verified in the shores of San Francisco Bay, CA, in 1904, likely introduced a few years earlier in the wooden hulls of transoceanic gold rush ships coming in from New Zealand, Australia, and Tasmania (the isopod's native range). Several bays in CA as well as Bajia San Quintin in Mexico now harbor large populations of the NZ Isopod, and it had made its way to OR by the mid-1990s. The NZ Isopod is likely to continue slowly expanding its range northward and southward, limited primarily by temperature (it would be unlikely to survive in freezing water). The species spreads rather slowly, over a period of decades, probably traveling to nearby bays in floating debris.

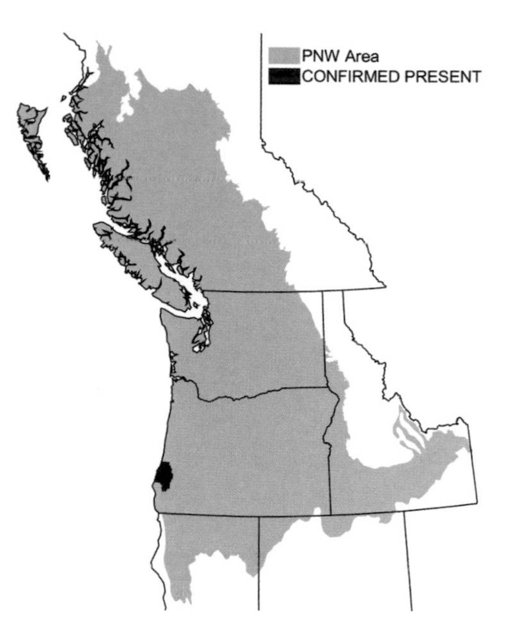

Acknowledgments: Many thanks to Theresa Sinicrope Talley at the University of California, San Diego, and to Dr. James Carlton at Williams College.

References: 272, 273, 301

Author: Stacey Solie

Asian Oyster Drill *Ocinebrellus inornatu*
Eastern Oyster Drill *Urosalpinx cinerea*

Species Description and Current Range

Oyster drills in the PNW include two species of gastropods (snails) in the family Muricidae. Both live intertidally on unconsolidated substrate such as oyster reefs, gravel, and sometimes mud in wave-protected areas or in estuaries. The Atlantic or Eastern Oyster Drill was introduced as a hitchhiker with eastern US oysters around the turn of the 20th century. It currently thrives in just one outer coastal estuary, Willapa Bay, WA, although a dwindling population was also reported in Boundary Bay, BC, in 1964, and it occurs in coastal estuaries south of the region. The Asian Oyster Drill was introduced as a hitchhiker with oysters from Japan in the 1920s. Its current range includes much of Puget Sound and Willapa Bay, WA, Netarts Bay, OR, and Ladysmith Harbor on the east side of Vancouver I., BC.

Drills get their name from their ability to make round holes in shells, allowing them to consume the organism inside. Drilling is a long process, requiring up to 3 days to drill and consume even a small (2.5 cm) oyster. The snails alternate between secreting an acid solution that dissolves the shell and rasping with their radula, which looks like a conveyor belt of teeth. This drilling habit is shared with several native species of Muricid snails, so it is important to distinguish the introduced species. Eastern Oyster Drills (to 30 mm) have dark, elongate shells with shallow grooves parallel to the whorls. Asian Drills (to 50 mm) have distinct varices—reinforced ridges produced periodically at the shell opening. Color ranges from beige to brown, occasionally orange or striped along the whorls. Because of their habitat, color at first glance may simply appear "muddy." The easily confused native Dogwhelk (*Nucella lamellosa*) tends to be larger, thicker-shelled, and more globose. Also, it does not eat oysters, preferring barnacles instead.

Left: *O. inornatu*; right: *U. cinerea*

Impact on Communities and Native Species

Most concern about oyster drills arises from their damage to aquaculture, because they feed on oysters, clams, and mussels. However, most of these cultured bivalves are also introduced. Ecological impacts are not well documented, although they may be high, given the generalist predatory habits of both introduced drill species and their high densities in some locations (to 100/m^2). In feeding experiments, individual drills often specialize on a single type of prey. Small oysters (1–2 cm) are drilled at rates of about 3/wk, and larger oysters (5 cm) require a full week to drill. A restoration effort for rare native oysters in south Puget Sound in 2000 completely failed because young oysters were eaten by drills. In Willapa Bay, native Dogwhelks were common prior to the introduction of Asian Drills but have apparently disappeared. Only shells inhabited by hermit crabs have been found recently. These ecologically similar native and introduced species still coexist in Puget Sound.

Control Methods and Management

The restricted distribution of Eastern Drills has prompted little management response. The introduction of Asian Drills, on the other hand, epitomizes a common pattern of response to invaders. *Step 1*, Ignorance: the Asian Oyster Drill was present in Willapa Bay for several years before its discovery in 1965. In the meantime, it had already been transferred from its site of introduction to at least three new locations. *Step 2*, Panic: State Fisheries biologists began weekly field inspections to ascertain the extent and density of the invasion. *Step 3*, Technological optimism: More than 60 chemicals were tested for their ability to selectively kill oyster drills. None of the safe options was entirely effective. *Step 4*, Accommodation: Today, shellfish growers pick up drills manually to reduce their impacts on aquaculture beds. Some beds where predation rates were particularly high have been abandoned altogether. The WA Dept. of Fish and Wildlife restricts the transfer of shellfish from areas where the Asian Oyster Drill has been found, to slow the spread of drills as hitchhikers. The Asian Oyster Drill epitomizes one other common pattern: malacologists enjoy renaming species. Its previous names include *Tritonalia japonica*, *Ocinebra japonica*, and *Ceratostoma inornatum*.

Life History and Species Overview

Both Eastern and Asian Oyster Drills gather in small aggregations to lay benthic egg capsules. Bright yellow in color, the capsules look like large grains of rice attached by one end to hard substrate such as rock or shell. Unlike many marine invertebrates, drills have no planktonic larval stage. Instead, the juveniles crawl out of the capsule in close proximity to their parents. About 10 small (2 mm) juveniles emerge from each capsule in midsummer. This direct development has probably slowed the spread of introduced drills. Adults crawl at a proverbial snail's pace, up to 1 m/day, but they tend not to leave areas where food is plentiful. Native predators play a large role in reducing drill densities in some areas. Annual survival rates are as low as 25% for adult drills, and fragments of crushed shell implicate native cancrid crabs as a major cause of mortality.

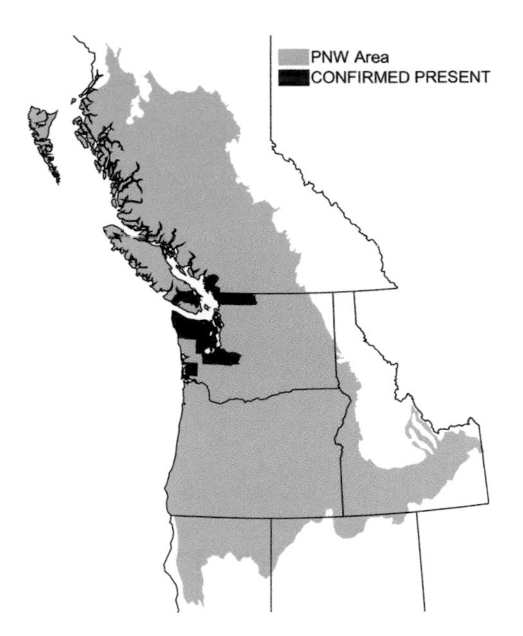

History of Invasiveness

Oysters have long been recognized as an exceptional vector for unplanned marine introductions. Oyster reefs harbor a high diversity of organisms that nestle among or live on shells. Given the global extent of oyster transfers, it is perhaps surprising that oyster drills have established in so few locations. The native range of the Eastern Oyster Drill extends along the East Coast of North America from Chesapeake Bay, MD, to FL. In addition to a few introduced populations on the West Coast, Eastern Drills also established in southeast England in 1920. Some populations subsequently disappeared due to pollution-induced imposex—when enough individuals are masculinized by exposure to antifouling paints such as tributyl tin, reproduction stops. The native range of the Asian Oyster Drill includes Japan and Korea. A new incursion was reported in French oyster-growing areas in 1995. Genetic studies indicate that these drills represent a secondary introduction from WA, rather than a direct transfer from the native range. New introductions should become less common, because most oysters are now transferred as larvae, where it's difficult for an oyster drill to hide.

Other Sources of Information: 26, 56
References: 255
Author: Jennifer L. Ruesink

Pacific Oyster *Crassostrea gigas*

Species Description and Current Range

The Pacific or Japanese Oyster is a bivalve mollusk up to 30 cm in length. The left shell, or valve, is larger and more deeply cupped than the right valve and is usually cemented to hard substrate. The valves are roughly oblong and often deeply fluted. The exterior is white to gray, often striped with purple or brown. The interior is coated with iridescent mother-of-pearl with a scar marking the attachment point of the adductor muscle, which closes the shell.

Pacific Oysters are found on hard and soft substrates in the intertidal and shallow subtidal zones of marine and estuarine areas. Although the species has been introduced from AK to central CA, self-sustaining populations occur only from BC to Willapa Bay, WA, and local distributions in Puget Sound and the Strait of Georgia have fluctuated.

Several other oyster species have been introduced to the PNW, including the Eastern Oyster (*Crassostrea virginica*), the European Flat Oyster (*Ostrea edulis*), and the Suminoe Oyster (*C. ariakensis*), but with the exception of a few remnant Eastern Oyster populations, none of these species has become established. The native Olympia Oyster (*O. conchaphila*) is small (to 5 cm), with a round, smooth shell that is often greenish inside.

Impact on Communities and Native Species

The most obvious impact of the Pacific Oyster has been as a vector; indeed, Pacific Oyster imports are a known or suspected mode of introduction for most of the 24 marine invertebrate and algal species covered in this book. Despite its conspicuous presence in many intertidal habitats in the PNW, very little is known about the Pacific Oyster's direct ecological impacts. The species is widely perceived as a beneficial seafood product rather than a potentially harmful invader. It supports a significant West Coast aquaculture industry, with 80% of the production based in WA. Nonetheless, the Pacific Oyster has probably substantially affected native communities. In WA and BC, it can be a dominant space-holder on rocky shores and likely competes for attachment space with native invertebrates and algae. Competition from the Pacific Oyster may impede the recovery of the Olympia Oyster, which was decimated by historical overharvest and pollution. On mudflats, reefs formed by Pacific Oysters are a striking topographic feature. These reefs may reduce oxygen levels and increase organic content in sediments, displace eelgrass and burrowing animals, and provide living space for species such as barnacles, mussels, clams, snails, sponges, and tunicates, many of which are invaders themselves. While the Pacific Oyster did not cause the introductions of these species per se, it may facilitate their establishment by providing habitat or food.

Control Methods and Management

Management of Pacific Oysters in the PNW is generally directed toward increasing aquaculture yields rather than mitigating impacts on native communities. No coordinated efforts exist to control or monitor naturalized populations or to determine the impact of the species outside of aquaculture settings. Large-scale eradication is unlikely, but care should be taken with live seafood products to avoid introducing oysters and other non-native species to locations where they are not already farmed or naturalized.

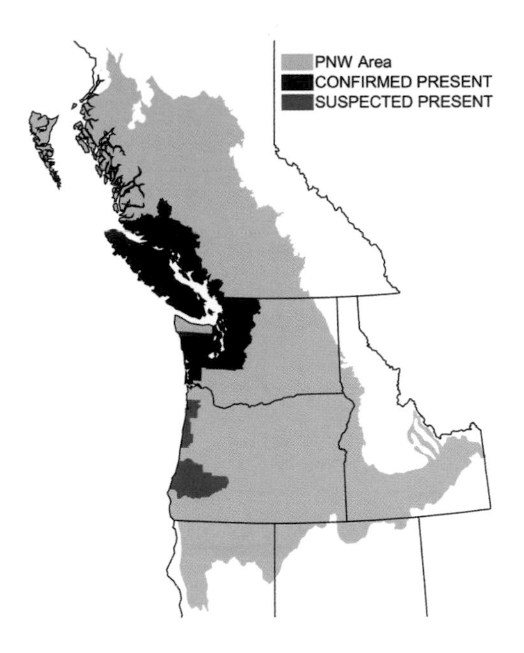

Life History and Species Overview

The Pacific Oyster is native to the Indo-Pacific from Pakistan to Japan, including Korea, China, and Southeast Asia. In the PNW it is most common in estuaries and sheltered areas to a depth of 6 m, but is occasionally found on exposed coastlines. Abundance ranges from a few scattered individuals in some outer coast localities to virtually 100% cover in rocky intertidal habitats of Pendrell Sound, BC. In WA, spawning takes place from June to August. Pacific Oysters are sequential hermaphrodites: each individual is either male or female at a given time but can change sex repeatedly. Gametes are released into the water column when temperatures reach at least 16°C, often during periods of warm, calm weather. Spawning is highly variable, with multiple bouts in some years and complete failure in others. The planktonic larvae feed and develop for 18–30 days before settling onto hard substrate and metamorphosing into juveniles, or "spat." In muddy estuarine habitats where hard surfaces are scarce, larvae settle gregariously on shells of older oysters. Many generations of settlement lead to the formation of solid, elevated reefs that can persist for decades or centuries if undisturbed. Oyster growers place old shell ("cultch") in intertidal areas to catch the settling larvae, which may then be transferred to areas of optimal growth. Growth rates depend on a variety of factors including food supply and temperature, but individuals can reach several cm in length in their first year and become sexually mature in their second year.

History of Invasiveness

British ecologist Charles Elton noted that "the greatest agency of all that spreads marine animals to new quarters of the world must be the business of oyster culture." Pacific Oysters have been widely introduced to regions outside their native range, including Australia, New Zealand, South Africa, Chile, Argentina, Europe, and eastern North America, and they are established in some of these areas. The importation of Pacific Oysters from the Miyagi prefecture in Japan to Samish Bay, WA, in 1902 initiated a long history of deliberate and accidental transfers of western Pacific marine species to the PNW. Pacific Oyster introductions helped revive the region's oyster industry following the destruction of natural Olympia Oyster stocks and a failed introduction of the Eastern Oyster. The Pacific Oyster's range expanded through continued imports of juveniles from Japan and Korea, transfers within the Northwest, and natural reproduction. Significant spawning and settlement began in the Strait of Georgia in 1932, Hood Canal in 1934, and Willapa Bay in 1936. Natural reproduction is currently the most important factor contributing to changes in the introduced range, as imports from Asia have mostly stopped. The distribution of Pacific Oysters outside of aquaculture settings may therefore be limited by the effects of temperature on reproduction mentioned above, suggesting that climate change could alter the species' range in the PNW.

Other Sources of Information: 66
References: 104, 298
Author: Eric R. Buhle

Asian and European Gypsy Moths *Lymantria dispar*

Species Description and Current Range

The Gypsy moth (GM) is one of the most devastating forest pests in the Northern Hemisphere. Native to Europe and Asia, GMs are decimating oaks in the eastern and midwestern US, where outbreaks occur every year. GM damage peaked in 1981, when over 5 million hectares were defoliated. GM are not established in the western US, but incipient populations are regularly eradicated. GM live in forest, woodland, and seminatural habitat reaching densities of 12.5 million larvae/ha. During nonoutbreak conditions, the moths are most visible in egg cases. A tan sheath covers 100–1,000 eggs over winter and remains after hatching in the spring. GMs lay egg masses in any sheltered location—on vehicles, houses, recreational items, under loose bark, or in leaf litter. Larvae grow to 6–7.5 cm, have long tufts of hairs and a double row of 5 blue and 6 red pairs of dots on their backs. Female adults are white, males are brown; both have brown scalloped lines on the wings, a 5-cm wingspan, and a characteristic zigzag flight pattern. European (EGM) and Asian Gypsy Moths (AGM) are different strains of the same species; genetic tests are necessary to tell them apart. Two other Lymantriids pose similar threats but are not yet established in North America: the Rosy Gypsy Moth, *L. mathura*, and the Nun Moth, *L. monacha*. Both seat broadleaved trees, especially oaks and fruit trees, but the Nun Moth also eats larch, fir, spruce, and pine. The Nun Moth is native to Europe and Asia; the Rosy Gypsy Moth is restricted to Asia, but includes India and Pakistan. Nun Moths cause more damage than any other forest defoliator in Europe.

Impact on Communities and Native Species

During outbreaks, GM can defoliate millions of hectares of forest and reduce food and habitat quality for many other species. Defoliation weakens trees, making them more likely to die from drought, disease, or other pests. As preferred GM hosts decline, less susceptible species increase. GMs eat over 600 species of trees and shrubs, including many ornamental and fruit trees. Preferred native hosts for both EGM and AGM in the PNW include Western Larch, Oregon White Oak, Alder, and Aspen. AGM grows much better than EGM on conifers, including Douglas-fir and Western Hemlock, which cannot survive repeated defoliation. Decline of any of these trees would alter food sources and habitat for many sensitive, threatened, and endangered woodland and forest species. Defoliation also changes water flow, chemistry, and temperature, altering habitat for sensitive riparian species such as salmon and trout. Once GMs establish, suppression methods likely would hurt native insects, especially butterflies and moths. Without suppression, the projected cost of GM and Nun Moth damage in OR and WA over 40 years is $2 billion for recreation, travel, and tourism industries, and $35–58 billion for timber harvests. Residential costs are much higher (~83% of total costs) because they include cleanup, removal, and replacement costs for defoliated trees and affect a large number of households.

Control Methods and Management

The GMs' defoliation potential was recognized soon after introduction, and it became the first species quarantined by the USDA in 1912. Control methods in the PNW for all Lymantriids concentrate on limiting human introductions and eradicating new colonies. The primary vectors for AGM are ships and wood products from Siberia. The moths are attracted to lights used when loading ships at night, so ports are now closely monitored and sprayed during outbreaks. EGM is carried across the US on personal items, vehicles, and some nursery plants. An extensive trapping network using sex pheromone to attract males helps detect new colonies. Colonies in the PNW have been eradicated with *Bacillus thuringiensis* (Bt). Bt kills nearly all moths and butterflies that eat foliage shortly after it is sprayed, but it is preferred over synthetic pesticides for several reasons: Bt kills fewer non-target organisms, has no known pathogenic effects on humans, and degrades within weeks instead of months to years. Once GM establishes, management focuses on keeping moth densities low with Bt, synthetic insecticides, a GM-specific nucleopolyhedrosis virus, and mating disruption. In 1990, 650,000 ha were sprayed with pesticide to control GM (350,000 ha Bt, 270,000 ha Dimilin). On average, $30 million/yr is spent nationwide on GM control. Over the past century, 60 natural enemies of GM were introduced; 15 established, but outbreaks still occur and GM keeps spreading. One species introduced for biocontrol in 1900 (the Tachinid Fly, *Compsilura concinnata*), combined with Bt spraying, may be responsible for the decline of native silkmoths in the eastern US.

Life History and Species Overview

GMs are native throughout Eurasia. Important physiological and behavioral differences between the European and Asian strains make AGM a greater threat in the PNW than EGM. The Asian strain typically comes from Siberia, is more cold tolerant, and has a greater preference for coniferous trees; females can fly >32 km before laying eggs. Female EGMs do not fly, slowing their spread in the eastern US. GM outbreaks are cyclical, depending on weather, host tree quality, and natural enemy abundance. The nucleopolyhedrosis virus and starvation usually end outbreaks. In 1989, a fungal pathogen, *Entomophaga maimaiga*

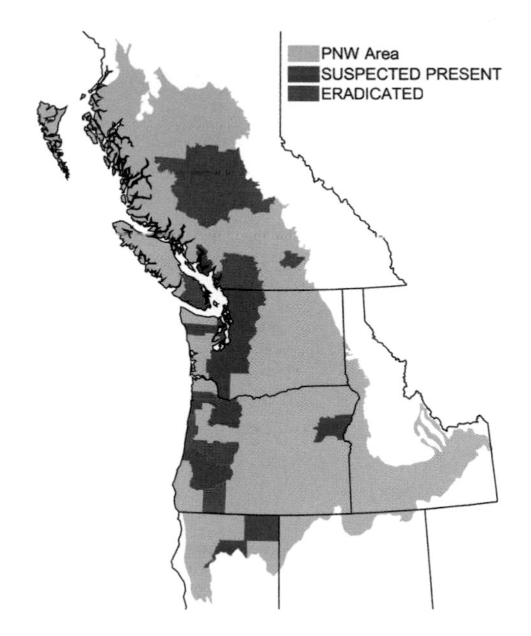

appeared in New England populations. It spread rapidly and is now a major cause of GM mortality throughout the moth's range. GMs complete a single generation each year. Adults lay eggs in June and July; the eggs overwinter and larvae hatch in April–May. Newly hatched larvae disperse to nearby trees by ballooning—sailing on the wind with a silk thread.

History of Invasiveness

The amateur entomologist Leopold Trouvelot brought GMs from France to Medford, MA, in 1868–69 to explore their potential for silk production. They escaped and within 15 years outbreaks started, at which time Mr. Trouvelot moved back to France. GMs' current North American range is from southern QC to NC and west to WI. The range is expanding slowly (21 km/yr), partly due to the "Slow the Spread" campaign and the flightlessness of EGM females. GMs spread short distances by larval ballooning and long distances by hitchhiking on cars and ships. Both strains appear frequently in the PNW, but control efforts have prevented establishment.

Other Sources of Information: 57, 72, 99, 105, 120
References: 165, 192, 198, 317, 333
Author: Lisa Crozier

Balsam Wooly Adelgid *Adelges picea*

Species Description and Current Range

The Balsam Wooly Adelgid (BWA) is a homopterans. This is a pestiferous order that includes aphids, leafhoppers, and many others that are problems to both agriculture and forestry. All stages of the BWA's life cycle are flightless, and the entire North American population consists only of females. Conspicuous to the human eye is the waxy, cottony fluff under which there is a single adult Adelgid. Groups of Adelgids appear as fuzzy white dots, each no bigger than a match head. Underneath its protective coating is a plump, round body about 1 mm in length. Like all homopterans, the BWA has specialized "piercing-sucking" mouthparts (stylets) designed for sucking the sap of living plants. The stylets of the BWA, used for feeding on the sap of fir trees, are about one and a half times longer than its body.

BWAs occur throughout the US and southern Canada. They follow the range of true firs but generally are limited to elevations below 1,800 m. The following are all signs of Adelgid infestation: swollen and deformed limbs (especially at the branch nodes), drooping tops, browning foliage, and in some cases death.

Impacts to Communities and Native Species

Unlike other homopterans, the BWA does not facilitate the introduction of pathogens into its hosts. When Adelgids feed, they pump fluids into the live tissues of their host tree causing abnormal reactions in the cells. This cellular damage is limited to an area of just a few millimeters from where the insect is feeding. If populations become very high (up to 75/6 cm^2), trees will die from the eventual failure of their water conducting system. Trees essentially die from physiological drought stress. Infested trees die over a period of 2–7 years depending on the susceptibility of the host and intensity of infestation.

Certain species as well as individuals show resistance to attack. Shasta Fir, Noble Fir, and Grand Fir above 300 m have the greatest resistance. Subalpine fir is the most susceptible of the PNW firs, with up to 90% mortality occurring in some stands 5 years after initial infestation. Alarmingly, BWAs have severely reduced subalpine fir populations from some harsh and disturbed environments (avalanche tracts, old lava flows). With warming temperatures, the BWA is capable of advancing into higher-elevation subalpine fir stands and causing serious damage. Grand Fir occurring at below 300 m is also sensitive to attack. Many trees have died and others are experiencing severely reduced cone crops. As a result, Grand Fir seems to be in a pattern of decline in low-elevation PNW valleys.

Control Methods and Management

The BWA's waxy coating helps shield the insect from insecticides. For this reason, insecticides are ineffective and expensive to use in a forest setting. However, intensive applications of insecticides can control BWA populations in nurseries and arboretums.

Sprays are most effective soon after nymphs hatch from eggs. Biological control efforts began in the 1950s with the introduction of 23 natural BWA predators into WA and OR. Of the 5 that survived, none has significantly reduced Adelgid populations. Most predator populations do not keep up with BWA populations, and host plants are harmed by even a small infestation. At present, biological control is limited to the ability of firs to develop resistance to the BWA. Other control measures involve restricting the spread of infested plant material and, in some cases, removing and destroying infested trees. If infested material must be transported, it is best done in winter months when Adelgids are dormant. Material must be used or destroyed before dormancy ends in the spring.

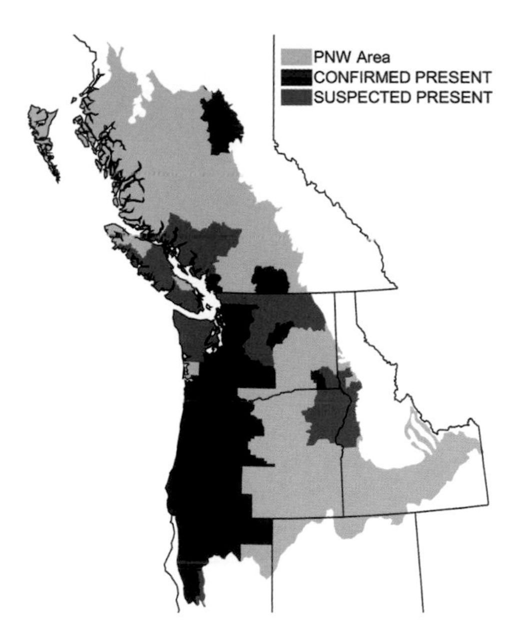

Life History and Species Overview

The BWA's native home is in the forests of central Europe. In some areas of Europe it relies on two different host species throughout its 2-year life cycle. This life cycle includes a sexual stage during which the primary host tree is spruce (*Picea* spp.) and the secondary host is a true fir. Throughout most of Europe, however, females have strayed from the range of spruce and are able to live and reproduce solely on fir. Introduced to North America about a century ago, the now very successful Adelgid population is made up entirely of flightless females. These produce 30–100 eggs by means of asexual reproduction and potentially produce 900–10,000 offspring during their 2-year life span. Each egg is genetically identical to its parent. After hatching, the first generation, or instar, are capable of crawling short distances before settling on a place to feed. These "crawlers" are also readily transported by wind, the primary method of BWA dispersal. Once on a suitable host, the insect will begin feeding by inserting its slender stylets into the tree, remaining until it molts three more times and becomes an adult. BWA overwinter as first larval instars, though all stages of development can successfully overwinter if temperatures do not fall below freezing. BWA typically produce two generations per year in the PNW; however, four generations have been observed at some low-elevation sites.

History of Invasiveness

Adelgids were first introduced to North America around 1900 and appeared in San Francisco, CA, on ornamental firs in 1928. By 1954, BWAs were causing serious damage to both silver and subalpine firs in southwest WA and OR. In a 160,000-ha area in southwestern WA, over 1.5 billion board feet of sawtimber died or was seriously damaged between 1950 and 1957. Since 1958, BWAs have been slowly expanding north into BC. By 1999, the BWA was found throughout the range of subalpine fir in ID.

Other Sources of Information: 104, 124
References: 26, 222, 223
Author: Carson Sprenger

Common Night Crawler *Lumbricus terrestris*
Red Worm *Lumbricus rubellus*

Species Description and Current Range

The European family of Lumbricidae earthworms is likely to be found in garden soil or used as fishing bait. These include the Common Night Crawler and the Red Worm. The adults of the most common invader, the Red Worm, reach 60–150 mm in length and are 4–6 mm wide. Their dorsal, or back, side is purplish or reddish-brown with an iridescent sheen, and the ventral side, or belly, is pale in color. Earthworms are generally considered beneficial. However, because late glaciations in WI wiped out most species in the northern US and Canada, forests in much of this region developed in the absence of earthworms. Remnant populations survived along the Pacific coast, and the native earthworm fauna of the PNW includes 28 species and another 80 species waiting to be described. There are 45 known invasive earthworms in North America, with at least 19 in BC alone. While it is unknown how many species of exotic earthworm inhabit the PNW, they likely occur in soils throughout the region.

L. terrestris

Impact on Communities and Native Species

Exotic earthworms change soil characteristics. By shrinking the duff, the top layer of soil composed of fallen leaves and debris not yet broken down, earthworm activity can increase the evaporation of moisture from the soil, making an area more susceptible to drought. Burrowing worms, such as the Common Night Crawler, increase erosion of the topsoil by removing and burying surface residues. Their feeding and burrowing reduce the ability of the soil to insulate from freezing temperatures and physically disrupt seed germination, lowering plant survival. Introduction of exotic earthworms to a mixed forest in MN led to a 75% loss in plant cover, decline of several sensitive plant species, and little or no tree seedlings compared with worm-free zones only a few meters away. As worms invade an area and consume organic materials, they respire carbon dioxide into the atmosphere. Earthworm activity modifies the carbon cycle by decreasing the soil's ability to store carbon. Invasive earthworms can disperse weedy seeds and plant and animal pathogens. Some species may be vectors for foot-and-mouth disease. A substantial threat to the integrity of native earthworm communities also exists, as exotic earthworms can outcompete and replace native fauna. European earthworms are thought to be partially responsible for the demise of the WA and OR giant earthworms (*Driloleirus americanus* and *Driloleirus macelfreshi*), which could grow to a meter in length.

Control Methods and Management

Exotic earthworms are unlikely to invade large, intact habitats. Consequently, preventing habitat fragmentation and the introduction of roads is a good way to reduce colonization by invasive earthworms. Beyond habitat protection, policy changes to ban or restrict imports may reduce the spread of harmful invaders. The US has no specific regulations concerning earthworm importation.

Canada regulates the import of cultivated species of earthworms, with the exception of those coming from the US. The only known way to rid soils of earthworms is to apply massive amounts of poisons or induce soil compaction. Prescribed burning in prairies led to increases in populations of native earthworms and decreases in exotics. Earthworms are killed by frost in soil, but their cocoons are more tolerant of low temperatures. Large-scale elimination of invasive earthworms is unlikely to be feasible but barriers to further expansion, such as buffer zones of unsuitable habitat to impede migration, warrant study. Most local introductions probably come from fishing activities and transportation of worms or cocoons on machinery or in tire treads. Priority should be given to the prevention of invasions from agricultural areas into surrounding ecosystems.

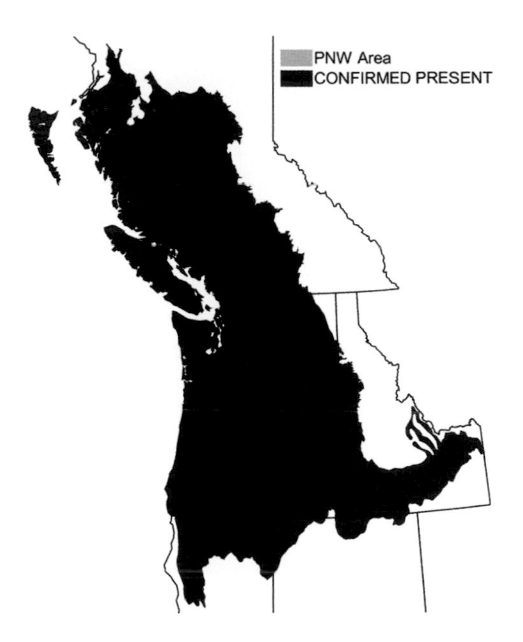

Life History and Species Overview

Red Worms, originally occupying mainland Europe and the British Isles, now occur throughout the US, Canada, New Zealand, and Australia. They are not yet documented in Asia, Africa, or South America, but likely occur in parts of these regions. In the PNW, European earthworms are the dominant earthworms in disturbed areas, including vegetable gardens. Earthworm densities range from less than 10 individuals and 1 g/m^2 to more than 1,000 individuals and 200 g/m^2 under favorable conditions.

Life histories are unknown for all but 20 of the world's more than 8,000 species of earthworms. Earthworms are hermaphroditic (having both male and female reproductive organs) and produce cocoons with fertilized eggs. Although most species are not self-fertilizing, some are able to reproduce without a mate, a characteristic that aids rapid expansion into new habitats. Life spans vary by species, ranging from less than 1 to more than 10 years. Earthworms mate underground or on the surface of the soil and produce, in the case of Red Worms, about 70 fertile cocoons/yr. There are 1–20 fertilized eggs in each cocoon but often only 1 or 2 survive to hatch. Cocoons are produced throughout the year, but seasonal fluctuations cause the number to vary. Earthworms grow throughout their lives and have great powers of regenerating injured or lost parts of their bodies. They can regenerate either end of their bodies, but the posterior part grows again more readily. The nerve system must remain intact for the worm to regenerate, which won't happen if too many segments are lost. Wound tissue forms in about 7 days but takes 2–3 months to become fully pigmented.

History of Invasiveness

European earthworms were first introduced by colonial settlers to the New England and Mid-Atlantic regions in ship ballast or in material adhering to imported plants. The worms spread throughout North America with the movement of people and goods. Exotic worms are included in home composting kits, shipped in agricultural and horticultural products, and used in commercial applications such as waste management. Local introduction to forested areas is recent, probably from logging, road-building activities, marijuana cultivation, and recreational pursuits such as fishing. Invasive earthworm densities in forested areas are closely associated with distance to roads.

Other Sources of Information: 65
References: 100, 102, 130, 149, 150
Author: Carly H. Vynne

European Fruit Fly *Drosophila subobscura*

Species Description and Current Range

Many people have seen the "fruit fly" *Drosophila melanogaster* in a biology lab. The European Fruit Fly (EFF) looks like a big, dark version of *D. melanogaster*, though it is still vastly smaller than a housefly. If you spot a "black" *Drosophila* in the PNW, it is probably an EFF. However, the only way to know for sure is to look at it carefully under a microscope, as an EFF closely resembles native members of its species group (the *obscura* group). Under magnification, the male EFF's large red testes will be visible.

In North America, these flies are found in shaded habitats from the northern tip of Vancouver I., BC, to Ojai, in southern CA. They have crossed the Cascade Mountains and spread north into the Okanogan in Canada, east to the foothills of the Rockies, and even south to Phoenix. Being relatively cold-adapted, these invaders are most abundant at northern sites, where they are now the most common *Drosophila*, at least in urban areas.

Impact on Communities and Native Species

Humans will most likely have a close encounter with an EFF only when one inadvertently drowns in a wine glass. (Although we don't appreciate them in our wine glass, they do add protein to our diet.) Plenty of indirect evidence suggests that the EFF has had a major impact on the native *Drosophila*. In fact, often 80% or more of all flies attracted to baits in northern sites (e.g., Seattle and Bellingham, WA) are EFFs; and the remaining percentage is split among 8 or more native species and at least 2 other invasive species. Biologists trying to collect native *obscura*-group flies in the north complain that EFFs just clog their traps. However, why EFFs seemingly outcompete native species in the north remains a mystery.

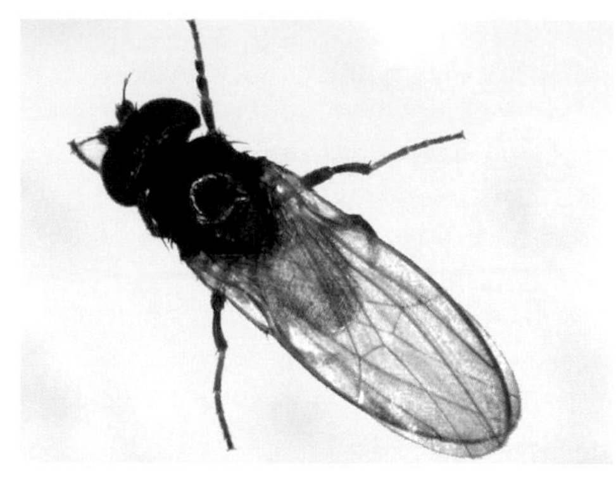

Control Methods and Management

Trying to control or manage the EFF is difficult. One option might be to introduce parasitoids and nematode parasites from the Old World that might drive down the populations. However, this approach might backfire, as those parasitoids and parasites would almost certainly attack native *obscura*-group flies, thereby only adding to the woes of those species.

Life History and Species Overview

The EFF lays eggs on various fruits, vegetables, and other plant materials. Larvae hatch a day or two later. They feed voraciously for several weeks, growing rapidly, before settling into a pupal stage. After a week or longer, they metamorphose into adults and begin to search for mates and egg-laying sites. During courtship, a male regurgitates a drop of food onto his proboscis while he dances in front of the female. If he successfully charms her, the female takes the meal and then lets him mate. Large males regurgitate large meals, but small males may be better dancers. So the choice a female makes may depend on her hunger level. In northern areas, these flies can have 4–6 generations/yr. Despite being called "fruit flies," they never attack ripe fruit. Instead, they eat the yeast and alcohol that accumulate in rotting fruit or other plant matter.

History of Invasiveness

The EFF is native to the Eastern Hemisphere and has a huge latitudinal range (>30°) from North Africa to Scandinavia. In 1978, these flies were discovered in Puerto Montt, Chile, and they almost instantly colonized much of Chile (>16°). In 1982, EFFs were found in Port Townsend, WA. Collections showed that they had already spread north into BC, and they were soon detected as far south as Ojai, near Santa Barbara, CA.

How EFFs made it to the New World is unknown, although they likely arrived on a cargo ship from Europe. Genetic studies reveal that fewer than 50 flies were involved in the invasion and that the North and South American flies come from a common source, probably Mediterranean Spain. No evidence of any additional invasion has been detected. This species is also highly invasive in South America, where it has crossed the Andes and is spreading across Argentina and northward. On all three continents, the EFF is most abundant in moderate to high latitude sites, where it vastly outnumbers native *Drosophila*.

Although the EFF is an ecological disaster for native flies, it is a boon to evolutionary biologists. By monitoring changes in these flies since the invasion, biologists have documented some of the fastest rates of evolution ever recorded. Unfortunately, such rapid adaptation to New World conditions suggests that EFFs will become an even more effective competitor of native species. Invading species are not just an ecological problem: they are also an evolutionary one.

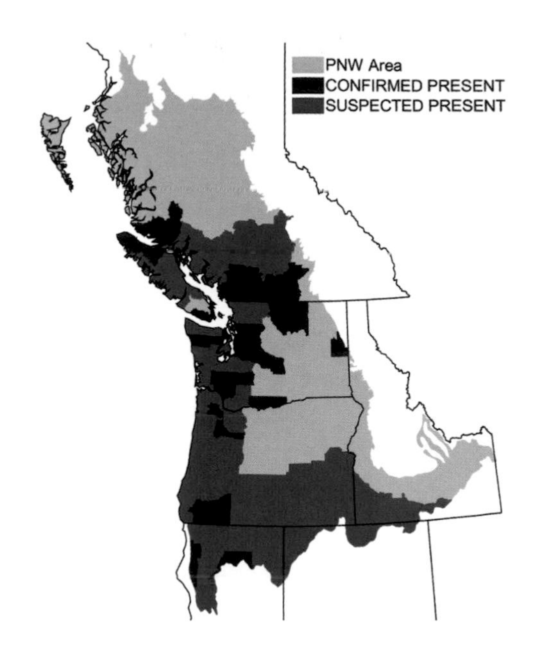

Acknowledgments: Research was supported by the Royalty Research Fund (University of Washington) and by the National Science Foundation.

Other Sources of Information: 9

References: 17, 167

Authors: Raymond B. Huey and George W. Gilchrist

European Yellowjacket *Paravespula germanica*

M / H

Species Description and Current Range

The European Yellowjacket, also called the German Yellowjacket, is one of 16 species of yellowjacket found in the US, of which 6 are found in the PNW. Workers are 10–16 mm long, with a stubby or stocky appearance and yellow and black (or white and black) stripes. The European Yellowjacket, an invasive wasp, can be distinguished from native wasps by the head shield, which has three black spots, and an incomplete yellow ring about the compound eye. Additionally, the first antennal segment is completely black. Native yellowjackets, such as the Western Yellowjacket (*V. pensylvanica*), have a complete yellow eye ring. Queens are half again as large as workers. When provoked, the species can give a painful sting. Unlike the Honeybee, which, because of its barbed stinger, only stings once and leaves the venom sac behind, wasps sting their victims multiple times with a smooth stinger. The chemical responsible for the wasp's painful sting is melittin, which stimulates the nerve endings of pain receptors in the skin. The sting results in an initial sharp, local sensation, which dissipates into a dull ache that can last for several days. In the search for antivenom for those allergic to wasp stings, pharmaceutical researchers can capture 500,000 wasps/yr. The European Yellowjacket is common throughout much of the US, except for the Deep South. In OR and WA, it is primarily found near urban areas, fields, gardens, and forests.

Impact on Communities and Native Species

As its range has steadily expanded into the western US, the European Yellowjacket has become the dominant foraging wasp, often at the expense of native wasp populations. Yellowjackets consume large numbers of beneficial insect pollinators. They compete with butterflies for nectar and displace hummingbirds from feeders. In the PNW, the European Yellowjacket competes with the Western Yellowjacket, in some cases outcompeting them, in others living sympatrically with them. Some researchers speculate that the two species may hybridize, if they do not already. European Yellowjackets are more successful in dry climates, and the PNW's mild climate may allow nests to survive for more than one season, although this has yet to be documented. Usually, colonies do not survive the winter, freezing in the cold temperatures. Unlike native PNW yellowjackets, the European Yellowjacket remains active into late fall, especially during unseasonably warm summers. Because of this, and also since it nests in such close proximity to people, the European Yellowjacket is considered a pest species. During the late summer, yellowjackets forage more on meats and sugars, and large numbers of yellowjackets come into close contact with humans at picnics, barbeques, and around houses. Defensive swarms deliver painful stings to passersby who accidentally come too close to a nest.

Control Methods and Management

Control of European Yellowjackets is complicated because they are beneficial in some ways, eating insect pests such as caterpillars that attack plants and crops. Large-scale management of European Yellowjackets does not exist; control, rather, is left to individuals who want to destroy nests. A variety of pesticides sprayed directly into the nests can kill yellowjackets, although this technique has some hazards, most notably the vigorous defense mounted by the yellowjackets. Setting of bait traps, such as hanging a dead fish over a bucket of water with detergent added, is effective when the wasps take chunks of fish that are too large, fall into the bucket of water, and drown.

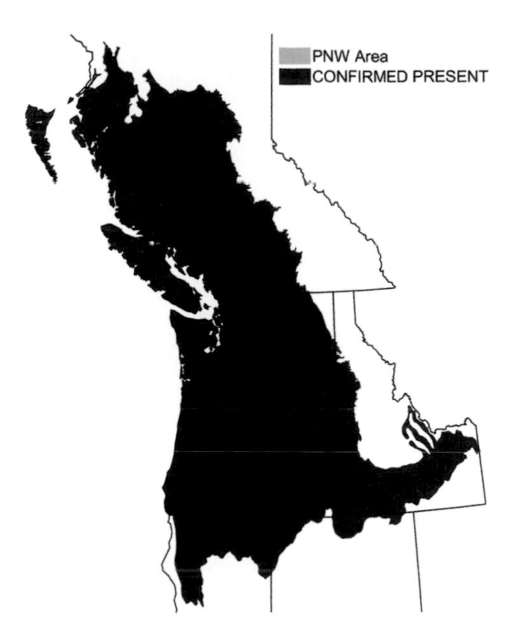

Life History and Species Overview

European Yellowjackets typically live less than a year. Only an inseminated queen overwinters, usually in a protected area such as under bark or in stacks of firewood. When the weather warms up in the spring, she emerges, selects a site, and begins building a small paper nest. The site is usually one of two types: below ground, in abandoned mouse burrows, under bushes or shrubs, and similar sites, such as building or house walls; or above ground, in aerial nests, suspended in trees, in woodsheds, under loose roof shingles, or under the eaves of houses. One of the primary criteria in site selection is whether it is suitably dry. Above-ground aerial nests are suspended by means of a suspensorium, which supports a small comb of cells. A paper envelope, made from salivary secretions and vegetable fibers chewed up by the queen, protects the combs.

The queen finds the nest site, builds the nest, lays eggs, and feeds the larvae for 3 weeks before they pupate. New larvae hold themselves in place in the cell with mucus excreted from the posterior. Mature larvae are white, elongate, and dorsoventrically flattened. When the larvae are ready to pupate, they spin a cell cap, then excrete accumulated wastes, called meconium, into the bottom of the cell. Sealed inside their cell, the larvae metamorphose into teneral adults, "teneral" meaning soft and unpigmented. Once they emerge, the workers (small, infertile female yellowjackets) take over the duties of feeding and rearing the brood. Males, which will leave the nest to mate with females, are haploid and come from unfertilized eggs. The queen normally stays in the nest once her first brood begins to forage, and the nest quickly grows to as many as 5,000 workers and 15,000 cells. Over her lifetime, a queen may lay upwards of 7,000 eggs. Combs of cells are stacked over one another in tiers, with mature nests usually having between 8 and 10 tiers. Yellowjacket nests can be over 30 cm in diameter.

Yellowjackets are most active from July through November. As the end of the yellowjacket year approaches, workers begin to build larger cells for the rearing of new queens and males (drones) to mate with the new queens. Once the queens mate, the annual season is effectively over—the hive slowly dies, the queen finds a spot in which to survive the winter, and the cycle starts again.

History of Invasiveness

European Yellowjackets are native to most of Europe, but are now found worldwide. They were first introduced in the northeastern US and have been steadily working their way westward. First recorded in San Francisco, CA, in 1987, they reached OR and WA by 1990. Their numbers have steadily increased since then.

Other Sources of Information: 64
Author: Eric Wagner

Multi-colored Asian Ladybeetle
Harmonia axyridis
Seven-spotted Ladybeetle
Coccinella septempunctata

Species Description and Current Range

Ladybeetles, also known as ladybugs, are in the family Coccinellidae ("little sphere"). Adult ladybeetles come in a variety of colors. North American populations range from orange, to yellow, to red, with or without spots on their wing covers. The number of spots varies from zero to 20, differing in size and intensity. Adults are 4.5–7 mm long and 5 mm wide, with a black-and-white head, black mouthparts, and short black antennae.

In the wild, Multicolored Asian Ladybeetles (MALs) are easily confused with native ones, especially the Convergent Ladybeetle (*Hippodamia convergens*) and the Transverse Ladybeetle (*Coccinella transversoguttata*), two of the most common Northwest species. Larvae are black, flattened, elongate, and covered with spines and tubercles. A voracious predator, Seven-spotted Ladybeetle (SSL) larvae consume 200–300 aphids and MAL larvae eat up to 1,200 aphids during development. The larva sheds its skin three times before metamorphosis; the last instar has a strikingly patterned red, orange, and black abdomen. Eggs are bright yellow. MALs lay their eggs in clusters on the bottom of leaves, while SSLs lay theirs in the middle of aphid colonies. Although females may lay up to 1,000 eggs, individual clumps that number around 20 ladybeetles are found throughout the US and Canada, including all of OR, WA, ID, and parts of southern BC.

Harmonia axyridis with larva

Impact on Communities and Native Species

MALs and SSLs have beneficial and detrimental impacts on local environments. The US Dept. of Agriculture (USDA) releases them as biocontrols for crop-pest species because they eat 90–270 aphids/day and consume other soft-bodied insects that damage trees, shrubs, and crops. Their impact on these pests limits the need for pesticides.

In the fall, they search for a place to overwinter. In their native Asia, this would normally be a cliff or some other sheer face. In the PNW, MALs do not migrate. (The native Northwestern Ladybeetle does migrate, spending the winter in the Cascade Range foothills.) They tend to swarm houses, office buildings, or any other reasonably warm and sheltered place. When ladybeetles are disturbed, they exude their own blood in a process known as "reflex bleeding." The noxious yellow-orange fluid permanently stains walls, clothes, or drapes.

Biologically, both SSLs and MALs may affect populations of native ladybeetles. In addition to a steady diet of aphids, Coccinellid larvae prey on one another. Introduced ladybeetles displace native ladybeetles in some areas of WA and OR through the combined effects of competition and predation. One advantage that exotic ladybeetles have is the absence of established predators; they also continue to be released as biocontrols.

Control Methods and Management

There is no organized management or control scheme for MAL, in part because they are beneficial for agriculture and used to control aphids. Commercial pesticides that are effective against ladybeetles include residual pyrethroid pesticides; research by the USDA indicates that camphor and menthol repell ladybeetles.

Life History and Species Overview

The complete cycle from egg to adult takes about 30 days. Female ladybeetles deposit their eggs in or near prey infestations in early to mid-spring. Eggs usually hatch within 5 days. Larvae then feed between 12 and 14 days. Their preferred prey is Crapemyrtle Aphids (*Tinocallis kahawaluokalani*), but they also consume scale insects and other soft-bodied insects. During the larval stage, ladybeetles molt three times before pupating. Pupation lasts 5–6 days, after which the adult beetle emerges. Adult ladybeetles can live up to 3 years. Upon metamorphosis into adults, ladybeetles devour aphids and other agricultural pests before cold weather compels them to seek shelter for overwintering.

The name "ladybeetle" originated in the Middle Ages. Several species of ladybeetle, including the SSL, were dedicated to the Virgin Mary and called "beetles of Our Lady." Eventually, the name was shortened to "ladybeetle." The MAL has a variety of common names, including Halloween Ladybeetle for its orange-with-black-eyes appearance and because the largest swarms tend to occur near the end of October.

History of Invasiveness

Over 170 species of ladybeetles were introduced in the US and Canada, compared with over 500 native species. Of the introduced species, 26 were able to establish. The first alien ladybeetle species were introduced into CA in the late 1880s. The MAL brought by the US government was released in CA in 1916 and the SSL in 1956. Scientists hoped that the ladybeetle, a native of arboreal regions across much of eastern Asia, would combat aphids, pear psylla, and other insects that harmed trees. Native ladybeetle species supposedly were not effective against such pests. Specimens of the exotic ladybeetle were imported from Russia, Japan, and Korea in an attempt to establish a viable population. When initial attempts at introduction failed, the ladybeetle was released again in the mid-1960s, again without notable success. Concerted introduction efforts did not begin again until the mid-1970s, when the USDA released thousands of ladybeetles into southern and eastern states. Large releases were made again in 1979–80, including more than 14,000 near Yakima, WA. Although efforts were discontinued because of perceived failure, by 1994 the ladybeetle had established populations in the southeast US. Soon after, it began spreading, and now is widely disseminated in the US and Canada.

Other Sources of Information: 64
Author: Eric Wagner

Asian Citrus Longhorned Beetle *Anoplophora chinensis*
Asian Longhorned Beetle *Anoplophora glabripennis*
Brown Spruce Longhorned Beetle *Tetropium fuscum*

Species Description and Current Range

The Asian Long-horned Beetle (ALB) and the Asian Citrus Longhorned Beetle (ACLB) had costly debuts, decimating trees around NY City, Chicago, and Seattle. A third beetle, the Brown Spruce Longhorned Beetle (BSLB), also made a costly arrival in Halifax, Canada, in early 2000. ALB and the ACLB are cousins and look nearly alike. Both are black with white spots and are 1.9–3.2 cm in length. The antennae are segmented, with black and white bands, and can be up to 2.5 times longer than the body. Less than 2.5 cm long, the BSLB has a black head, brown thorax, and two long reddish antennae.

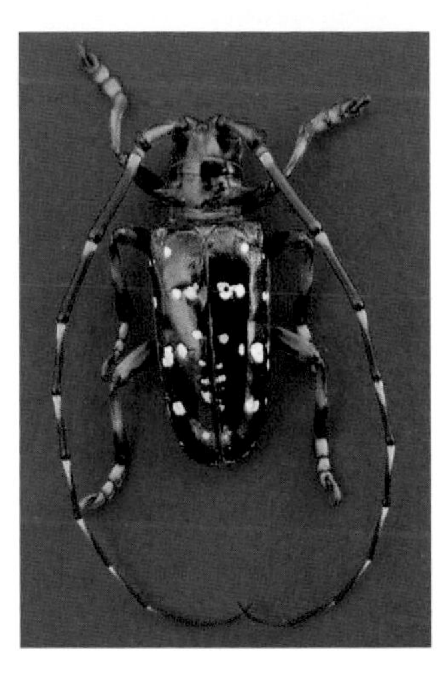

Anoplophora glabripennis

Impact on Communities and Native Species

Many hardwoods like maples, elms, and willows are susceptible to the ALB, which has attacked 29 different species of hardwood trees and infected over 7,000 trees in New York, Chicago, and NJ. The estimated cost to eradicate ALBs by 2009 is $365 million. Just as destructive, ACLBs attack 26 different families of plants, which include pear, apple, and citrus trees. If ACLBs become established in the PNW, the economic cost to fruit farmers in WA is estimated by the USDA at $1 billion/yr. One thousand trees have been removed from the infected area in Tukwila, WA. Conifer trees, including larches, spruce, firs, and pines, are the primary targets for BSLB. When BSLBs arrived in Halifax, authorities considered destroying 10,000 trees. Since May of 2000, 3,500 trees have been removed. BSLB attacks dead or weak trees, but healthy trees are also attacked during population outbreaks.

Control Methods and Management

The only known method to prevent ALB infestations from spreading is cutting and removing infested trees. The insecticide imidacloprid was field tested in Chicago in 2003. The results were promising enough to prompt the USDA to implement plans to treat 89,000 susceptible trees starting in April 2004. Imidacloprid is injected either into the soil or directly into uninfested trees. It kills adult beetles when they feed on twigs and newly hatched larvae when they feed under the bark. The effectiveness of imidacloprid is still unknown. Other control techniques for ALBs include surveying trees for evidence of infestation and planting resistant species such as oak, Basswood, and Honeylocust. Biological controls are also being researched. The ACLB is controlled and managed similarly to the ALB, including surveying, the use of imidacloprid, and replanting with resistant species like cedars, firs, and hemlocks. The BSLB is controlled and managed by biological methods including natural predators such aswasps and woodpeckers, and imidacloprid was effective in test treatments. Cutting, chipping, and incinerating infected trees are also effective.

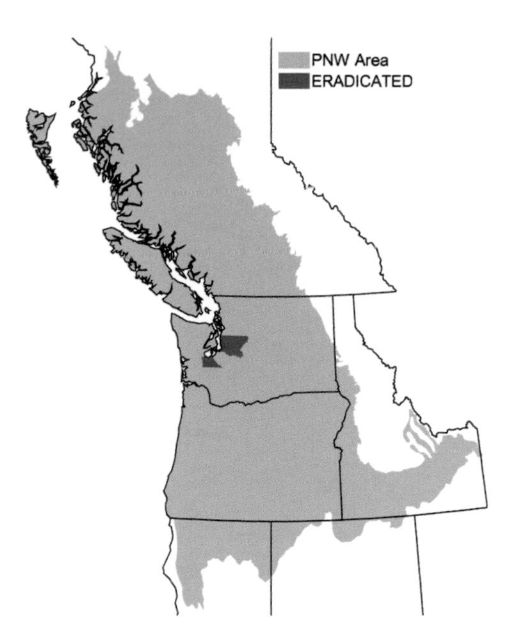

Life History and Species Overview

ALB adults begin to emerge in spring. For 2–3 days, the adults feed on twigs before mating. After hatching, the larvae bore into the wood to overwinter and feed on the xylem and cambium, the food- and water-conducting vessels within a tree, and can kill the tree by girdling (i.e., preventing the transport of water and nutrients). ACLB adults emerge from trees in mid to late summer and feed on small leaves and twigs for up to 3 months. Females are more fecund than ALB females. After reproduction is completed, the adults begin to die. Like ALB, ACLB grubs chew deep into the tree during the fall and winter months to feed on the xylem and cambium. BSLBs emerge, sexually mature, from late May to late August. BSLB larvae also bore into the wood to overwinter and feed on cambium and phloem tissues before emerging in spring as adults. Adults are strong fliers, but dispersal range is believed to be limited by the fact that BSLBs are sexually mature after emergence.

History of Invasiveness

New invaders of North America, these three beetles are potentially catastrophic to urban greenbelts, forest ecosystems, wetlands, and orchards. The main method of entrance is via imported cargo containing infested wood packaging and wood products like logs, tree trimmings, firewood, ornamental trees, and untreated lumber. ALBs can be transported as eggs, larvae, and pupae. Based on its range in China, the ALB could become established from southern Mexico to the Great Lakes region wherever there is suitable habitat. Because it is difficult to identify ALB as eggs or larvae, initial infestations may have occurred 10 years before detection. Infestations spread at a rate of 300 m/yr. Since being detected in NY City, ALB has been identified in Suffolk and Nassau counties on Long Island and the Toronto metro area in Canada and in shipments in at least 16 other states. Two ACLB beetles, one a female, were detected in Lacey, WA, in 2000. ACLB cannot be considered officially eradicated from Tukwila, WA, until beetles are absent for 4 years after tree removal. BSLB has been found at approximately 50 sites in the Halifax area, most located within 10 miles of the initial infestation area.

Acknowledgment: James Marra, Managing Entomologist, WA Dept. of Agriculture
Other Sources of Information: 21, 107, 120
Author: Demetrius Fletcher

Table 1. A comparison of biological characteristics among the three beetles

Species	Life Cycle	Adult Life Span (days)	No. of Eggs	Incubation Period (days)	Larval Coloration	Larval Length (cm)
ALB	1 year	3–66	35–80	10–15	White	~5
ACLB	1 year	78–98	200	30–90	Cream	~5
BSLB	1 year	~ 21	80	10–14	Cream-White	~2.5

Asian Bush Mosquito *Ochlerotatus japonicus*
Asian Tiger Mosquito *Aedes albopictus*

Species Description and Current Range

The translation of the Latin *Aedes albopictus* —"annoying, white-painted"—aptly describes the behavior and appearance of an Asian Tiger (or Forest Day) Mosquito. When the Asian Bush (or East Asian) Mosquito, (*Ochlerotatus japonicus*) was reclassified from *Aedes japonicus* in 2002, the new translation upgraded it from "annoying" to "horrible" (from Japan). Stowaway Asian Bush Mosquitoes were found and exterminated from cargo shipments in Portland, OR; Seattle, WA; and Vancouver, BC. Single specimens of Asian Bush Mosquitoes were sighted and eradicated in King and Thurston counties in WA. Vigilance prevented both from establishing residence in the PNW.

Glossy black scales enhanced by distinctive white stripes distinguish the Asian Tiger and Asian Bush Mosquitoes from other species. Asian Tiger Mosquitoes have a silver stripe on the upper surface of the head and body. Asian Bush Mosquitoes have belly stripes. Both have silver-white banded legs and feed with their hind legs held in the air. They rest with abdomens parallel to the surface and heads lowered at an angle. The 5–8 mm bodies of the Asian Tiger and Asian Bush Mosquitoes are slightly larger than the 3–5 mm native *Culex pipiens.*

Both Asian Tiger and Asian Bush Mosquitoes are found in urban, suburban, rural, and forested settings, small lakes, ponds, and wetlands. Mosquitoes, true parasites, belong to the order Diptera, characterized by a single pair of wings. Halteres, organs of equilibrium, replaced their hind wings and provide them with in-flight gyroscopic balancers. Mosquitoes "taste" with their feet and detect sounds with their antennae. Their rapid wing-beats generate their buzzing sound, with smaller-bodied females producing a higher frequency sound than males. Dragonflies are mosquitoes' natural predators. Their deep wing-beat frequency signals danger to the mosquito. Female mosquitoes can live weeks longer than males. The labium, a flexible sheath, protects the proboscis, but bends to allow the enclosed, needlelike, two-channeled stylet to pierce the victim's skin. Digestive enzymes and anticoagulants injected through the salivary channel cause skin irritation in sensitive individuals when blood is sucked up through the insect's feeding channel.

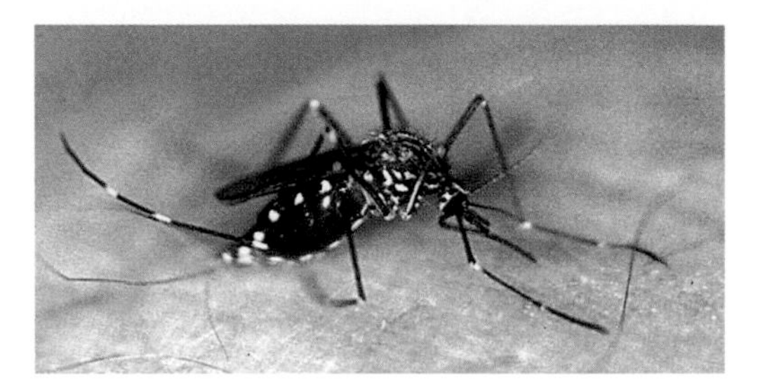

Ochlerotatus japonicus

Impact on Communities and Native Species

Asian Tiger and Asian Bush Mosquitoes displace other mosquito species and act as a transmission vector of potentially lethal parasites and viruses, including West Nile Virus (WNV). In Houston, TX, the Asian Tiger Mosquito's attempts at hybridization eliminated the native *Aedes aegypti* mosquito in less than 10 years. Mammals and amphibians can carry WNV, but birds, especially corvids such as crows, jays, and magpies, are particularly susceptible. They act as sufficient amplifier hosts, prior to succumbing to the virus.

Control Methods and Management

The key to control is early detection, reduction of breeding sites, and extermination of larvae. Larvacides such as *Bacillus thuringeinsis israelinsis* (Bti) target only mosquitoes and blackflies and do not harm other species. Methoprene regulates larval growth, while petroleum and mineral oils coat water surfaces, blocking larval breathing siphons. Once adults are present in large numbers or have spread, pesticides and chemical prevention are necessary. Organophosphates, natural pyrethrins, and synthetic pyrethroids are used with some success but harm other species. Individuals can help by eliminating possible breeding sites. Remove containers that collect standing water from your yard. Toys and discarded food containers offer sufficient temporary water storage for breeding mosquitoes. Change birdbath water and animal watering containers every few days.

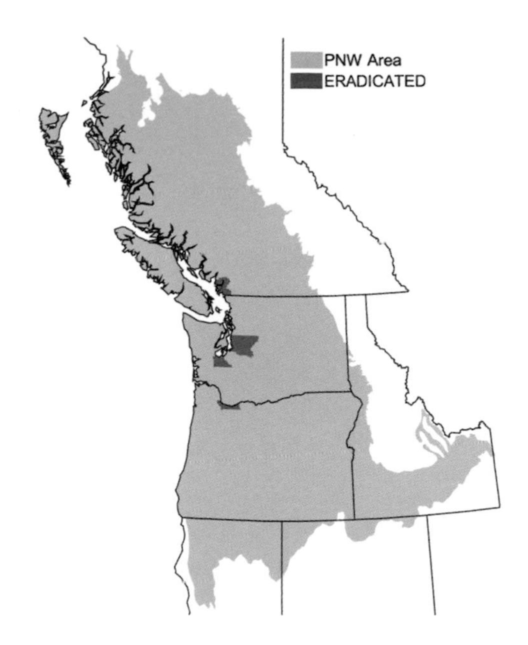

Life History and Species Overview

Native to Japan, Korea, China, Taiwan, and eastern Russia, Asian Tiger and Asian Bush Mosquitoes have spread throughout the world. Asian Tiger Mosquitoes are more aggressive toward humans than Asian Bush Mosquitoes. Generally mosquitoes are most active at dawn and dusk, but Asian Tiger and Asian Bush Mosquitoes are day biters, giving them more opportunities to feed and transmit disease, and making avoidance more difficult. Male Asian Tiger and Asian Bush mosquitoes dine on plant nectar. Only females require blood meals to provide nutrients for eggs. After feeding, the female finds a resting place and eliminates the liquid from the blood to lighten her load. She mates only once, storing the unused sperm for future eggs.

From April through November, females deposit hundreds of eggs in moist forests with black, red, or woody substrates such as treeholes, in rock pools, and small, shallow pools of standing water, or in damp soil subject to flooding. The eggs don't hatch until flooded, and can remain viable for years. Larvae live in standing water and are rarely found in deep, clean lakes and ponds or flowing water. Asian Tiger and Asian Bush Mosquito larvae require water just 6 mm deep to complete the first life cycle, breathing through posterior siphons at the surface and pupating within 10 days. The adult emerges from the pupal stage after 1–10 days. Asian Tiger Mosquito eggs can survive cold winters, but Asian Bush Mosquito adults are better adapted to cold northern winters. Both species have increased in northeastern US states.

History of Invasiveness

A recapped-tire shipment from Japan is thought to have delivered the first recorded US specimen of an Asian Tiger Mosquito to Memphis, TN, in 1983. Sightings throughout the South and Midwest followed. The Asian Bush Mosquito is suspected of sneaking WNV into the country when it arrived in NY in 1998. Asian Tiger Mosquito specimens were found and subsequently eliminated in cargo containers carrying shipments of ornamental bamboo in Seattle, WA (1986) and Portland, OR (2001), and in a shipment of the popular houseplant Draecena in Vancouver, BC (1996). Bamboo and used-tire shipments from Asia were the most likely invasion vectors for the single Asian Bush Mosquito specimens found in WA in 2002, in King and Thurston counties.

Other Sources of Information: 69, 120, 122
References: 7, 25, 214, 295, 349
Author: Rebecca Gamboa

Spiders *Philodromus dispar*

Overview of Spiders

Spiders are the earth's leading land predators. They range in size from tarantulas down to half-mm microspiders, and in abundance from hundreds/m^2 in meadows and forests to thousands in bogs. Of 842 (and counting) known spider species in WA, 47 (5.6%) are introduced, up from 33 known in 1980. On average, another new species arrives every 2 years. The situation in BC, where almost 600 spider species are known, is similar. We know very little about the species present in OR and ID or their ranges. The introduced spiders of WA vary greatly in their rate of spread and ecological effects. A species first found in 1994 (*Xysticus cristatus*) is already as widespread as the first introduced spider found in WA (*Theridion melanurum*, 1897), which spread much more slowly. Of the 47 species, 24 live mainly in and around houses; 11 are widespread only in human-inhabited areas; and the remaining 12 are widespread in natural areas and appear to be displacing native species. Introduced spiders have no economic impact on agriculture or forestry, but two of the three medically harmful spiders in WA are introduced (Hobo Spider, *Tegenaria agrestis*; Yellow Sac Spider, *Cheiracanthium mildei*). This account deals with *Philodromus dispar* (no species common name), one of our 12 ecologically invasive spiders.

Species Description and Current Range

Philodromus dispar is a 4–5 mm long crab spider of the subfamily Philodrominae, the "running crab spiders." Crab spiders have laterigrade legs that are twisted backward at the base so the whole leg lies more nearly in the plane of the body, which tends to be flatter than in typical spiders. Adult male *P. dispar* appear distinctively black from above, with a narrow white rim. Females and juveniles are mottled tan and are indistinguishable from many other species without a microscope. Species level identification of spiders almost always requires examining the fine structure of genitalia, and takes training and experience.

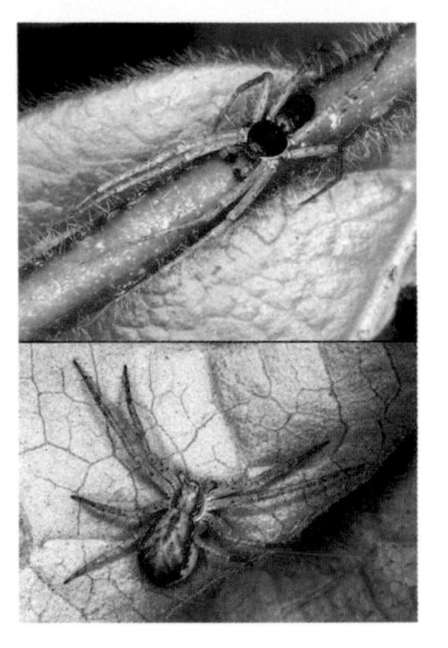

male

female

P. dispar occurs throughout the Puget Sound lowlands, north to Vancouver and Victoria, BC. It must occur in northwest OR, but no one has studied spiders there. Abundant in many cities and towns, it is increasingly common in lowland forests generally and occurs even in remote nature preserves. It is not known at elevations above 260 m or in the wetter coastal areas of WA. Its main microhabitat is foliage of shrubs and other forest understory vegetation; another important habitat is conifer canopy foliage. The species seems to require shade and is never found far from trees. An occasional population becomes established inside a house. Macrohabitats are urban-suburban residential and natural areas, rural disturbed areas, lowland moist forest, and oak and alder woodland.

Impact on Communities and Native Species

In urban and suburban areas of Puget Sound, *P. dispar* is almost always one of the two most abundant spiders on shrubs, ferns, and similar vegetation near trees or other shade. Native Philodromus species (about 11 exist in the area) are absent from these habitats when *P. dispar* is present. In its secondary habitat of conifer canopy, *P. dispar* has not completely excluded native species but has reduced their numbers. Replacement of native species is not as far advanced in areas far from human habitation but seems to be ongoing. In time, native *Philodromus* species (and perhaps some other spiders) could be largely eliminated from large tracts of western WA and nearby areas. A similar story could be told about several of the other invasive spiders. The mechanics of how the non-natives displace the natives are unknown. Presumably they are able to outcompete the natives for prey and physical space because they either tolerate wider climate extremes or resist native parasites.

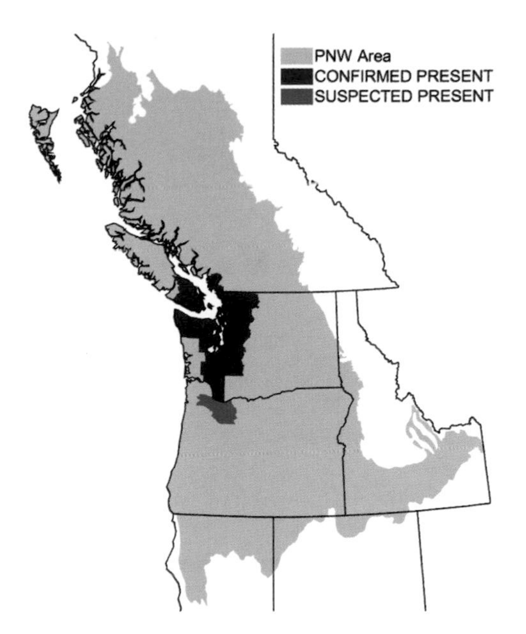

Control Methods and Management

It is not possible to manage or control specific spider species. Once non-native spiders are established, they will stay unless the habitat changes or they are outcompeted by another introduced spider. Biological control by imported parasites or pathogens has never been tried for spiders, might further harm native species, and would require funding that does not exist. Restoring native vegetation on sites overgrown with invasive plants might help a little. Prevention, not control, is the only action likely to do much good. In the case of vegetation-living spiders like *P. dispar*, this would mean much tighter controls on importation and transport of nursery plant stock.

Life History and Species Overview

Like most introduced spiders in the PNW, *P. dispar* is native to Europe. It is a very successful ambush predator, making no web and detecting its insect prey mainly by touch. It has a 1-year life cycle. Juveniles overwinter 1–2 molts from maturity and mature and mate in April. Females make their egg sacs in a curled leaf and guard them until the spiderlings emerge a month later. This species, like many crab spiders, practices "sexual bondage," the male fastening a female down with silk threads before copulating with her.

History of Invasiveness

The first North American record of *Philodromus dispar* is from Seattle, WA, in 1950. The mode of transport for this or any other non-native spider is not known, but *P. dispar* was probably an unintended passenger on imported trees or shrubs. Future spider introductions could be sharply reduced by restricting or eliminating live plant importation. By 1980 the species was only found within a 30-mile radius of Seattle, WA, and Victoria, BC. It has spread more quickly since then, and its range is continuous from Vancouver, BC, to Vancouver, WA, and an unknown distance into OR. Some spiders disperse on the wind by ballooning, but *P. dispar* seems to balloon only occasionally. Another spider, *Lepthyphantes tenuis*, introduced to Seattle in the same year, balloons massively and unlike *P. dispar* is now found throughout the PNW. *P. dispar* seems to spread first by transport on shrubs and trees within the PNW, especially to new construction, and then by short-range dispersal from these inhabited sites into surrounding natural habitat. Nothing can be done about natural dispersal, but cleaning or fumigating nursery plants before transport would help a great deal.

Other Sources of Information: 16, 83
References: 76, 77, 95, 338
Author: Rod Crawford

Black Arion *Arion ater*
Black Garlic Glass-snail *Oxychilus alliarius*
Brown Gardensnail *Cornu aspersum*
Gray Fieldslug *Deroceras reticulatum*
Leopard Slug *Limax maximus*

Species Description and Current Range

Introduced slugs vary in size from 1.3 to 25.5 cm long, and in color from yellow to black. Non-native snails are mostly smaller than 2.5 cm and their shells vary in color from white, yellow, and orange, to brown. Slugs and snails are usually nocturnal and are more active during wet weather. Mollusks are found in almost all terrestrial habitats. Introduced species are the most abundant mollusks in urban and suburban areas. They are spreading into wild lands and may become more common than natives. There are more than 70 native and 29 introduced terrestrial mollusk species in the PNW, with 23 native and 16 introduced slug species. The PNW's worst slug invader is the Black Arion, a voracious herbivore that damages gardens, agriculture crops, and native plants. It varies in color from black in more rural areas to brownish with an orange skirt fringe around the edge of the body in cities, and young animals are striped. Adults can reach up to 15 cm. The upper surface of this slug is deeply grooved and the entire animal is firm to the touch. The 5 cm Gray Fieldslug or Milky Slug, another common invasive slug, eats garden and flowering plants including iris, tulips, and other vegetation. The species tends to be light brown to gray in color, and releases a distinctive milky-white mucus when irritated. The Brown Gardensnail is a common garden snail with a 2.5 cm streaked, brown shell. The glossy, flat, brown shell of the Black Garlic Glass-snail is no longer than 0.8 cm, and the species releases a garlicky smell when irritated.

Impact on Communities and Native Species

Introduced mollusks compete with natives for food, and some, such as the Leopard Slug or Great Gray Garden Slug (*Limax maximus*), prey on native slugs such as the Banana Slug (*Ariolimax columbianus*). In 1961, the Banana Slug was the most abundant slug species in Stanley Park in Vancouver, BC, but by 1975 Black Arion was more abundant. The interactions between the species that displace the Banana Slug are not known. Introduced slugs cause major damage to large-scale agriculture and home gardens. They have damaged up to 75% of the strawberry, sweet corn, and bush bean crops in WA and up to 50% of the potato crops, 25% of the lettuce crops, and 10% of the tomato crops in BC. In BC, 26 different crops are damaged by slugs.

Control Methods and Management

There are no good control methods for introduced slugs and snails. The most effective time for finding slugs is at night when they are active and exposed. Given the wide distribution and sometimes high densities of terrestrial mollusks in the PNW, picking off slugs or snails from vegetation is not an effective

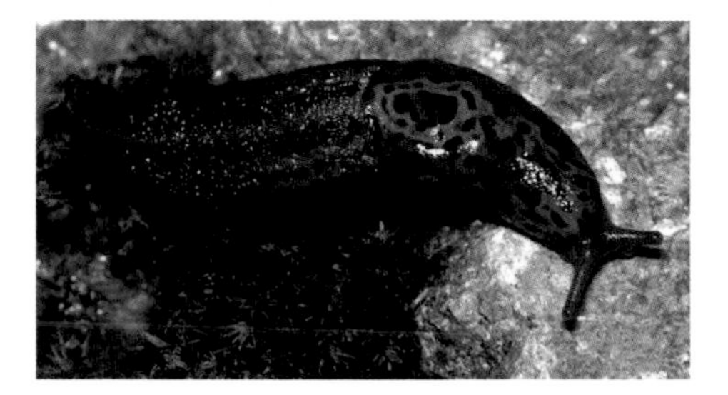

Limax maximus

extermination technique. Traps baited with beer or apple cider are good techniques for ridding a small garden of slugs, though they also attract and kill native species. Gardening with plants that slugs do not like—such as ferns, rhododendrons, and Bleeding Heart—or running a copper strip, which slugs dislike crossing, around your garden keeps slugs out of a small local area. Cats, raccoons, ducks, geese, garter snakes, and chickens eat slugs. Some chemicals and poisons kill slugs but are toxic to the environment.

Life History and Species Overview

Slugs and snails like moist, humid areas with leafy vegetation and places for retreat from the sun, such as logs and rocks. Most slugs live above ground. Slugs and snails mostly eat herbaceous material but will also eat conifer seedlings, fungus, worms, insects, dog scat, and other slugs. They taste, smell, and detect light using sensory cells that are scattered all over their bodies. Slugs and snails travel farther and faster than most people realize. The native Banana Slug can travel 10 m/hr, and the introduced Gray Fieldslug travels up to 12 m a night. Slime mucus of slugs and snails enables them to stick to vertical surfaces, glide unscathed across surfaces as sharp as razor blades, and seals water in the body to prevent dehydration. Slime protects them from predators by making them harder to grasp and swallow, and may seal shut the mouths of snakes and shrews. Slugs and snails are hermaphrodites, possessing both male and female reproductive organs, and are capable of mating with themselves. Mating behavior between slugs is elaborate, and may reduce the chance of two separate species mating. Leopard Slugs have a penis about half their body length and they copulate in midair while dangling from mucus thread up to 45 cm long. After mating, slugs deposit up to 50 eggs in soil or vegetation or under rocks that shelter the eggs from desiccation. Some slugs mate many times in a year and lay up to 500 eggs annually. Most slug and snail eggs are round, white to clear in color, and between 0.25–0.4 cm in diameter. Depending on the species, terrestrial mollusks can live 1–6 years, and survive in temperatures as low as –8°C. Slug densities can reach up to 50/m^2 in moist, heavily vegetated coastal areas.

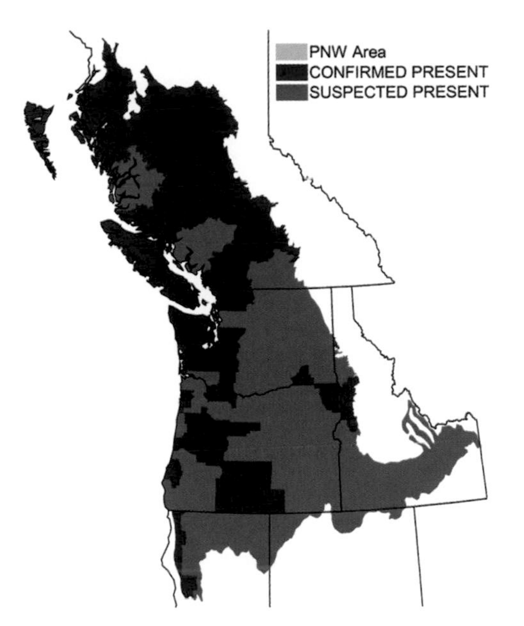

History of Invasiveness

Slugs and snails are hitchhikers and are regularly intercepted at ports of entry in baggage and mail or on horticultural and agricultural products. Most introduced terrestrial mollusks in the PNW are of European origin. Some snail species, such as the Brown Gardensnail, were intentionally introduced to gardens as early as the late 1800s as curios or delicacies for later consumption. It is suspected that Black Arion was first introduced into North America from Europe in the Puget Sound region, and by the mid-1900s it was recognized as an economic problem. Once introduced, snails and slugs readily move into wild lands and invade most terrestrial habitats. For example, slugs were introduced to forests in hay bales used to prevent erosion. In a survey conducted in 1973, Black Arion was the most common slug encountered in coastal, humid WA. Currently in BC, about 25% of the terrestrial mollusk species are exotic. Many populations of exotic slug and snail species are expanding.

Acknowledgments: Data for the map were provided by Robert Forsyth and the Royal BC Museum.
Other Sources of Information: 15, 73, 116
References: 28, 123, 137, 185, 268
Author: D. Shallin Busch

American Bullfrog *Rana catesbeiana*

Species Description and Current Range

The American Bullfrog, North America's largest frog species (0.5 kg), is found around permanent freshwater ponds and lakes in every life-history stage and generally prefers lowland, warm, and still bodies of water. American Bullfrogs are similar in appearance to Green Frogs (another PNW invader), with olive to bright green skin and a yellow throat, but are larger (180–200 mm) and lack raised folds along the edges of the back. Adult females have tympanic membranes (external eardrums) the size of their eyes, whereas the male tympanum is several times larger. American Bullfrog tadpoles are also large (150 mm long) and often occur at extremely high densities. Originally introduced to the PNW to produce frog legs (current market price $12.99/lb), American Bullfrogs quickly colonized most lowland habitat west of the Cascade Range in WA, OR, and northern CA, but are also present in the Great Basin and Central Valley. American Bullfrogs are native to eastern North America, but are now present in every western US state (except AK), HI, and many areas of Europe, Asia, the Caribbean, and South America. While no one factor limits American Bullfrogs' range, they generally avoid extremely cold or dry areas and appear to require unshaded waters.

Impact on Communities and Native Species

American Bullfrogs contribute to the decline of many native PNW species through consumption or competition. Affected species are primarily other amphibians such as Northern Leopard Frogs, Spotted Frogs, Red-legged Frogs, Foothills Yellow-legged Frogs, and possibly Tiger Salamanders. They have also been implicated in the decline of the Western Pond Turtle (by consuming hatchlings and juveniles), a species of conservation concern, and may even influence nesting waterfowl.

While American Bullfrogs are a significant threat to native PNW pond communities, their impact on species is often confounded with their preference for heavily disturbed areas, which is another key contributor to the decline of most native taxa. Though little studied in the PNW, American Bullfrogs may alter wetland ecosystems by consuming crustaceans, insects, fish, garter snakes, baby turtles, ducklings, bats, blackbirds, and especially other American Bullfrogs. Due to their large size and voracious appetite, American Bullfrogs consume anything they can catch and fit in their mouths. Alternatively, American Bullfrogs serve as prey for a variety of wading birds, raccoons, mink, and snakes and as a food source for humans (frog legs).

Control Methods and Management

Local American Bullfrog eradication programs involve the collection and destruction of adults and eggs, and occasionally the application of general pesticides such as Rotenone. However, because individual bullfrogs can travel relatively long distances overland in periods of rainy weather or when aided by humans, eradication successes are often short-lived as recolonization occurs quickly. Therefore, American Bullfrog populations are extremely difficult to control and nearly impossible to eliminate, posing a very serious challenge to restoration and conservation efforts.

Life History and Species Overview

The American Bullfrog's complex life history requires both land and water. Adult males are highly territorial and can be heard making their distinctive and booming "jug-o'-rum" calls in the shallow margins of wetlands from late spring through the summer. Both male and female American Bullfrogs have a wide repertoire of vocalizations that act to attract potential mates, warn potential competitors, and startle predators. Adults can be easily captured during the breeding season as individuals are abundant and will often freeze when caught in a bright spotlight at night. Individual females can lay up to 20,000 eggs, allowing populations to expand rapidly. Eggs are laid in a film at the surface of the water throughout the summer and sink to the bottom over the 4–7 day incubation period.

Tadpoles hatch into free-swimming individuals and usually spend 2 years in the water before growing large enough to undergo metamorphosis into terrestrial juveniles. Tadpoles are primarily herbivorous but can graze on a variety of biofilms and invertebrates that collect on the surfaces of plants, rocks, and sediment. Juveniles and adults are carnivorous, and individuals can live up to 9 years. Juvenile and adult American Bullfrogs from northern latitudes generally hibernate over winter in the mud and debris at the edge of ponds, and those from warmer climates can be active all year. As with many amphibians, American Bullfrogs produce a toxin in their skin that makes them distasteful to most vertebrate predators.

History of Invasiveness

The earliest record of an American Bullfrog introduction in the PNW is from CA in 1895. Introductions occurred through the early part of the 1900s across WA and OR, and as late as the 1940s in BC. American Bullfrogs have a higher tolerance than many native taxa for the elevated water temperatures that are often found in wetlands modified by humans, and this likely aids their spread throughout the region. Research shows that the presence of other exotic species like sunfish and bass can facilitate American Bullfrog colonization by reducing predatory dragonfly larvae that limit American Bullfrogs in their native range.

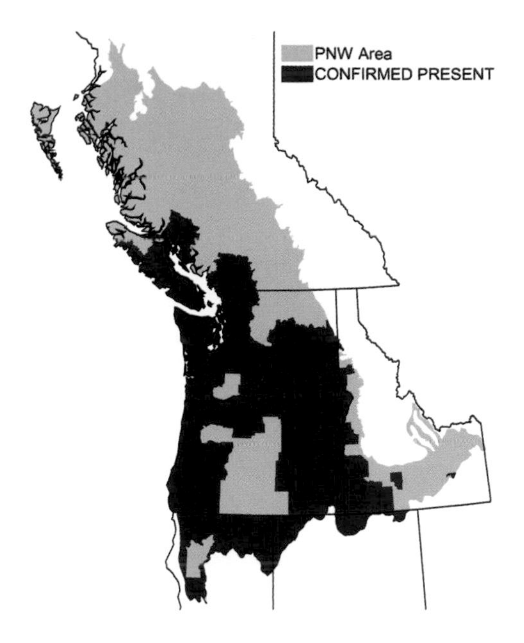

References: 2, 44, 145, 180
Author: Wendy J. Palen

American Shad *Alosa sapidissima*

M / H

Species Description and Current Range

American Shad, named after the Saxon word *allis*, an old name for European Shad, and *sapidissima*, meaning "most delicious," are the largest members of the herring family. Shad males (bucks) can weigh as much as 1.2 kg and females (hens or roe fish) weigh up to 2.2 kg. On the West Coast, schools of American Shad inhabit the Pacific Ocean, coastal rivers (Chehalis, Coos), and larger inland rivers and their tributaries (Columbia, Fraser, Willamette) from CA to AK.

Adult Shad are 300–635 mm in length and predominantly silver, with some blue-green along the top of their backs, a row of 3–23 dark spots on their sides, and a deeply forked tail. A distinct row of modified scales called "scutes" form a sharp edge or ridge along their belly. The coloration of Shad deepens when they return to fresh water to spawn.

Impact on Communities and Native Species

The May–July migration of adult Shad overlaps with that of Sockeye and Chinook Salmon. The less agile but more abundant Shad block entrances to Columbia River fish ladders, delaying salmon passage over dams. In 2003, half a million Shad passed Bonneville Dam on the Columbia River in a single day. Some fish ladders have been modified, eliminating submerged entrances and decreasing the height of pool steps to aid Shad passage. Other impacts of Shad on native communities are not well known; however, the presence of millions of Shad may change food-web interactions. The diet of juvenile Shad may overlap with that of native salmon species in the estuary, and the abundance of young Shad greatly increases the prey available to larger predator species.

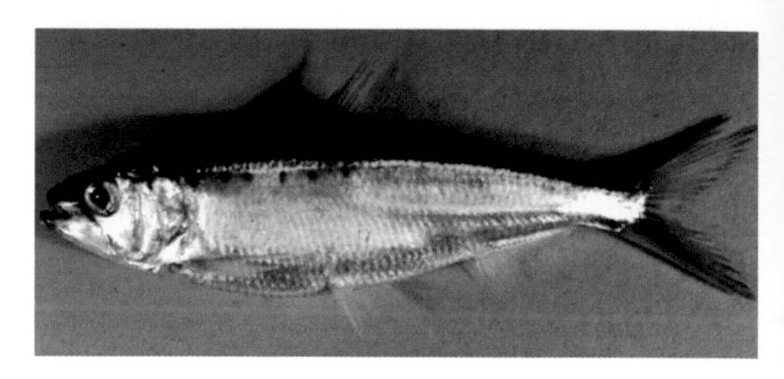

Control Methods and Management

There is minimal management of Shad as a game fish or as a food fish, and no current management of Shad as an invasive species. Though highly valued as a food fish on the East Coast, only a small commercial market for Shad developed in the west, and it has had little impact on the Shad population. BC currently has no limits on Shad fishing. However, OR and WA generally restrict the commercial and recreational seasons and gear to correspond with the salmon fishing season, thereby minimizing the impact on salmon by Shad fishers. Studies are under way to evaluate if the availability of American Shad in the lower Columbia River causes the Northern Pike-minnow, a native predator with a huge appetite for salmon and Shad, to grow larger faster. These results could affect the management of American Shad and Northern Pike-minnow in the Columbia River.

Life History and Species Overview

American Shad begin their lives in fresh water and spend most of their adult lives in salt water, only returning to the freshwater place of their birth to spawn. Shad spawn in estuaries and rivers in the late spring to early summer in shallow water areas of sand, gravel, or silt. Spawning is usually at night. One female can lay more than 100,000 eggs. Shad fry hatch in 5–10 days, remain in fresh water until the late summer or early fall, and then migrate to the ocean. They feast on plankton at sea for 2–5 years before returning to fresh water to spawn. Shad, like salmon, fast while migrating upstream, relying on fat reserves to survive. Most fish die after spawning only once, although some Shad complete the spawning cycle two or three times.

The native range of American Shad on the East Coast of North America stretches from the Gulf of St. Lawrence, Canada, to the St. Johns River, FL. Atlantic Coast Shad grow up to 125 mm longer and 2 kg heavier than Shad introduced to the Pacific. American Shad, which had been harvested by Native Americans for centuries, quickly became a staple of the colonial diet. Shad was consumed fresh, dried, and salted for the winter and was even used as fertilizer. The settlers were so dependent on Shad that in 1724 the colonial legislature passed a law forbidding waterway obstructions that deprived upstream settlements of fish (but the law was not strictly enforced). In the spring of 1778, the arrival of the Shad run at Valley Forge saved George Washington's army from starvation. American Shad constituted the largest fishery in the mid-Atlantic region throughout the 1800s and gained the name "poor-man's salmon" because it was cheap and readily available. Like other fish in the herring family, Shad is mild, oily, and rich and cooks to a creamy white. Today, the roe or eggs are the most valued part of the fish, considered a seasonal delicacy, but prior to the early 1800s the roe was discarded. Shad meat is laced with around 900 small bones in multiple rows and many bones that branch into a Y shape. The Shad's numerous bones led the Lenni Lenape native people of the mid-Atlantic States to refer to Shad as a "porcupine fish turned inside out." For those with dexterity, boning Shad was a good business. As many as 11 separate cuts are required, taking a skilled deboner a few minutes to finish a single fish. As Shad populations have declined, so has the skill of Shad boning.

In 1900, Chesapeake fishermen caught 7,700 metric tons of Shad but by 1940 the catch had dropped to 1,400 metric tons. Over-fishing, spawning barriers (dams), and water pollution caused Shad populations to plummet. VA, once home to the largest Shad fishery, in 1994 banned Shad fishing in the Chesapeake Bay and surrounding waters to help the population rebuild.

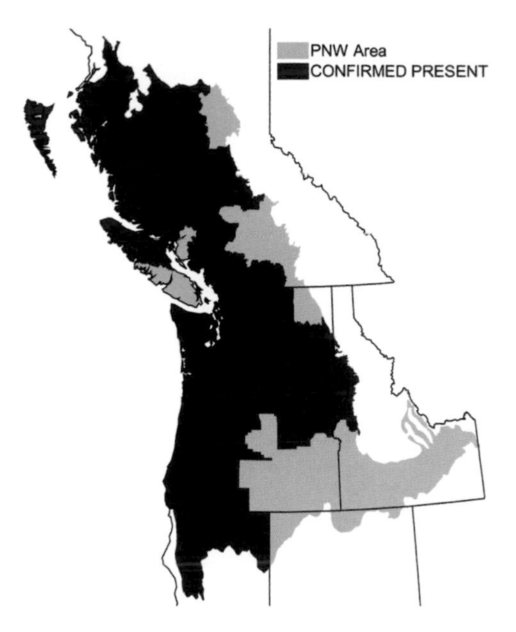

History of Invasiveness

In 1871, at the request of the CA Fisheries Commission, 10,000 American Shad fry were released into the Sacramento River, after a 7-day cross-country trip from the Hudson River, NY, on the newly completed transcontinental railroad. Nobody expected the introduction to succeed. Shad were thought to spend the oceanic portion of their life cycle close to the rivers of their birth and therefore would not disperse up or down the coast. As is now known, a small percentage of Shad stray into rivers other than those of their birth. Helped along by coastal currents, Shad were caught off WA and BC, Canada, 5 years after their introduction. By 1880 Shad had been found in the Columbia River. The completion of the Dalles Dam on the Columbia River in 1957 covered Celilo Falls, a passage barrier to Shad, and made accessible hundreds of kilometers of prime spawning habitat. As a result, the Shad population in the Columbia River exploded to 5 million, the largest population of American Shad in the world.

References: 101, 111, 171, 217, 244
Author: Katie Barnas

Largemouth Bass *Micropterus salmonides*
Rock Bass *Ambloplites rupestris*
Smallmouth Bass *Micropterus dolomiu*

M / H

Species Description and Current Range

Smallmouth Bass, Largemouth Bass, and Rock Bass belong to the sunfish family and aren't really bass at all. They are closely related to other large sunfish including Bluegill and Crappie. Smallmouth, Largemouth, and Rock Bass have been introduced into freshwater lakes and rivers in all regions of the continental US and parts of Canada. In the PNW they are among the most widely dispersed non-native fishes.

Like all sunfish, these three species have a single large dorsal fin (the fin on the top of the back) with two distinctly different parts: one portion has sharp spines and the other has fin rays. Despite this, black bass, a group of sunfish that includes Smallmouth and Largemouth, in 1803 were erroneously given the name *Micropterus*, Greek for "small fin," based on a single specimen with a torn dorsal fin. The Largemouth has a mouth that extends past the center of the eye and a horizontal line of dark spots running from gill to tail. The mouth of the Smallmouth does not extend past the center of the eye, and these fish are more mottled, with dark vertical bars along their sides. In the PNW, Smallmouth generally weigh 2–3 kg, with the bigger Largemouth reaching 3–4 kg.

Rock Bass look more like the Lepomis sunfish than like Largemouth or Smallmouth Bass, are smaller (rarely weighing more than half a kilogram), and have a stocky build and red eyes. Rock Bass have five or more spines on their anal fin (on the belly by the tail), while the rest of the Lepomis sunfish, including Warmouth and Green Sunfish, have only three.

Ambloplites rupestris

Impact on Communities and Native Species

Freshwater bass are aggressive predators, consuming any prey smaller than the size of their head, including fish, rats, mice, ducklings, frogs, snakes, and salamanders. This healthy appetite has contributed to the decline of some native fishes, frogs, and salamanders where freshwater bass are introduced. Smallmouth and Largemouth Bass prey on juvenile salmon. In the John Day Reservoir of the Columbia River, each Smallmouth consumes an average of 1 salmon/day during the summer months, or 170,000–300,000 juvenile salmonids a year.

Control Methods and Management

Smallmouth and Largemouth Bass continue to be intentionally stocked and managed as game fish in WA, OR, and ID. These states limit the number of bass caught to 3, 5, or 6 per day and have restrictions on length to ensure a stable population for sportfishing. BC has never stocked bass but still limits anglers to 4 bass per day.

Life History and Species Overview

Smallmouth, Largemouth, and Rock Bass natively inhabit the fresh waters of central and eastern North America stretching from the Great Lakes and St. Lawrence River south through the Mississippi River and its tributaries. Smallmouth and Rock Bass inhabit clear streams, rivers, and lakes with gravel or rocky substrate and little vegetation. They tend to congregate in bottom habitat with ledges and boulders. Largemouth Bass prefer slightly warmer water with vegetation and submerged debris where they can hide. All three species are warm water fish favoring streams with summer water temperatures of 15–20°C. Spawning occurs in the spring when water temperatures rise above 15°C. Feeding activity slows at water temperatures below 10°C and stops at water temperatures around 5°C. All three species move to deeper waters with less current in the winter.

Like other sunfish, males build and fiercely defend large circular nests visible from the surface. The male uses his head and tail to clear the area of debris and create a depression near shore in areas with slow-moving water and ample cover. Females produce thousands of eggs and then leave the nest immediately after spawning and may spawn several times in one season. After mating, the male hovers over the nest, fanning the eggs to keep oxygen-rich water moving over the eggs and prevent debris from smothering them. The eggs hatch out in 2–10 days depending on water temperature. The males continue to guard the fry on the nest for 5–7 days and will swim with them for 1–2 weeks after they leave the nest. At first the young eat mainly invertebrates but later include larger insects and fish. All three species prefer to feed in low light, either at dusk or on cloudy days, and often feed by ambushing unsuspecting prey. Smallmouth and Largemouth mature in 3–5 years and have a life span of up to 15 years. Rock Bass have a shorter life span, maturing by age 3, with few living beyond age 6.

History of Invasiveness

Largemouth and Smallmouth Bass arrived in the PNW over 100 years ago upon the completion of the transcontinental railroad in 1869. Smallmouth were introduced to

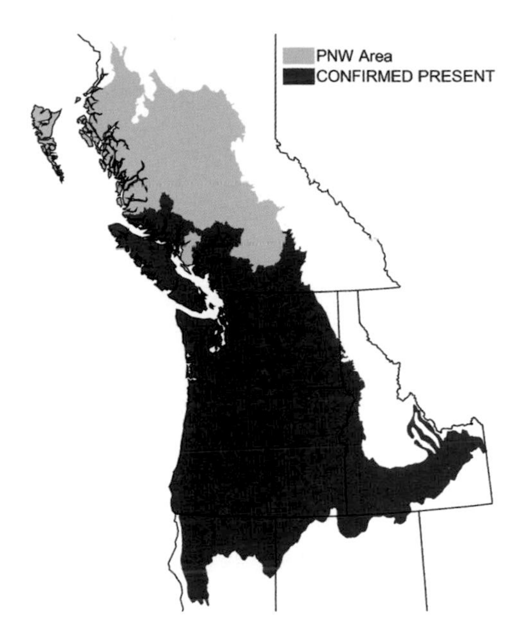

the West Coast as game fish in 1874 (CA). Subsequent early introductions occurred in Kettle River, BC (1901), in waters deemed too warm to support native trout, followed by Blakely Island, WA (1920), Willamette River, OR (1924), and Yakima River, WA (1925). Only 4 years after introduction into the Yakima River, E. J. Farley of Prosser, WA, caught a 6-pound Smallmouth, to take second place in a national *Field and Stream* contest.

Private citizens introduced Largemouth Bass into the Boise River, ID, and Willamette River, OR, in 1887 and 1888. Two years later, the US Fish Commission released the first 1,220 fish in WA. Over the next 10 years another 4,000 were planted in lakes throughout the state. Largemouth Bass in BC likely descended from the escapees of a private pond near the Kootenai River, ID, in 1916.

Rock Bass arrived slightly after Smallmouth and Largemouth. The first documented introduction of Rock Bass was in the lower Columbia River in 1893. Rock Bass are not as popular for sportfishing as other freshwater basses so they have been introduced less frequently; however, Rock Bass populations exist in WA, OR, and BC.

References: 111, 171, 264, 309
Author: Katie Barnas

Bighead Carp *Aristhythys nobilis*
Black Carp *Mylopharyngodon piceus*
Common Carp *Cyprinus carpio*
Grass Carp *Ctenopharyngodon idella*
Silver Carp *Hypothalmichthys molitrix*

Species Description and Current Range

This group includes the Common Carp and four species generally falling under the name of Asian carp: Bighead Carp, Silver Carp, Black Carp, and Grass Carp. Carp are large members of the minnow family. Bighead weigh up to 50 kg, reaching lengths of 1.5 m. The head has a protruding lower jaw and is abnormally large for the green to olive-colored, deep, laterally compressed body. Because the eyes are below the midline of the body, the fish appears to look downward. Long, closely spaced gill rakers efficiently strain plankton. Silver Carp are typically smaller, with fused gill rakers. Bighead and Silvers are common in the Mississippi and Missouri river basins in slowly moving water at least 8 ft deep. Both species are very excitable, and lights and engine noise cause them to leap as much as 3 m out of the water. They often land in boats, injuring boaters, breaking windshields, and knocking people overboard. Some wildlife agents consider the fish a danger to public safety because of these accidents.

Black Carp are distinguished by their black color, a more cylindrical body, and strong pharyngeal teeth in the throat. They typically grow to 1 m and average 15 kg but live up to 15 years or more, reaching up to 1.5 m and 68 kg. Common Carp are brownish, have two long, whiskerlike sensory barbels on each side of the upper jaw, and may grow to 0.5 m in length and weigh up to 25 kg. Grass Carp are golden brown or silvery and without barbels. They are typically about 30 cm long when stocked, but they grow rapidly, up to 1 m, and weigh up to 30 kg. Only Grass and Common Carp are found throughout the PNW. Common Carp live anywhere in shallow lakes and streams.

Cyprinus carpio

Impact on Communities and Native Species

Large size, high fecundity, and voracious appetite make carp a competitive threat to native, sport, and commercial fish. Feeding at the base of the food chain, Bighead and Silver Carp compete directly with other filter feeders, including other fish, insects, and mussels. As filter feeders become less abundant, carnivorous fish decline. Although not competing directly with native fish, Grass Carp significantly change the community composition of large plants, phytoplankton, and invertebrates; decrease refuge and modify habitat; disrupt food webs; and introduce parasites and diseases. Rapid reduction of plants may cause an abrupt input of nutrients via feces, causing algal blooms. Carp could potentially acclimatize to brackish water and establish in coastal marshes, denuding vegetation used by overwintering ducks. Grass Carp are usually stocked in lakes having large nuisance-plant populations to control plant growth. Both Grass and Common Carp uproot vegetation, stirring up the bottom, muddying the water, and deteriorating aquatic habitat. A mollusk eater, the Black Carp impacts native freshwater snail and mussel populations.

Control Methods and Management

Bighead and Silver Carp are filter feeders, so traditional fishing methods are not very effective. The best control method is preventing their spread. The US FWS has proposed making transport of live Bighead, Silver, and Black Carp across state lines illegal. In the Great Lakes, to prevent ecological and financial impacts to the $4.5 billion sport and commercial fisheries, a new $8.5 million electrical barrier across the Chicago Sanitary and Ship Canal deters Mississippi River fish from entering Lake Michigan. As with invisible dog fences, electrical current induces fish swimming upstream to turn back. Grass Carp are controlled by stocking only sterile, triploid fish, released only by permit in enclosed systems and irrigation canals. Sterilization may not always be effective; in the wild, even sterile carp feed voraciously for years.

Life History and Species Overview

Asian carp originate in eastern Russia and China; the Common Carp is a Eurasian relative. The group tolerates a wide range of water quality and weather conditions. Asian carp are native to large rivers, and spawning generally tends to occur in large rivers during spring flooding, over 2–3 months. Grass Carp reproduce in flowing water, but Common Carp breed easily in ponds. Common Carp may produce up to 2 million eggs per fish. Bighead and Silver Carp are filter feeders, consuming small algae, zooplankton, and aquatic insects. Lacking a true stomach, the Bighead eats constantly, consuming nearly half its weight in plankton daily. Grass Carp are nonselective feeders, known to eat at least 175 species of submerged aquatic plants. Young fish can eat 2–3 times their weight daily and are used to control aquatic weed infestations. Black Carp consume snails and mussels almost exclusively, but omnivorous

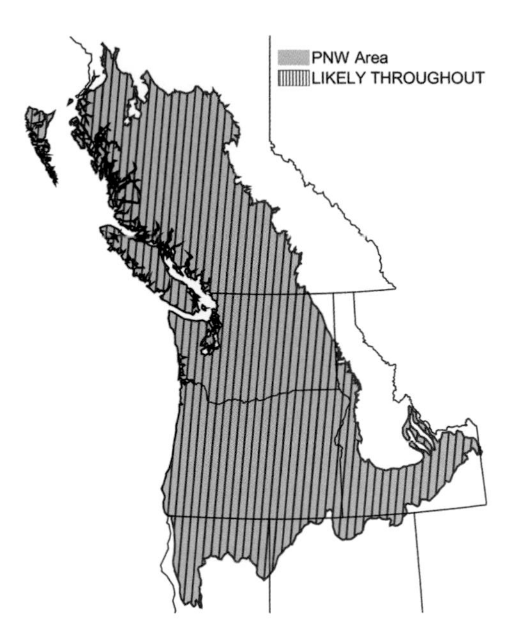

Common Carp eat plants, insects, annelids, mollusks, and other species.

History of Invasiveness

Carp are prolific and very invasive. Common Carp were introduced in the US in the 1850s as a food fish and into WA, OR, and ID in 1882 and are widespread. In 1963, AR imported Grass Carp for aquaculture and research. Accidental release occurred as early as 1966, and lakes were intentionally stocked in 1971. By 1972, they were in 40 states and now occur in at least 46 states. Sterile Grass Carp were developed in the 1980s. In 1972, AR fish farmers introduced Bighead and Silver Carp to control vegetation in aquaculture ponds. The fish escaped into the Mississippi River during the 1994 floods, spreading rapidly to at least 19 states. Black Carp first arrived in AR in the early 1970s, as a "contaminant" in Grass Carp shipments, and were later purposefully introduced in the 1980s by AR catfish aquaculture to control snails that harbor trematode parasites infecting catfish. Their only known escape was into the Mississippi River during the 1994 floods, but predictably, this species is now spreading as well. Bighead and Silver Carp are imported live for Asian community consumption, providing another potential introduction pathway.

References: 4, 111, 291
Author: Joan Cabreza

Black Bullhead *Ameiurus melas*
Brown Bullhead *Ameiurus nebulosus*
Channel Catfish *Ictalurus punctatus*

Species Description and Current Range

Brown Bullhead, Black Bullhead, and Channel Catfish are members of the Catfish family. All three species are found in lakes, rivers, and streams of the PNW. Brown Bullhead are the most common in the region, followed by Channel Catfish and Black Bullhead. Yellow Bullhead are present though rare.

All species have 8 whiskers or barbels, characteristic of catfish, on their chin and mouth, a scaleless body, and spines on their dorsal (top of the back) and pectoral (side of the body) fins. Channel Catfish are silver or slate in color, grow much larger than Bullheads (up to 915 mm and 13 kg), and have a forked tail. Bullheads, named for the bullish behavior of breeding males, grow to 300 mm in length with an average weight of 1 kg. Brown Bullhead can grow slightly larger than Black, up to 500 mm and 1.3 kg. Black Bullheads and Brown Bullheads have a dark and mottled appearance with a rounded tail and look identical at first glance. The only differences are the presence of thorns on the end of the pectoral spines of Brown Bullhead and the more plump body of Black Bullhead.

Ameiurus melas

Impact on Communities and Native Species

The unlawful introduction of Brown Bullhead to Hadley Lake, BC, caused the extinction of two endemic sticklebacks by 1999. In Columbia River reservoirs, salmon are a favorite staple of the Channel Catfish diet. Channel Catfish 466–674 mm in length have a diet containing approximately 50% salmon, and large Channel Catfish (over 674 mm) feed almost exclusively on salmon. A single catfish eats an average of one juvenile salmon every 3 days in summer months. No estimates have been made of the catfish population or the number of salmon consumed.

Control Methods and Management

WA, OR, and ID stock Channel Catfish in lakes. It is assumed that most of these lakes do not reach the 21–27°C required for catfish reproduction by late spring. OR, ID, and BC have no statewide limits on catch; WA limits anglers to 5/day.

There are no current attempts to control Bullheads; they are extremely difficult to eradicate. Bullheads are capable of surviving in low oxygen and high temperatures above 30°C that would kill most fish. Brown and Yellow Bullhead have the highest rating of "tolerant" on the US EPA scale for non-specific pollution tolerance.

Life History and Species Overview

These catfish species are native to the Midwest and eastern North America. Brown Bullhead's native range extends from ND and OK to the Atlantic Ocean. Channel Catfish and Black Bullhead extend farther west from southern

SK, Canada, to NM. Catfish species are adapted to live in murky waters and usually feed at night. Their barbels are loaded with taste buds and olfactory sensors to help them locate food. Catfish are omnivorous and will eat just about anything, including frogs, snakes, birds, crustaceans, snails, worms, leeches, fish, plants, and seeds. Channel Catfish are more piscivorous (fish eating) and are more visual hunters than most catfish species. Channel Catfish prefer large streams and rivers with swift water, while Bullheads are most often found in slowly moving water like backwater areas in streams and in lakes. Spawning occurs May–July when water temperatures reach 18–21°C for Bullhead and 21–27°C for Channel Catfish. Since spawning-water temperatures for Channel Catfish are so warm, only the Snake River, ID, lower Yakima River, WA, Walla Walla River, WA, and Columbia River have naturally reproducing populations.

During the breeding season, males, sometimes with the help of females, make nests in natural shelters such as undercut banks and logs or in litter such as tires and buckets. The nest is then cleared of silt by the mating pair. Spawning occurs in a head-to-tail position characteristic of the Catfish family, in which the male wraps his tail around the snout of the female. Female Bullhead spawn up to 5 times in an hour. After mating, both Bullhead parents fan the nest but only male Channel Catfish stay on guard. Bullhead and Channel Catfish males continue to guard the young after the fry hatch out. Bullheads mature at 3 years and live 7–9 years. The larger Channel Catfish mature later, at 4–6 yrs, and live longer, 15–20 years.

Brown Bullheads are the larger and more abundant bullhead species in the PNW, though where they coexist with Black Bullheads in their native habitat, Blacks often dominate. Because they are smaller, Black Bullheads are not as popular with anglers as Brown and neither are a particularly popular game fish in the PNW, despite an abundance of healthy populations and their reputation for being an excellent table fish.

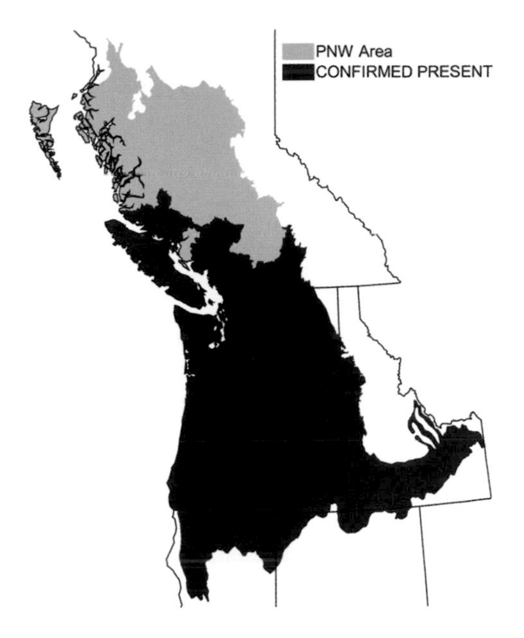

History of Invasiveness

Brown Bullheads were first introduced in Silver Lake, WA, in 1883. They apparently moved out of the lake and reached the Columbia and Willamette rivers. In spring 1890 the *Oregonian* newspaper reported, "The ponds and lakes of Sauvies Island [where the Willamette River flows into the Columbia] are literally alive with catfish which have been carried in by the late flood waters. By every appearance our waters will soon be swarming with these fish, as they increase at an appalling rate." There are also reports that private citizens introduced Black and Brown Bullheads into the Willamette River in 1880.

In 1892, 250 Channel Catfish were released into three lakes in WA. A year later, the US Fish Commission released 100 adult Channel Catfish in the Boise River, ID, and the Willamette River, OR. Black and Yellow Bullheads were likely released from an exhibit at the closing of the Lewis and Clark 100-year exposition in Portland, OR, in 1905. Black Bullheads may also have been accidentally mixed in with Brown Bullheads or Channel Catfish introduced to ID.

Other Sources of Information: 128
References: 111, 171, 187, 287, 329
Author: Katie Barnas

Nutria *Myocastor coypus*

Species Description and Current Range

Nutria are large aquatic rodents with thick, yellowish-brown to dark-brown fur and white whiskers on their chin. Their arched, ratlike body is 0.45 m long, and their scaly tail is about 0.3 m long. They have prominent front teeth that are pigmented with orange. The front paws of the Nutria are clawed, and their back feet are webbed. Nutria weigh 4.5–11.3 kg, with males weighing about 15% more than females.

Though native to southern South America, Nutria now live on all continents except Australia and Antarctica. They live in 15 states in the US, and are causing major destruction to wetlands in LA and MD. Nutria are found in OR and WA in wetlands surrounding lakes, ponds, and rivers and in brackish or salt-water marine estuaries. Nutria populations were expanding in OR and WA in the mid-1990s, though are most likely stable over the long term. In October 2004, a Nutria was seen feeding in Union Bay, Seattle, WA. Nutria were once found in BC, but it is unknown if any populations still persist in the province.

Impact on Communities and Native Species

At low densities, Nutria have a small impact on ecosystems. However, at high densities, foraging and burrowing by Nutria can be very destructive to wild and agricultural lands. For example, they have damaged an estimated 40,500 ha of coastal wetlands in LA and convert 65–91 km^2 from coastal wetlands to open water every year. Nutria can raid farmlands, causing serious damage to sugar, rice, alfalfa, corn, clover, and root crops. In addition, they damage dikes and irrigation facilities, weaken and erode banks, and obstruct wetland rehabilitation projects. Nutria carry a variety of parasites and diseases, such as a Strogyloides nematode, rabies, coccidiosis, and salmonella, that can infect humans, livestock, and wildlife. They also compete with and replace

the native Muskrat, damage native plants, reduce food and cover available to migratory birds and waterfowl, and reduce wetland habitat available as nurseries for finfish and shellfish.

Control Methods and Management

Once established, Nutria are difficult to eradicate. However, when Nutria populations are small and contained in a discrete area, eradication programs can be successful. For example, Nutria were eliminated from East Anglia in England after a 7-year, 24-person-strong trapping campaign. Organized trapping in the region began in 1962 in response to damage from Nutria whose numbers might have been as high as 200,000 in the late 1950s.

The best ways to eliminate Nutria are trapping, poisoning, and shooting. Eradication programs are most successful in years of very high or low temperatures, when Nutria have lower reproductive success and survival. In areas where Nutria populations are causing severe habitat damage, programs promoting the use of Nutria for their fur and as a source of low-fat, nutritious meat can be helpful.

Life History and Species Overview

Nutria are native to South America, occurring from central Bolivia and southern Brazil to Tierra del Fuego. Although primarily nocturnal, they shift to more diurnal activity to avoid cold nighttime temperatures. Nutria usually live 2–3 years in the wild. About 80% of young die in the first year and 60–80% of adults die each year. Nutria breed throughout the year but have reproductive peaks in OR in January, March, May, and October. Both sexes become sexually active at 4–5 months of age. The female gestation period is approximately 130 days. Litter size ranges from 1 to 13, with most litters having 3–6 individuals. Females produce 2–3 litters/yr, with annual productivity ranging from 8.1 young/female in MD to 15 young/female in FL. Female productivity is limited by food type and availability, weather conditions, predators, and disease.

Nutria are highly aquatic and can remain submerged for over 10 minutes. Though they are tied to water throughout most of their range, they have been observed breeding without access to water when temporary water sources were dry. Nutria consume about 25% of their body weight/day, but can survive for up to 29 days without food. They feed primarily on aquatic vegetation (stems, leaves, roots, and bark), though also feed on terrestrial vegetation. Nutria dig 1–7 m long burrows and tunnel complexes into steep banks adjacent to waterways, and use local vegetation to build nests and feeding platforms. Nutria family groups have 2–13 animals (adult females, their young, and an adult male), and adults are territorial. The average home range of females is 2.5 ha and of males is 5.7 ha, with a population density of 0.2–25 Nutria/ha.

The body temperature of Nutria is quite variable and is positively correlated with air temperature. Weather extremes, not food, seem to limit Nutria populations. For example, 60–70% of the Nutria population in the marshes of midwestern France died during one harsh winter. Heat stroke caused the deaths of 45,000–50,000 Nutria during one summer in Transcaucasia when ambient temperatures were above 35°C. The wet and temperate climate of the PNW may limit Nutria survival and reproductive success.

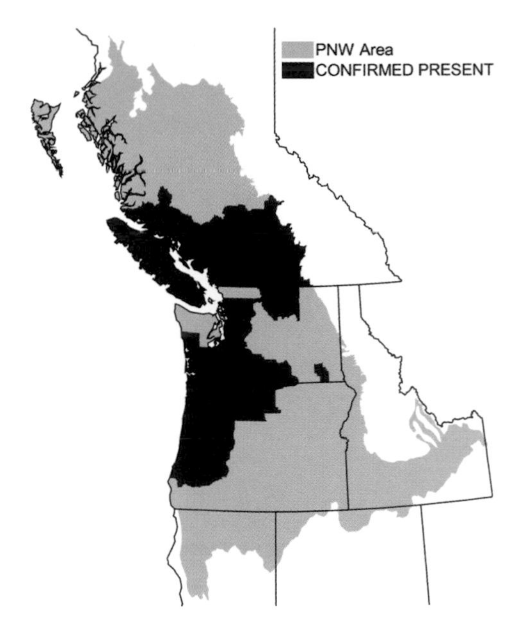

History of Invasiveness

Nutria were valued for their fur in the 1800s. They were first introduced into CA in 1899 for fur and were released to control weeds in the mid 1900s in the southeast. Fur farmers imported Nutria to WA in the late 1930s and to OR in 1937, and by 1941 both states had established feral populations. Nutria were found in BC by 1943, but now might be extinct in the province. Nutria populations were expanding in WA and OR in the mid-1990s, potentially recovering from high mortality due to severe winter weather in the late 1980s. The natural spread of the species is limited by waterways, which it uses to colonize new areas. The population explosion of the Nutria and resultant major ecological damage that has occurred in the southeast US are not expected in the PNW.

Other Sources of Information: 73, 116
References: 27, 55, 235, 320, 348
Author: D. Shallin Busch

Muskellunge *Esox masquinongy*
Northern Pike *Esox lucius*
Pickerel *Esox americanus*

Species Description and Current Range

Northern Pike, Muskellunge, and Pickerel are all "pikes," distinguished by long, cylindrical, spearlike bodies and a duck-billed snout filled with sharp teeth. The species name for Northern Pike, *lucius*, means "pike," the spearlike weapon. The similarly shaped Northern Pike and Muskellunge are distinguished by coloration, tail fins, and numbers of pores under each side of their lower jaw. Northern Pike have light markings on a dark body, rounded fins, and five or fewer pores. Muskellunge have dark markings on a light body, pointed fins, and six or more pores. Pickerel are a small pike, rarely reaching 30 cm, with dark bars over a light body and rounded fins. Where distributions overlap, Northern Pike and Muskellunge produce a slightly smaller sterile hybrid, the Tiger Muskie, with the coloration of the Muskellunge, the rounded fins of the Northern Pike, and the fighting spirit of both. Tiger Muskies are artificially propagated in fish hatcheries for release into reservoirs and lakes.

Pike, introduced into the PNW since at least the late 1800s, are an important sport fish and are used to control nuisance fish. Unsanctioned or illegal introductions are not uncommon; hence Northern Pike occur in isolated lakes from CA to AK. Fishermen illegally introduced Northern Pike into the Chain Lakes of the Coeur d'Alene drainage in ID, where they spread throughout the system and into WA. Now in the upper reaches of the Columbia drainage, Northern Pike are moving down through the Columbia River and probably inhabit sloughs bordering the river. WA Dept. of Fish and Wildlife (WDFW) manages Tiger Muskies as part of its sportfishing and biological control program.

E. americanus

Although female hybrids are considered safe for introduction because they are sterile, some are fertile. WDFW hopes the voracious appetite of the Tiger Muskie will help control populations of other introduced species, the Northern Squawfish and large-scale suckers. Pickerel were introduced in northeastern OR in the early 1900s and were found in southeastern WA by 1935. These smaller pike are collected for aquariums. Unfortunately, aquarium contents are sometimes dumped into places where the fish, snails, and weeds survive and reproduce.

Pike live in freshwater lakes, rivers, and streams. Northern Pike occupy the most diverse habitats, extending into brackish water in lagoons. Muskellunge are found primarily in lakes and creek backwaters and Pickerel mostly in streams. Newly hatched pike eat invertebrates, but these insatiable predators soon switch to a diet of fish.

Impact on Communities and Native Species

Pike are fearsome top predators, primarily eating other fish. When introduced into undisturbed systems, pike radically disrupt communities either directly through predation (and eating other fish's prey) or indirectly by causing behavior changes in other fish. Introduced pike reduce or eliminate other fish species except those with natural defenses such as stiff spines. Fish managers in the PNW are concerned about illegal pike introductions into salmonid habitat: introduced Northern Pike have reduced numbers of native Cutthroat and Bull Trouts in MT and decimated endemic Sticklebacks in AK and northern BC. Although Tiger Muskies could have a similar effect, their introduction into disturbed systems may help to control other introduced species.

Control Methods and Management

OR's policy is eradication, but if pike do not eliminate all their prey, eradication is difficult. CA wildlife officials tried for a year to eliminate Northern Pike in Davis Lake through electrofishing, nets, traps, and explosives. Pike thought eradicated in 1997 were back by 1999. WA manages hybrid pike, Tiger Muskies, as game fish and uses them for biological controls. Most hybrids are infertile, so their numbers are controlled through stocking.

Life History and Species Overview

In spring, Northern Pike move inshore or upstream to release sticky eggs in weedy areas. Muskellunge spawn in open areas and their eggs settle to the bottom. Pickerel spawn in vegetated areas including flooded plains, sloughs, and ditches. Pike spawn once or twice within 2 weeks and leave their eggs unguarded. Eggs hatch in around 2 weeks, and juveniles grow rapidly, becoming piscivorous when around 4 cm long and maturing at 2–4 yrs. Pike are solitary, visual predators that lurk in vegetation and ambush prey. This hiding behavior makes them difficult to find and eradicate. In warmer water, pike live around 10 yrs but live nearly 30 yrs in cooler water: the world record Northern Pike, captured in Germany, was nearly 150 cm long and weighed 25 kg. When hooked, pike put up a fight and the bigger the pike, the bigger the fight, making them an important sport fish. Although rarely fished commercially, their flesh is firm, white, and flaky.

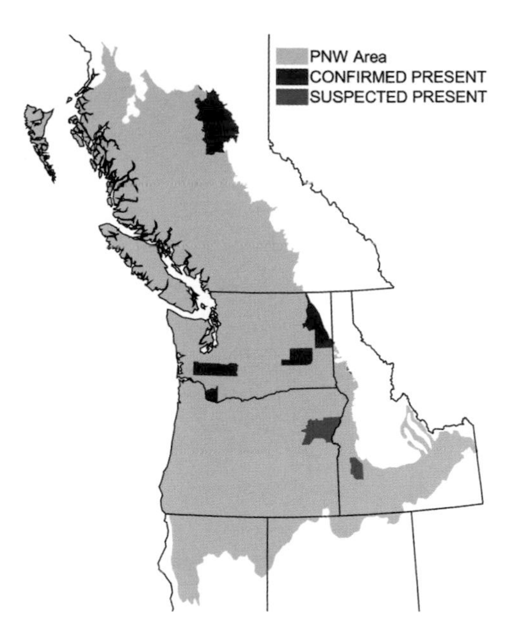

Northern Pike are found in northern waters in Europe, Asia, and North America. In North America, their native distribution extends east of the Rocky Mountains throughout Canada and as far south as IL and MO. Muskellunge are restricted to North America east of the Great Plains, from Canada south to TX. Pickerel are native to states bordering and to the east of the Mississippi River and from QC to the Gulf of Mexico.

History of Invasiveness

Initial introductions were poorly documented. Northern Pike were stocked in CA and ID during the 1800s for sportfishing. Although by the late 1990s most states and Canadian provinces had restrictions against transporting live fish, boats with live-wells and determined fishermen with buckets make control of illegal introductions difficult. Pike illegally introduced into northern BC, MT, and ID colonized the whole area by using connecting waterways. The only sanctioned introductions are Tiger Muskies in WA and ID. Similar in size and appetite to other predatory pike, Tiger Muskies eat organisms up to a third of their body length, including other fish, frogs, mice, ducklings, and muskrats.

Other Sources of Information: 30, 35, 111
References: 146, 171, 215, 239
Author: Maureen P. Small

Bluegill *Lepomis macrochirus*
Green Sunfish *Lepomis cyanellus*
Pumpkinseed *Lepomis gibbosus*
Red-ear Sunfish *Lepomis microlophus*
Warmouth *Lepomis gulosus*

Species Description and Current Range

The Sunfish family is a diverse taxonomic group of spiny-rayed fishes. The genus name *Lepomis* comes from the Greek for "scaled gill cover." The common name "sunfish" alludes to the compressed, deep-bodied profile, to the colorful appearance of many species, and to the propensity of some species to "sun" in open waters. The sunfish family is native only east of the Rocky Mountains from southern Canada to northern Mexico. Of the 11 sunfish species, 5 are widely distributed in the PNW. Sunfishes characteristically inhabit clear warmwater ponds, lakes, reservoirs with dense aquatic vegetation, and slow-moving streams, but are occasionally found in turbid and coldwater lakes. Because of their aggressive behavior and high productivity they are abundant in suitable habitats (e.g., one small 10 ha lake had 3,000 fish).

The sunfish species are characterized by a united spinous and soft dorsal fin with 9–11 spines at the front part. The anal fin has 3 spines and the pelvic fin 1 spine. The scales are ctenoid, which means that they have rough edges. Bluegills are distinguished from other sunfishes by the blue-black flaps on the rear of the gill cover and a black spot on the rear of the dorsal fin, and Pumpkinseeds by the black gill cover with a bright orange-red spot at the edge. Red-ears have a red-orange edge ahead of the dark spot of the gill cover. Warmouths resemble Green Sunfish but differ from them in their brown color, by having teeth on their tongue, and by lacking a dark spot at the base of the last 3 soft dorsal rays. Sunfish are indeterminate growers, meaning they may shrink in size when food is scarce but may grow to 20–40 cm in length. Bluegill and Pumpkinseed are the largest in the family.

L. gibbosus

Impact on Communities and Native Species

Most sunfishes are aggressive toward other fishes and their own kind, chasing and nipping them. Predation by introduced populations of Green Sunfish can seriously reduce population size and growth of native fishes such as Minnow, Pikeminnow, Sticklebacks, and Roach. The Green and Bluegill Sunfish outcompete native perch by chasing them from their spawning areas. These two species and Warmouth are responsible for the decline of native frogs and salamanders in CA. Adult Red-ears and Pumpkinseeds feed on snails and clams and Red-ear can eliminate snails in small ponds. Predation by sunfish possibly limits the dispersion or population size of Zebra Mussels.

Control Methods and Management

Because sunfishes are important sport fishes, they are managed as game species. The major management consideration for Bluegill and Green Sunfish is population control. Because of their high reproductive rates, they form large populations, resulting in stunted fish. Introduction of large predators, removal by trapping or netting of abundant young, intensive angling (e.g., bag limit of 25 fish per angler per day), and the use of chemicals to kill the eggs can reduce their numbers, resulting in better growth and larger fish. In some lakes, removal of up to 90% may be necessary to achieve a Bluegill population with some large fish. The Green Sunfish is often considered a nuisance and is removed from lakes by chemical treatment; this is usually unsuccessful and surviving fish replenish their numbers within a few years.

Life History and Species Overview

Male sunfish excavate a shallow circular nest 20–60 cm in diameter and 5–15 cm in depth in the gravel or sand by fanning their caudal fin from a nearly vertical position in shallow waters. In crowded ponds, nests are often close together and may be visible from shore. Spawning occurs in mid spring to early summer. The owner of the nest, a territorial male, is approached by a female, and they circle over the nest repeatedly until the female swoops down and releases eggs. The male follows and releases a cloud of sperm. Several females may spawn in the same nest, so hybridization may take place when a female deposits her eggs in the wrong male's nest. Each female lays 600–80,000 eggs; larger and older females lay the most. Territorial males guard the nest for about a week or at least until the eggs hatch. Fry enter the open water where they feed on zooplankton and later settle into beds of aquatic plankton, feeding on insects, small fish, fish eggs, crayfish, or snails.

Among Bluegills, intermediate-sized males in female colors (female mimics) may gain admission to a nest by posing as a receptive female. Small, inconspicuous males, called "sneakers," which do not make nests, also may dart in and add sperm to the mix. Parental males are constantly occupied with chasing satellite and sneaker males away from their nests. These alternative male strategies

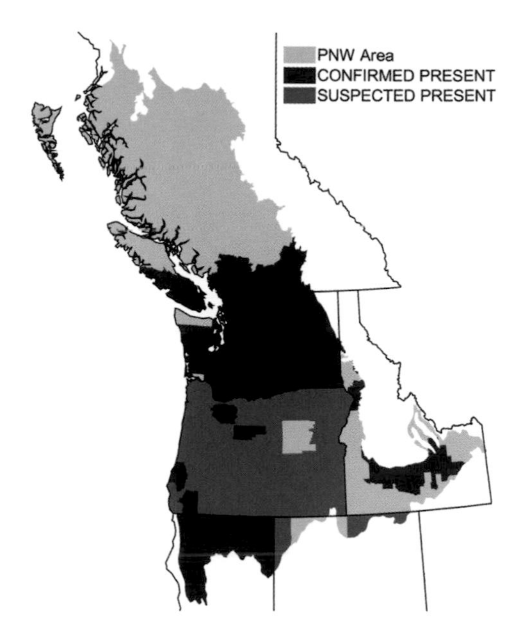

are important because otherwise only a few dominant parental males would do all the spawning.

Adult sunfish males are some of the most colorful fishes in North America. Color is important in courtship and species recognition in this genus. Sunfishes mature at 1–2 yrs and have a maximum life span of 6–8 yrs. Hybridization is relatively common and most frequently occurs in overpopulated, turbid, or polluted waters. The results are fast-growing and sterile male hybrids. Sunfishes can tolerate a wide range of water temperature (2–40°C) and salinity gradients (1–5 parts/thousand) and can survive in waters of low oxygen content (<1 mg/l).

History of Invasiveness

Sunfishes were stocked as sport fish in the PNW by the US Fish Commission. Mixed sunfish species were first introduced in Deer, Sprague, and Loon lakes in eastern WA and in the lower Columbia River during 1890 and 1891 and became widely distributed throughout the western US. The Bluegill is the most widely introduced species of sunfish in the PNW. The Green Sunfish was accidentally stocked as Bluegill. Sunfishes are established in most locations where they have been introduced.

Other Sources of Information: 34
References: 111, 171, 216, 227, 350
Author: Monika Winder

Striped Bass *Morone saxatilis*
White Bass *Morone chrysops*

Species Description and Current Range

Striped Bass and White Bass are true bass and members of the temperate Bass family. The species name for White Bass, *chrysops*, means "golden eye," their defining characteristic. The name for Striped Bass, *saxatilis*, means "among the rocks," the preferred Striped Bass spawning habitat. Striped Bass were introduced to the Pacific coast over a century ago. Their range now extends from Northern Mexico to BC, but populations are concentrated in San Francisco Bay, CA, and Coos Bay and the Umpqua River, OR. Smaller populations exist in the Columbia River, WA, and along Vancouver I., BC. The only documented population of White Bass is in southwest WA, near Vancouver.

White Bass live in large freshwater lakes and rivers. Striped Bass are naturally anadromous, beginning their life in fresh water, migrating out to coastal waters, and then returning to fresh water to spawn, but they can spend their whole lives in fresh water when trapped in landlocked lakes. Landlocked adult Striped Bass reach weights of 7–13 kg, with saltwater fish getting even bigger, up to 140 cm and 40 kg. Females grow much larger than males, so any oceangoing Striped Bass over 120 cm or 14 kg are likely female. White Bass are tiny in comparison, weighing less than 1 kg as full-grown adults. Juvenile Striped Bass and adult White Bass look almost identical: both are silver with longitudinal dark stripes and a double dorsal fin (the fin on the top of their back) made up of a spiny portion and a soft portion. The only certain way to distinguish the two species is to look in their mouths. White Bass have a single, heart-shaped tooth patch on the back of their tongues while Striped Bass have two distinct and parallel tooth patches. These two species also closely resemble the Wiper, a White Bass–Striped Bass infertile hybrid that has

M. saxatilis

been artificially propagated since 1965 and introduced throughout the US.

Impact on Communities and Native Species

True bass prey on small fish and are often blamed for the decline of small native fish throughout their introduced range. Juvenile salmon have been found in the stomachs of Striped Bass; however, the impact of this predation on salmon populations is unknown.

Control Methods and Management

In OR, the Striped Bass stocking program is being phased out to lessen predation on native fish, but there are still strict daily catch limits of two fish (72 cm minimum) to ensure some future for the Striped Bass population. Because of their limited distribution, White Bass are not managed as a game fish or as an invasive species.

Life History and Species Overview

Striped Bass are native to coastal regions from the Gulf of Mexico up to the St. Lawrence estuary. They spawn in early summer in rivers with rocky habitat and swiftly moving water. The eggs must stay suspended in the water column to survive and then hatch out in 2–3 days. Larval Striped Bass float downstream, seeking out the nursery habitat of estuaries for 2 years before moving into the coastal ocean and later returning to streams to spawn. They feed on zooplankton, progressing to a diet of fish, crabs, and invertebrates as they grow larger. Females mature at age 4 and males at age 2. Striped Bass spawn yearly, migrating from sea to river throughout their 10–12 yr average life span. As females grow larger, their reproductive capacity increases from thousands of eggs in young smaller fish to millions in older adults. The damming of rivers along the East Coast preceded the discovery that populations of Striped Bass trapped in the newly formed landlocked reservoirs could complete their entire life cycle in fresh water (although with less reproductive success). This prompted the introduction of Striped Bass into inland reservoirs and lakes for sportfishing. Striped Bass are fierce fighters, which makes them an exciting catch. The 35 kg world angling record Striped Bass fought for 1 hour and 40 minutes before it was landed.

Striped Bass have been commercially important since colonial times and were consumed by the residents of Plymouth as early as 1623. They were in such high demand that in 1639, in one of the earliest conservation measures, the Massachusetts Bay Colony banned the sale of Striped Bass as fertilizer. Further legislation in 1776 outlawed the sale of Striped Bass in winter months. Striped Bass populations declined throughout the next two centuries due to overfishing and habitat destruction, prompting the passage of the 1984 Striped Bass Conservation Act. Striped Bass populations are rebuilding, and active management of the species continues.

White Bass are a fast-growing, short-lived, schooling fish found in large lakes and rivers. They are native to the Great Lakes, Hudson Bay, the Mississippi River basin, and to areas of TX and NM. White Bass spawn in lakes and rivers over sandy or rocky bottoms. The eggs sink, adhere to the bottom, and then hatch out in two days. White Bass grow quickly, up to 200 mm in the first year, with a full-grown size of 330–380 mm. Their quick growth is a result of their rapacious appetite

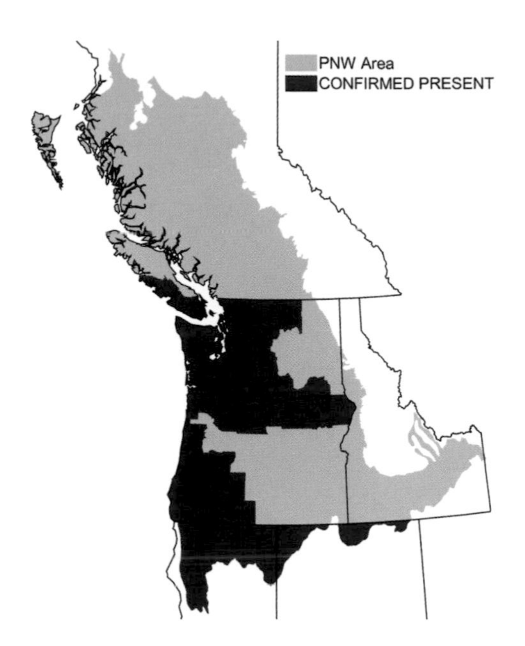

for fish, but they will settle for invertebrates when fish are not available. White Bass mature at age 2 and live only 3 or 4 years. Like their larger cousin, White Bass put up an intense fight and are considered good eating, making them popular among anglers.

History of Invasiveness

In 1879, 132 Striped Bass packed in milk jugs traveled on the newly completed transcontinental railroad from NJ to CA. Those that survived the journey were let loose into San Francisco Bay. Three years later, 300 more fish made the same trip. After release, the Striped Bass migrated north, and by 1914, some were caught in Coos Bay, OR. A commercial fishery opened in OR in 1922 and peaked in the 1940s at Coos Bay and in 1971 on the Umpqua River. Declining Striped Bass populations and pressure from sportfishers prompted the permanent closure of the commercial fishery in 1975. The White Bass introduction in WA is not well documented, possibly occurring in the 1960s.

References: 90, 111, 173, 190, 309
Author: Katie Barnas

Brook Trout *Salvelinus fontinalis*
Brown Trout *Salmo trutta*
Cutthroat Trout *Oncorhynchus clarki*
Golden Trout *Oncorhynchus mykiss roosevelt*
Lake Trout *Salvelinus namaycush*
Rainbow Trout *Oncorhynchus mykiss*

Species Description and Current Range

Trout inhabit streams, rivers, lakes, and ponds with relatively cold and clean water. The term "trout" refers informally to a group of species of both trout and char in the Salmonidae family. Two species of foreign trout arrived in the PNW within the last 200 years: Brown Trout and Brook Trout. They have been introduced into rivers and lakes from CA north to BC. Several species of trout, including Rainbow, Cutthroat, Golden, and Lake, native to restricted regions of the PNW, were aggressively stocked in thousands of water bodies since the 1800s and have ranges that now encompass the entire PNW. Trout can live in virtually any permanent freshwater lake or stream that is not too hot. Trout are found from alpine to lowland regions. Most trout in streams are relatively small, averaging 15–30 cm in length. Trout that live in larger rivers and lakes or migrate to and from the ocean can grow large—Brown and Rainbow Trout can be 75 cm long and weigh 9 kg.

Because trout have similar appearances and variable coloration and often hybridize, they can be difficult to identify. Brown Trout have pale brown to yellow sides, with red and dark-brown spots on their sides and back. A light ring surrounds the darker spots, giving the Browns a leopard-like appearance. Brown Trout can tolerate warmer temperatures than other PNW trout. Brook Trout have bright orange bellies and golden, wormlike squiggles on their dark green backs and are most common in alpine lakes and streams. Their sides are dotted with pale gold and blue-ringed red spots. Lake Trout have the irregular pale spots of the Brook Trout but lack an orange belly and red spots. Rainbow Trout are named for the band of red that usually brightens their silvery sides. Their dark green backs are speckled with small black spots. Cutthroat Trout, depending on the subspecies, often appear fairly similar to Rainbow Trout but have two red slashes under the throat, hence their name. Golden Trout, originally native to CA, are often considered the most beautiful trout, displaying sides that are gaudy with orange and gold.

Oncorhynchus clarki

Impact on Communities and Native Species

Trout are voracious predators that eat everything from invertebrates to frogs, salamanders, and fishes. As a result, when trout are introduced to a lake that is naturally barren of fish, they quickly eat the native invertebrates and amphibians, causing population crashes or extinctions. When introduced to areas with native fish, trout eat not only the the food that the native fish eat, but the native fish themselves. Introductions of trout are blamed for the local exterminations of rare native fish such as Lahontan Cutthroat (*Oncorhynchus clarki henshawi*), and may contribute to the decline of populations of threatened or endangered salmon. For example, survival of juvenile Chinook Salmon in streams without Brook Trout was twice as high as survival of juvenile Chinook in streams with Brook Trout.

Control Methods and Management

Trout are managed as sport fish throughout the PNW, and populations of invasive trout are often protected by strict regulations. In addition, trout stocking is still common on many private lands and in wilderness areas managed by the US Forest Service. Impacts of non-native trout on native communities are minimized through regulations and changes in stocking. For example, there has been no trout-stocking recently in most western national parks, and fishing regulations allow anglers to heavily harvest invasive trout while protecting native trout in some areas. However, once a population of trout is self-sustaining, it is often difficult to remove completely, necessitating years of intense netting, or possibly catastrophic, lakewide poisonings with rotenone.

Life History and Species Overview

Trout spawn in either spring (Cutthroat, Rainbow, Golden) or fall (Lake, Brown, Brook). Trout dig nests, or redds, in lake or stream gravel, where they lay and then bury their hundreds of eggs. Developing eggs need clean gravel with subsurface flow of cold, oxygenated water. Eggs develop into tiny juveniles in a period of 1–4 months, when they subsequently emerge from the gravel. Juvenile trout feed upon invertebrates and insects, while adults tend to feed more on small fish. Most trout spend all of their lives in fresh water, but some migrate out to the ocean to feed and grow before returning to their natal waters to spawn. Depending on the species and population, trout reach maturity at 2 or more years of age and can live more than 15 years. Brown Trout are native to Europe and Asia, Brook Trout are native to eastern North America, and Lake Trout originally were distributed throughout

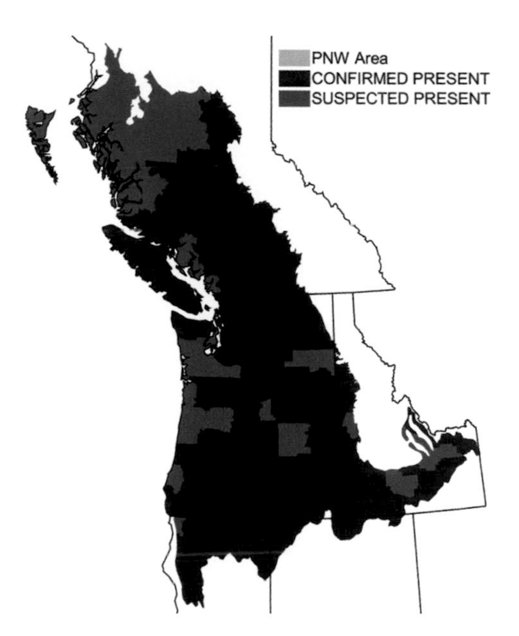

northern North America. The various subspecies of Cutthroat and Rainbow Trout are from select drainages west of the Mississippi. Golden Trout are from remote drainages in the Sierra Nevada, CA.

History of Invasiveness

Euro-American settlers began to introduce trout to lakes and streams in the PNW in the late 1800s. Historically, most of the lakes in the PNW had been fishless since the last glaciation, 10,000 years ago. Since then, governmental agencies as well as private individuals have stocked thousands of lakes and rivers. Now, 95% of larger, deeper mountain lakes have non-native trout, and amphibian and invertebrate populations have plummeted. Although often illegal, citizen plantings were popular among mountaineering and sportsmen's clubs. After World War II, trout were often dropped into remote lakes by helicopters and airplanes or were transported to more accessible locations by train or truck. Because trout have been stocked in such a haphazard fashion, fresh water bodies in the PNW now hold a patchwork of introduced non-native Brown and Brook Trout, introduced trout stocked to regions outside their native PNW range, and remote relic populations of native trout.

References: 11, 253, 257, 292
Author: Jonathan W. Moore

Common Snapping Turtle *Chelydra serpentina*
Red-eared Slider *Trachemys scripta*

Species Description and Current Range

The Red-eared Slider and the Common Snapping Turtle occur in a variety of freshwater habitats throughout the PNW, but they prefer shallow, muddy-bottomed lakes with dense aquatic vegetation. In the PNW, the Red-eared Slider and the Snapping Turtle live from CA north to BC, usually around urban areas. Populations are documented in CA, WA, OR, and BC. The Red-eared Slider is a medium-sized turtle with a shell reaching up to 29 cm in length. Defining features include a bright yellow underside, yellow stripes on the legs and neck, a dark green outer shell, and red stripes on each side of the head behind the eyes. The native Western Painted Turtle, the largest of the painted turtles with which it could be confused, has red on the margin of the carapace and red on its yellow-streaked skin. Snapping Turtles are gray, have long necks, and can strike quickly. Their scientific name translates to "serpent turtle." Their large head, a long jagged-edged tail, and a small, serrated shell make them look prehistoric. The Snapping Turtle can grow to 49 cm in length and weigh 39 kg. Coloring varies from gray to dark brown or olive. The Snapping Turtle's powerful pointed beak is its most distinguishing characteristic.

Trachemys scripta

Impact on Communities and Native Species

Disease and competition for resources are two threats that the Red-eared Slider and the Snapping Turtle pose to native PNW turtle populations. High demand for the pet trade motivated breeders to farm Red-eared Sliders in extremely unsanitary conditions. These turtles are often sold with preexisting respiratory and other diseases. Sliders usually live less than a year after purchase, even though their life span in captivity can be 20+ years. Unfortunately, when pet owners are unable to care for their sick turtle or handle the size of their growing turtles, they often release the animals into the wild, thereby introducing diseases into native turtle populations that have no inherent defenses. Released turtles compete with native turtles for basking sites, nesting sites, and food.

Control Methods and Management

People are still releasing pets into the wild despite efforts to educate against it. Turtle adoption and relocation organizations, such as the California Turtle and Tortoise Club, help protect native turtle populations from introduced pet turtles. Methods to control established invasive turtles include a program to catch, sterilize, and release turtles in order to reduce breeding in some areas. Check with your local wildlife agency to see if there are any adoption or relocation programs in your area.

Life History and Species Overview

Red-eared Sliders' native range is from eastern North America to northern South America. They bask daily for several hours to maintain their optimal body temperature of 28°C. When water temperatures drop below

10–15°C, Sliders and Snapping Turtles hibernate in muskrat burrows and hollow stumps, under logs, or in underwater shelters. When the turtle hibernates underwater, its metabolism slows, allowing it to survive on oxygen absorbed through its skin from the flowing water. Female Sliders produce 2–23 eggs per clutch depending on their body size and may lay 5 clutches/yr. When a young Red-eared Slider is swallowed by a fish, it holds its breath and chews on the lining of the fish's stomach until the fish regurgitates it.

The Snapping Turtle is native throughout most of eastern North America. Females produce 6–104 eggs with an average of 20–40, laying 1 clutch/yr. When hibernating, Snapping Turtles' body temperature can reach as low as 1°C. These turtles can crawl beneath the ice and remain underwater over 14 days, depending on the oxygen content of the water. When turtles hibernate, they excrete water, and because their cells are dehydrated they do not freeze until temperatures are well below 0°C. Turtles lack sex chromosomes. Sex determination of young for both the Red-eared Slider and the Common Snapping Turtle depends upon incubation temperature. In the Snapping Turtle, eggs incubated at 20°C and 29–31°C produce all females, temperatures of 23–24°C produce all males, and temperatures in between produce a mixture of males and females. In Red-eared Sliders, temperatures of 30°C produce females and temperatures of 22.5–27°C produce males. This is the opposite of alligators, which produce all males at high temperatures and all females at low temperatures.

History of Invasiveness

Red-eared Sliders taken from the wild were widely available in flea markets and dime stores starting in the early 1900s. Unsanitary farming began in the 1950s and by 1975 the sale of Red-eared Sliders less than 10.2 cm long was banned in the US after 300,000 children were diagnosed with turtle-transmitted salmonellosis. In 1997, an upper respiratory disease, suspected to have been transmitted from an introduced turtle, wiped out approximately 20% of the endangered Western Pond Turtle population in WA. A relative of

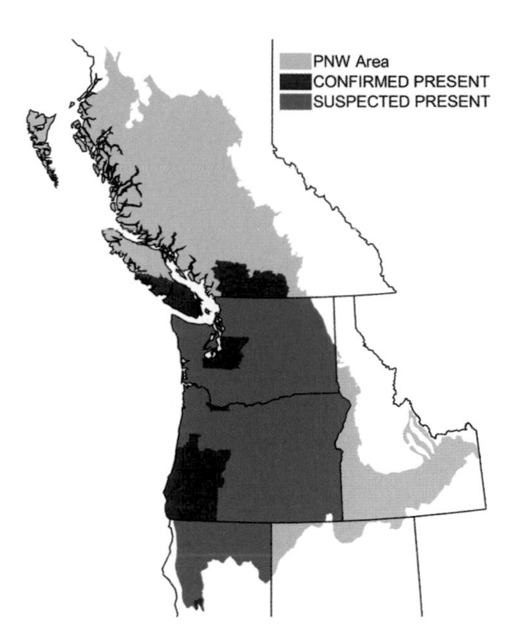

the Western Pond Turtle, the European Pond Turtle, experienced fierce competition with introduced Red-eared Sliders. As a result there is a ban on the importation of Red-eared Sliders in the European Union. Importation is also banned in South Africa and Canada. It is illegal to possess or transport Red-eared Sliders in OR. Snapping Turtles were introduced into the wild by pet owners. The Common Snapping Turtle grows quickly and at just 15 cm, less than a third of its adult size, can break a child's finger in one bite. Snapping Turtles can eat young ducks and young turtles, and can outcompete native turtle populations so severely that the possession and transport of the Snapping Turtle is banned in CA and OR. Other introduced turtles in the PNW include the Reeve's Turtle, several different species of soft-shelled turtles, various Asian turtle species, and several painted turtle species.

Acknowledgments: Bruce Bury, USGS Forest and Rangeland Ecosystem Science Center
References: 107, 340, 345
Author: Carmen M. Albert

Yellow Perch *Perca flavescens*

Species Description and Current Range

Yellow Perch is a medium-size fish that inhabits many lowland lakes, ponds, and rivers throughout the PNW. This species is recognized by a series of 6–9 dark bands or saddles on its dorsal surface, covering a yellowish or greenish, scaly body. Perch have two distinct dorsal fins, two anal fins, and a forked tail. Its eyes are yellow-green in color. Body depth is about one-third of body length and head size is relatively large (about one-quarter the body length). Perch are quite prickly, with sharp spines on each operculum, 13–15 spines on the first dorsal fin, 2–3 on the second dorsal fin, and 1 spine on each pelvic fin. The pelvic and pectoral fins are usually yellow but can be bright orange/red in spawning males. Small perch can be confused with the closely related Walleye (also exotic in the PNW). A distinguishing characteristic of the Walleye is its strong canine-like teeth, which perch lack.

Impact on Communities and Native Species

Perch are both competitors with and predators of native fishes. Because perch are highly omnivorous and their diets change substantially over the course of their lives, they can compete with a wide range of native fishes. Perch have been observed to prey on native juvenile salmonids and other small fishes (e.g., endangered suckers in the Klamath River, CA), and recent introductions to AK have raised concern regarding their predation on juvenile salmon there. Despite a wealth of information about the effects of perch on other species in its native range, there is little specific information regarding its effects on native species in its introduced range. However, because of its wide distribution and documented importance as a competitor and predator of fishes and invertebrates (from zooplankton to benthic insects and mollusks) in its native range, it is highly likely to have an important effect on native aquatic communities in the PNW.

Control Methods and Management

Yellow Perch is a very popular fish among anglers because of its ease of capture and excellent taste. Thus, most management for perch in North America promotes large, fast-growing populations of this species. Like most other exotic fishes, the most effective method for control of perch is to use a piscicide (e.g., rotenone) to eradicate it from lakes or ponds where it has been transplanted. Unfortunately, current piscicides are not species-specific and eradication of perch would also eliminate most other fishes in a control site. Thus, the best management practice is to control further distribution of this species by better control on bait-bucket introductions and stocking for angling opportunities. Stocking of predators may reduce perch population sizes but is unlikely to eliminate perch from lakes. Stout bodies and spiny fins allow perch to coexist with most piscivores in North America.

Life History and Species Overview

Yellow Perch are often found in loosely aggregated schools and are most common in lakes with extensive aquatic plant cover and relatively clear water. Perch usually spawn early in the spring, from mid-April to mid-May when water temperatures are between 9 and 12°C. Female perch deposit a transparent and gelatinous, accordion-like string of eggs that is usually draped over aquatic vegetation or dead trees in shallow water. This string can be up to 2 m long and hold more than 100,000 eggs. Most perch begin spawning in their second year of life. Upon hatching, usually within 2 weeks, larval perch (about 5 mm long) are distributed throughout the open water of lakes, where they begin feeding on small zooplankton within about 5 days post hatching. By mid to late summer, young-of-the-year perch move back into nearshore habitats, where they switch to feeding primarily on benthic insects. Perch are omnivorous as adults, feeding on prey ranging from large zooplankton to benthic insects, mollusks, and small fishes. Growth of perch is highly variable among lakes. Individuals can be as long as 30 cm in large lakes but stunting of perch populations is common in small lakes where they can reach especially high densities. In these lakes, predatory impacts on zooplankton can be substantial, shifting communities from large-bodied species to small-bodied species. Perch typically live up to about 10 years in lakes where they are not exploited heavily. Because perch are a highly desirable game fish, exploitation rates can be high, resulting in substantially younger average ages in many systems. Perch are generally considered a warm-water species and therefore flourish in lowland lakes and rivers. However, they are tolerant of cold-water habitats and can overlap considerably with trout populations.

History of Invasiveness

Yellow Perch are native to the northern half of North America, extending from the Arctic as far south as GA, and are found inland in the Great Lakes and Mississippi River basin as far west as the Rocky Mountains. Since the late 1800s, this species has been introduced into every state and eradicated only in AR. The first recorded introductions in WA were by the US Fish Commission between 1890 and 1895, when Yellow Perch were stocked as a sport/food fish in lakes near Spokane

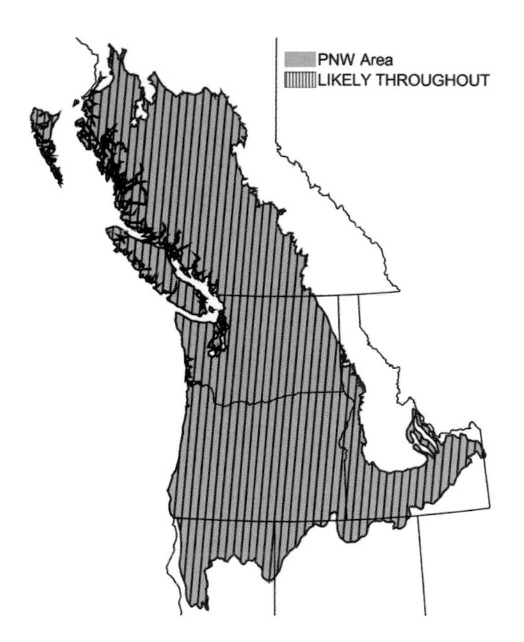

and probably introduced into the lower Columbia River. In 1891, 6,000 yearlings were brought into southern CA from IL, and 406 of these perch were transplanted into northern CA in 1896. Yellow Perch was thought to have been eradicated in CA until it was found again in 1946. The range of this fish is expected to expand via illegal releases by anglers, accidental release from bait-buckets, and natural migration between drainages. Once introduced, the species usually establishes quickly due to its high fecundity and early age at reproduction.

References: 18, 90, 227, 283, 350
Authors: Daniel E. Schindler and Jackie L. Carter

Atlantic Salmon *Salmo salar*

Species Description and Current Range

Atlantic Salmon are anadromous fish (living in marine environments and reproducing in fresh waters). They were transplanted to the PNW, primarily for aquaculture, where they are preferred over native Pacific salmon for their high growth rates and tolerance for captivity. Atlantic Salmon can be found in any freshwater or marine habitat including the open ocean, estuaries, large rivers, small streams, and lakes. While quite similar in appearance to Brown Trout (also introduced to the PNW) and Steelhead, they can be distinguished from native Pacific salmon by the presence of black spots on the gill cover of ocean-phase fish, a hooked lower jaw on breeding males, and the absence of pigment on the fins. Atlantic Salmon are propagated and raised for aquaculture at over 100 locations in WA and BC, primarily in sheltered marine bays, and now represent ~75% of all salmon produced in the region. Individual Atlantic Salmon found outside aquaculture operations in the PNW are likely recent escapees, and they often have eroded fins and tails as a result of rubbing against the net-pens in which they are raised. Outside of artificial propagation and aquaculture, Atlantic Salmon have reproduced at three locations on Vancouver I., BC—Amor de Cosmos Creek, Adam and Eve River, and Tsitika River—though low survey efforts likely make this an underestimate of establishment in the wild.

Impact on Communities and Native Species

Very little is known about the impact of Atlantic Salmon on native communities in the PNW, but their potential effects are enormous. The primary delivery of Atlantic Salmon to PNW waters is through aquaculture, as both captive individuals in marine net-pens and escapees from these operations. Evaluating the extent of Atlantic Salmon's

impact on PNW ecosystems is made more difficult by the failure of aquaculture programs to carefully monitor the number of fish that escape. For example, Atlantic salmon were found in BC commercial fishery catches in 1984, 4 years before the aquaculture industry first reported any escaped fish. It is believed that prior to 1999 there were no established breeding populations of Atlantic Salmon in the PNW. However, in WA, between 1996 and 1999, over 600,000 Atlantic Salmon were reported as accidentally released from aquaculture operations (industry statistics), emphasizing that even in the absence of "naturalized" populations, Atlantic Salmon could be having substantial effects on PNW ecosystems. The greatest concerns are that escapees will establish self-sustaining populations and harm native Pacific salmon through competition for prey and breeding sites, disease transfer to and from wild salmon, and potential erosion of the genetic composition of native stocks through hybridization. Even those Atlantic Salmon successfully caged can have considerable impacts, including acting as sources of nutrient and antibiotic pollution and as reservoirs for disease and parasites. Atlantic Salmon operations in Norway, for instance, each year produce sewage equivalent to 3.9 and 1.7 million people in nitrogen and phosphorous.

Because Atlantic Salmon are predators near the top of the food chain, growing them in captivity requires a tremendous amount of wild fish as feed. Every kg of salmon produced requires 2.8 kg of wild fish, representing a net loss of marine productivity. Farmed Atlantic Salmon also carry higher levels of organic contaminants than wild Pacific salmon. Many of these ecological impacts are not factored into the market price of farm-raised Atlantic Salmon, which encourages the continued deterioration of marine resources. Despite the uncertainty surrounding the specific impacts of Atlantic Salmon, AK preemptively banned net-pen aquaculture in 1990, fearing that its robust Pacific salmon industry would be jeopardized by salmon farming.

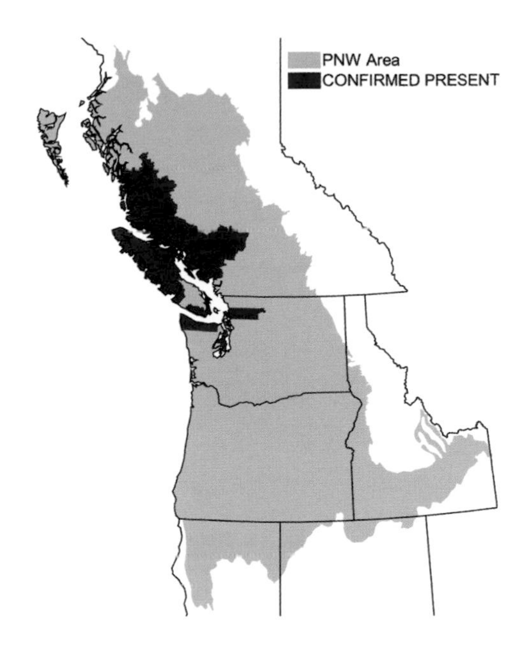

Control Methods and Management

Fisheries and Oceans Canada in BC maintains an Atlantic Salmon watch program to document the distribution of escaped aquaculture fish and monitor the establishment of "wild" Atlantic Salmon populations, though recent evaluations of this program have cast considerable doubt on the accuracy of its statistics. Conservation groups are currently lobbying for changes in the licensing of Atlantic Salmon aquaculture to include only land based, fully contained aquaculture operations.

Life History and Species Overview

Atlantic Salmon are possibly the most ancient of salmon species, commonly referred to by anglers as "leapers" for their acrobatic displays and their ability to overcome tall waterfalls (up to 3.7 m). The native range of Atlantic Salmon is the North Atlantic basin, from New England to the Arctic Circle and south to Portugal. Many native Atlantic Salmon stocks were depleted or exterminated as a result of overexploitation and habitat degradation. Their life cycle is similar to Pacific salmon, whose eggs are laid in fresh water followed by a juvenile stage of 1–8 years, after which individuals travel to the ocean for one or more years before maturing and returning to fresh water to reproduce. Unlike Pacific salmon species that return to fresh water and spawn only once before dying, Atlantic Salmon are capable of breeding in multiple years, with females laying 8,000–26,000 eggs (1,150–3,050 eggs/kg female). Although salmon have high fecundity, at least some populations of all six anadromous Pacific salmon species are listed as threatened or endangered under the US Endangered Species Act.

History of Invasiveness

The history of Atlantic Salmon introductions to the PNW parallels that of other exotic fishes. Hundreds of thousands of Atlantics were transplanted to WA, OR, and BC by federal, state, and private agencies with the goal of establishing sustaining populations for sport and commercial fishing. Despite the huge magnitude of released fish, none of these early attempts resulted in "naturalized" Atlantic Salmon populations. This history of failed introduction has disproportionately influenced opinions surrounding Atlantic Salmon threats and understates the potential ecological impacts of current and future escapees on PNW ecosystems.

Acknowledgments: John P. Volpe, University of Alberta

Other Sources of Information: 8

References: 37, 231, 232, 330, 331

Author: Wendy J. Palen

Domestic Cat *Felis catus*

Species Description and Current Range

This is the familiar house cat. Size and color are highly variable. Weight ranges from about 2.5 to 13.5 kg. The PNW has populations of owned (pet) cats and feral (abandoned, stray, or wild-born) cats, which have similar impacts on native wildlife but different management issues. Pet cats live wherever humans live. Feral cats are concentrated near humans but also occur in remote areas. *Felis* and *catus* are both Latin words meaning "cat."

Impact on Communities and Native Species

The approximately 77.7 million owned cats and 30–100 million feral cats in the US kill several hundred million birds and about a billion small mammals every year. Researchers at Cornell University estimated that cats kill about $17 billion worth of birds annually in the US, based on spending for bird-watching and hunting. Cats kill some pest mice and rats, but they also kill native mammals, birds, reptiles, amphibians, insects, and other invertebrates. Endangered or at-risk species that cats harm include Least Terns, Piping Plovers, Loggerhead Shrikes, Key West Marsh Rabbits, Brown Pelicans, California Clapper Rails, Burrowing Owls, Snowy Plovers, and Saltmarsh Harvest Mice. Animals that escape after being captured by cats often die later. In WA 17% of animals brought to wildlife rehabilitators were attacked by cats (compared to 2% hit by cars and 1% poisoned). One cat in rural WI killed 1,690 animals in 18 months, but 5–365 animals/yr per cat is more typical. Both pet and feral cats that are fed continue to kill small animals. Rural cats kill more wildlife than urban cats do, but cats kill many birds at feeders in urban/suburban yards.

Cats directly caused the extinction of 3 bird species and contributed to the extinction of more than 30 additional species and subspecies on islands worldwide. Even in the PNW, cats negatively impact native animals. Much of our native habitat is fragmented into small "islands" surrounded by humans and domestic cats. Humans maintain cats at high densities by protecting them from starvation, disease, and predation, which limit numbers of other predators. Cats outnumber all other similar-sized predators combined. Cats do not exclude other cats from their hunting range. Cats indirectly affect native predators by reducing prey populations. Domestic cats have alternate food sources (from humans) when prey becomes scarce, but native predators do not. Diseases that feral cats transmit to humans include rabies, plague (especially in the Southwest), and toxoplasmosis, which is dangerous for pregnant women and people with compromised immune systems. Feral cats also transmit disease to native animals.

Control Methods and Management

Cats make wonderful pets, but they should not be allowed to kill native wildlife. These recommendations save wildlife.

1. *Pet Cats.* Keeping pet cats indoors is the only way to prevent them from hunting.

Bells and electronic sonic collars sometimes reduce hunting success, but cannot eliminate predation. Prey species may not associate the sounds with danger and very young animals cannot escape. Some animals such as amphibians cannot hear the sounds. Cats with bells may switch prey species to animals that do not respond to the tones. Declawing does not always reduce hunting success. Spay or neuter your cats. Outdoor cats can learn to be happy indoors.

2. *Feral Cats.* Never feed stray or feral cats. Do not maintain cat "colonies" or leave pet food or garbage where it will attract stray cats. Cat colonies do not form without an artificial food source. TNR (trap, neuter, release), also called TTVAR, or managed cat colonies, is not the answer for feral cats. People dump cats at colonies, even if dumping is illegal. The food attracts other cats and problem wildlife such as raccoons, skunks, and rats, increasing their densities above natural levels. Colony cats that are too wary to be caught are not neutered or spayed. Colony cats do not exclude other cats; colonies may grow to hundreds of cats. These cats kill wildlife, even if they are well fed. Colonies spread disease to humans and wildlife.

3. *Protect Native Wildlife in Your Yard.* Locate bird feeders away from cover, where cats can hide in ambush. Stop feeding birds if they are being killed by cats. Put animal guards on trees to prevent cats from reaching bird nests.

4. *Protect Native Wildlife on Farms.* Do not keep more cats than are necessary to control rodents. Neutered, well-fed females will stay closer to buildings where rodent control is needed. Traps or rodent-proof storage or construction may be more effective than cats in controlling rodents.

Life History and Species Overview

Domestic cats reproduce at 7–12 months old. Cats have 1–2 litters/yr, often fathered by multiple males, with a gestation period of 63–65 days. Feral cats seldom live longer than 4–5 years while 15–17 years is typical for pet cats. Adult cats eat 5–8% of their body weight per day, but females nursing kittens can eat 20% of their body weight. Cats can kill and eat animals as large as they are. Feral cats use a variety of habitats including rock outcrops, burrows, trees, shrubs, culverts, abandoned buildings, and farm outbuildings. They are generally opportunistic in habitat and diet. Home ranges of cats in rural areas can be as large as 5.5 km.

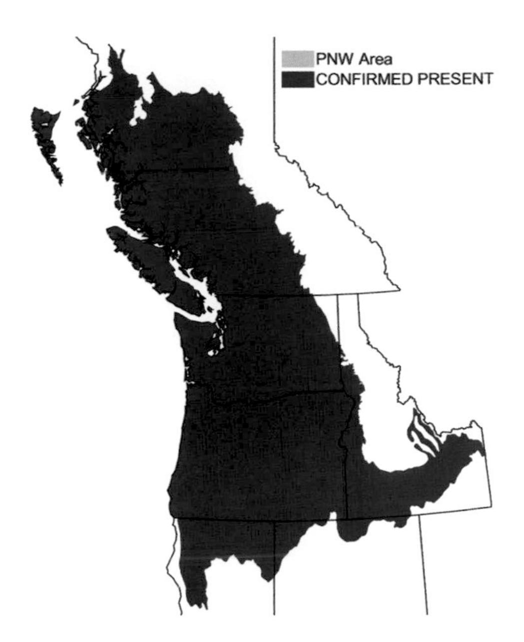

History of Invasiveness

The domestic cat evolved from wild cats of Europe and Africa. Mediterranean peoples may have kept cats as early as 9,500 years ago. Cats were sacred in ancient Egypt, where it was a capital offense to kill a cat, even by accident. It was also illegal to export cats. Special agents traveled around neighboring countries to buy and repatriate cats that had been illegally exported. In spite of this, domestic cats had spread around the Mediterranean by 500 B.C. and throughout Europe and Asia by the 10th century A.D. People transported cats on ships, as the spread of some color mutations followed shipping routes. Early colonists brought cats to North America from Europe.

Other Sources of Information: 2, 23, 80
References: 5, 81, 199, 252, 312
Author: Ginger A. Rebstock

Domestic Pig *Sus scrofa*

Species Description and Current Range

Pigs—relatives of sheep, deer, and cows—are descended from the wild boar. Domestication selected for a variety of breeds, from Mini Maialino pigs that weigh 9 kg at maturity to the Poland-China pig, the largest weighing in at 1,160 kg. Feral pigs in the US have long black, brown, blond, white, or red hair, can be a meter in height, and can weigh over 140 kg, although the average pig weighs closer to 50–80 kg. Pigs have an excellent sense of smell, several times better than humans and similar to that of dogs. They are used to track game, sniff out drugs, and find truffles (a fungus the size of a walnut) that grow half a meter underground. Pigs can learn their names when they are 2–3 weeks old, are excellent problem solvers, have great memories, and are easily trained. They are thought to be more intelligent than any breed of dog. Because they roll in mud, people often think they are dirty animals. They actually roll in mud to prevent sunburn and control their body temperature, as they have no sweat glands and use the mud to keep cool.

Feral pigs in the US may number close to 5 million, and TX may have had a million in the early 1990s. Pigs were introduced in the 16th and 17th centuries for food, and wild boars were introduced later for sport. Pigs do best in warm climates, but wild boars are more adaptable and do well in much colder conditions. Despite subzero temperatures and heavy winter snow, escaped wild boar established small but viable populations in MB and SK, Canada. The Spanish brought pigs to CA in the 1500s and by the 1980s their population may have numbered 100,000. Wild boar were introduced to CA as a game animal in 1925 or 1926. As of 2001, the only known population of feral pigs within the PNW, according to the OR Fish and Wildlife Service, could be found in central OR and consisted of several hundred animals most likely descended from escaped domestic stock. There are no known populations of feral pigs in WA, ID, MT, or BC.

Impact on Communities and Native Species

Feral pigs are omnivorous, opportunistic feeders that eat mostly seeds, plant roots, and tubers, but they will eat amphibians, reptiles, bird eggs, small mammals, carrion, and almost anything they find. Feral pigs trample and remove vegetation and compete for food with native animals such as deer, bear, small mammals, and even birds. When searching for food, feral pigs root, pushing their snouts through the soil. They clear vegetation and leave bare soil in their wake. This bare soil makes a great seedbed for invasive plants, and pigs are helping to spread weedy species, such as Kahili Ginger and Banana Poke, throughout HI. Pigs also actively disperse the seeds of invasive plants by eating their fruits and excreting the seeds with fertilizer directly into the freshly disturbed ground. Pigs are responsible for the dispersal and continued invasion of Strawberry Guava in HI. They are so integral to its spread that managers have identified the eradication of feral pigs as the first step in the eradication of this plant. Rooting not only destroys plants, it also changes water and mineral cycles, disrupts plant succession, and impacts ground-living species. The pig's keen sense of smell and tendency to root for food makes it a devastating force for ground-nesting species.

Pigs have had a drastic effect on seabirds and reptiles in the Galápagos Is., most notably on the Galápagos Tortoise. In the 1970s, in under two months, a single pair of pigs destroyed the nests and ate the young of 23 tortoises. Pig wallows, where pigs root and roll, often fill with water, creating habitat for mosquitoes that can carry avian malaria and other diseases. Avian malaria in HI has contributed to the extinction of over half of the 100 native bird species. Pigs transmit diseases directly as well, both to native species (brucellosis, salmonella, and foot-and-mouth disease) and humans (influenza, encephalitis, and tuberculosis among others). The debilitating flu pandemic of 1918, which killed 21 million people, came from pigs.

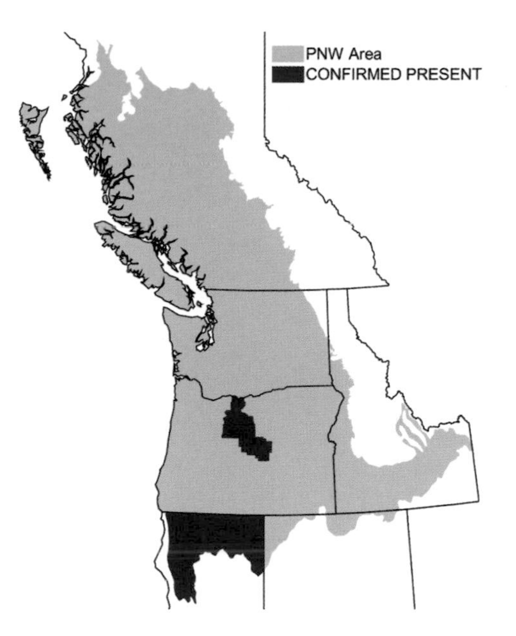

Control Methods and Management

Trapping, hunting, poisoning, and fencing are the most common methods used to control pigs. Trapping can be used year-round to capture groups of animals. Hunting is effective and most states manage pigs as a game animal. Poison, although effective, affects non-target species such as deer, sheep, and scavenging predators. Fencing prevents pigs from entering extremely sensitive regions, but does not affect the population. Large predators, such as pumas, can also help to control populations by preying on juvenile animals. If you find a group of feral pigs, contact your state department of fish and game. Pigs are not generally aggressive, but they can be dangerous when cornered.

Life History and Species Overview

Pigs are very social and live in large groups (sounders) of females and their young. Communication within and between sounders is by a rudimentary language of grunts, squeals, and whistles (so far more than 20 specific sounds have been identified). Males, solitary at maturity, seek out females in the spring in temperate regions but year-round in warmer climates. Males fight viciously to control groups of females (usually 1–3 but sometimes up to 8). When the females are no longer receptive, males become solitary again. About 115 days after mating, females give birth to 5–12 piglets, which they (and the other females within the group) ferociously defend. A feral pig's home range depends on climate and food supply and can vary from 0.9 to 19 mi. Pigs prefer moist woodland but can survive anywhere as long as water is available.

History of Invasiveness

Pigs, native to Europe, Asia, northern Africa, and the Malay Archipelago, were domesticated as early as 6,000 B.C. in China, and in Europe and the Middle East soon after. Selective breeding created over 377 breeds of pig. China still has the largest population of domestic pigs, with about 40% of the world's population. Pigs are closely associated with humans. Their tendency to eat almost anything allows them to turn refuse into protein at a ratio of 0.86:1, which means that pigs can turn food inedible to humans into meat protein with remarkable efficiency. Likely because of this and their general adaptability, they almost always accompanied human colonists, from early Polynesians to explorers of the New World. From initial widespread colonies, pigs, whether released or escaped, have adapted to their new environments, in many cases with drastic negative impacts on native species. They are number 5 on the IUCN's 100 worst invasive alien species list (see Appendix 1).

Other Sources of Information: 17
References: 147, 161, 207, 235, 339
Author: Sacha N. Vignieri

European House Sparrow *Passer domesticus*

Species Description and Current Range

The European House Sparrow is a stocky passerine, or perching bird, about 15–16 cm long. It is found in urban and agricultural habitats, where it thrives in close proximity to humans. Males have brown and black striping on the wings and mantle, a gray crown, black mask and bib, and a chestnut stripe behind a white face. Females are dusky brown above and gray beneath. The House Sparrow is likely the most common sparrow in North America and is distinguished from native North American birds by its stout appearance, owing in part to its thick bill and short legs. It is found throughout the PNW. The largest concentrations are in the Columbia River basin of eastern WA. House Sparrows breed throughout the US and Canada, and live as far north as southern AK. They are rare or absent in unaltered habitats such as forests, deserts, or prairies, but are abundant in cities, suburbs, and farms, where they are permanent residents. Their opportunistic feeding habits, which include eating grass-seed from newly planted lawns, birdseed from backyard feeders, pieces of crackers or bread dropped from the tables of outdoor restaurants, and crushed insects off car windshields, make them dependent on habitats modified for people.

Impact on Communities and Native Species

Owing to its aggressive behavior, the House Sparrow competes with native birds such as the Purple Martin (*Progne subis*) and House Finch (*Carpodacus mexicanus*) and forcibly evicts birds from their nests, sometimes puncturing the eggs, or killing the chicks or adults. For example, Purple Martin populations in BC, already low, were reduced when House Sparrows arrived in the early 1900s; in one area of southern BC, as a result of competition for nest sites with House Sparrows and European Starlings (*Sturnis vulgaris*), the Martin population declined steadily until only six breeding pairs remained. Only with the use of artificial nest boxes, and the House Sparrow's own declining population, have Martin numbers rebounded. However, not all nest-box interventions are so successful, as House Sparrows sometimes outcompete the birds for which the boxes are intended. In IL, efforts to halt declining numbers of Eastern Bluebirds were hindered when House Sparrows evicted or killed nesting Bluebirds. House Sparrows also have effects beyond competition with other bird species, and can be vectors for several viruses and diseases, including West Nile Virus and foot-and-mouth disease. Finally, House Sparrows may disperse non-native seeds common in bird feeders such as the Niger Thistle (*Guizotia abyssinicia*), the only imported birdseed.

Control Methods and Management

House Sparrows are not systematically controlled by any government agency but are considered a pest, meaning they can be killed. Controls are usually implemented by agriculturalists worried about their fields, urbanites who dislike noise or roosting birds in ivy or trees, or bird lovers concerned over the displacement of native species. Control techniques include scare tactics such as scarecrows or noise devices, chemical repellents such as methyl anthranilate, trapping, or destruction of active nests and egg-piercing.

Life History and Species Overview

Males court females by bowing and chirping. Females prefer males with large bibs, the size of which is correlated with the size of the testes. The pair selects a nest site, usually in an enclosed or protected area such as rain gutters, eaves of houses, natural tree cavities, or nesting boxes. Both parents line the cavity with twigs, grass, weeds, or other types of debris. In the absence of suitable cavities or eaves, House Sparrows will build large, globular nests in the branches of trees, which clearly distinguishes them from cup-nest-building, native New World sparrows (family Fringillidae). The female lays 3–6 eggs, which are white or green with brown speckles. Clutch size varies depending on latitude and is larger at higher latitudes. Both parents, but primarily the female, incubate the eggs for approximately 2 weeks. Once the young have hatched, both parents feed them for 2 weeks until they leave the nest, or fledge. House Sparrows can produce 2–3 broods per year, but at high latitudes, 5 broods are possible. North American populations increased until 1900 and have since declined. Although it is not clear why, possibilities include better agricultural practices, increased urbanization, and the decline in horses, all factors that may have reduced access to seeds.

The House Sparrow, not actually a sparrow, belongs to the weaver finch family. House Sparrows are capable of rapidly adapting to their surroundings, and several races, or subspecies, have been documented in North America. A close relative of the House Sparrow, the Eurasian Tree Sparrow (*Passer montanus*), was introduced into North America in 1870 near St. Louis, MO. However, it was not nearly as successful as the House Sparrow and currently its population is restricted to a small area in MO and IL.

History of Invasiveness

The House Sparrow was introduced to the US for two reasons: European immigrants wished to see the familiar birds of their homeland, the same argument that brought the European Starling (*Sturnis vulgaris*); and farmers hoped that the bird would help control agricultural

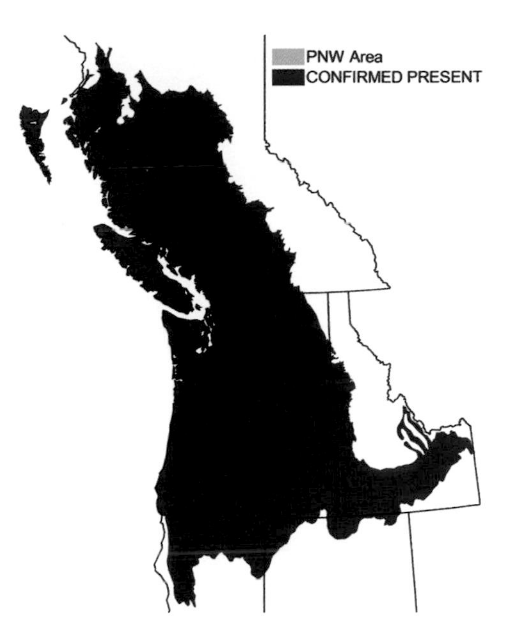

pests. (The most frequently mentioned is the larva of the Snow-white Linden Moth [*Ennomos subsignarius*].) The House Sparrow first arrived in the US in 1850, when eight pairs came from England to the Brooklyn Institute in NY. After overwintering in a large cage, they were released in the spring of 1851. They did not establish a population, so 100 more birds were brought over and released in 1852. During the 1850s, House Sparrows were released at several other sites along the Eastern Seaboard, and by 1870 populations were as far west as the Mississippi River border of IL. Introductions continued throughout the 1870s, in Salt Lake City, UT (1873–74), and San Francisco, CA (1871–72). The birds flourished on the partially digested seeds found in horse manure, and their spread may have relied on freight trains, since reports from western cities noted that House Sparrows were particularly well established in cities along railroads. By 1900, House Sparrows had populations in Portland, OR, and Seattle, WA, and by 1910, they had invaded every state except central NV, southern CA, and the northern Rockies. By 1969, House Sparrows had expanded into northern Canada and southern Mexico.

Other Sources of Information: 27, 73, 121
References: 71, 158, 179, 230, 322
Author: Eric Wagner

European Rabbit *Oryctolagus cuniculus*

Species Description and Current Range

This is the familiar domestic rabbit, used as food and, more commonly today, as pets. Derived from domestic stock and not directly from the true wild form, European Rabbits in the PNW are larger than their wild cousins and come in many colors and patterns, from black to white, shades of brown, and even pinto. Coloration in long-established populations may revert to the wild type, an overall sandy brown color. They may be confused with some PNW native hares and rabbits, of which there are seven species. European Rabbits are much larger than native rabbits, which include the cottontails, brush rabbits, and the endangered Pygmy Rabbit, and lack the prominent black ear-tips of native hares, namely the Snowshoe Hare and the jack-rabbits.

Although small populations of European Rabbits can be found in many city parks and residential neighborhoods in the PNW, they often disappear over time as predators, disease, or direct control methods decimate their numbers. Self-sustaining populations occur only on islands where natural predators such as coyotes and bobcats are absent. European Rabbits are established on most of the San Juan Is., WA; on Gedney I. in Puget Sound, WA; Destruction I., WA, off the coast of the Olympic Peninsula; on Triangle I., BC; and in the Victoria, BC, area. Introductions on the Gulf Is. and Queen Charlotte Is., BC, do not appear to have established persistent populations. Rabbits prefer coastal grasslands on offshore islands, and oak woodland habitats on the drier San Juan Is. and Vancouver I., BC.

Impacts to Communities and Native Species

Rabbits eat a variety of succulent plants and can damage native plant communities, particularly on islands where species may have evolved without grazers. In recent decades, populations on many islands in the PNW suffered from a variety of diseases and reduced numbers, awhich were then followed by population rebounds. The destructiveness of introduced rabbits is best known from Australia and New Zealand. Introduced to Australia in 1859, rabbits had within 50 years invaded the entire continent with an estimated population of 200–300 million. Vast areas of potential grazing lands were damaged, forcing ranchers to use expensive control methods. In New Zealand, predators introduced to control the rabbits preyed upon native flightless birds.

Besides overgrazing, rabbits increase erosion through their extensive warren systems, which can undermine buildings or cause injuries to livestock. Before extermination from Smith I. in Puget Sound, WA, in the early 1970s, the upheaved ground and lack of lush plant growth brought about by rabbit activities caused the bluffs to cave into the sea, gradually decreasing the size of the island each year. Tunneling by rabbits on oceanic islands is suspected of causing smaller, burrow-nesting seabirds such as Storm-petrels, Cassin's Auklets, and Ancient Murrelets to abandon whole islands. Larger burrow-nesters, such as Tufted Puffins and Rhinoceros Auklets, appear to be able to exclude rabbits from their burrows. Even in city parks and neighborhoods rabbits cause extensive vegetation damage, promote erosion through overgrazing, and can wreak havoc in gardens and flower beds.

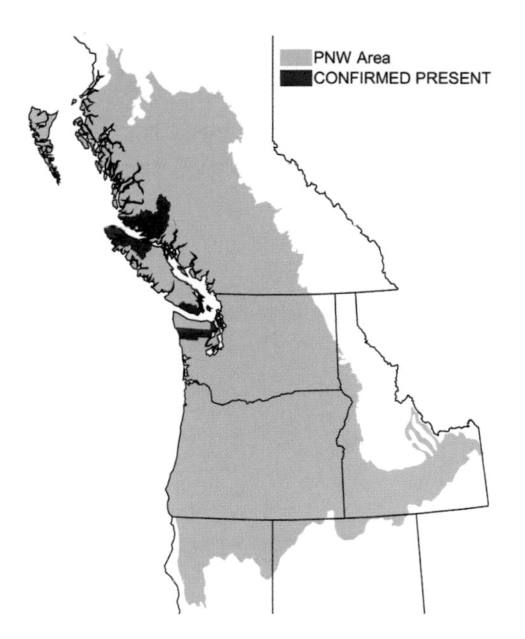

Control Methods and Management

Probably the most effective control method for rabbits on small islands, where they are likely to have their most negative impacts, is fumigation of burrows with calcium cyanide, which reacts with moisture in the burrows to form cyanogen, a suffocating gas. This method was successful on Smith I., but may have eliminated the Tufted Puffins there as well. Direct poisoning is not recommended because rabbit carcasses can then poison scavenging birds. Introduction of Red Foxes to San Juan I. to control rabbits failed to eliminate them all. Introduced pathogens such as myxoma virus can also reduce rabbit numbers but usually do not eliminate all of the individuals. Naturally, such methods should not be used on small populations in city parks, where live-trapping followed by adoption or humane disposal is a safer alternative. Establishment of rabbit populations can be prevented by not releasing pet rabbits into the wild or feeding feral rabbits. Wild rabbits should be reported to local animal control agents to prevent them from becoming a problem. Fences extending underground at least 0.5 m will protect gardens.

Life History and Species Overview

The European Rabbit is native only to southern France, Spain, Portugal, and northwest Africa. This is the only commonly domesticated rabbit or hare, and through direct release it now occurs over most of Europe. It has been introduced throughout the world, with large populations now found in Australia, New Zealand, Chile, and Argentina and on over 550 oceanic islands. European Rabbits in the wild prefer open habitats, shunning the deep forest, and do best in warm, dry climates similar to the Mediterranean region where they originated. The domesticated form released in the PNW is known as the "Belgian Hare," but it is a true rabbit, not a hare, meaning its young are born naked and helpless in underground burrows. By contrast, newborn hares are fully furred and can follow the mother shortly after birth. Typical wild adults weigh 1.5–2 kg. Baby European Rabbits open their eyes after 10 days, venture outside the burrow after 3 weeks, and are weaned at 4 weeks. A female typically has a litter of 5–7 young and may mate soon after giving birth, so she may produce 5–7 litters or more than 30 young/yr.

History of Invasiveness

Feral populations in the PNW are derived from farmed livestock that have gone wild, or from deliberately released pairs placed on islands as a source of food for settlers. It is believed that Hudson's Bay Company employees released rabbits on many of the San Juan and Gulf Is. during early exploration in the late 1800s. Farm operations released rabbits in the early 1900s. As recently as 1970, a lighthouse keeper released rabbits on Destruction I., WA. Small populations in city parks and neighborhoods are usually former family pets. European Rabbits breed prolifically, so one pair in a predator-free area can quickly populate an entire island or city park.

Other Sources of Information: 109, 116
References: 10, 134, 174, 229, 235
Author: Dale R. Herter

European Starling *Sturnus vulgaris*

Species Description and Current Range

The European Starling is an introduction success story. In less than a century the European Starling population grew from approximately 100 birds released in New York City's Central Park to over 200 million starlings spanning the continent. Now one of the most abundant and widespread bird species in North America, starlings are a familiar sight almost anywhere there is human development including throughout the PNW. The European Starling is a small, glossy, black bird with a short tail and pointed wings. In nonbreeding plumage it has a dark bill and feathers with pale spots at the tips, giving the bird an overall spotted appearance. In breeding plumage, the bill turns yellow and the pale tips wear away, leaving black feathers with a purple-green iridescence. Males and females have a similar appearance, but juveniles are a dull brown-gray color with a dark bill. Starlings prefer open areas for foraging and live in grasslands, seminatural areas, agricultural areas, and most habitat types where there is some human development. They avoid large areas of undisturbed forest as well as intact scrub and desert.

Impact on Communities and Native Species

Starlings evict woodpeckers from their nesting cavities, destroying the eggs and replacing them with their own. They also evict other birds, such as swallows, bluebirds, and wood ducks, from nest boxes or natural cavities. Disagreement as to whether this affects the reproductive success of the evicted birds continues. Some species of displaced woodpeckers nest successfully later in the breeding season; however, a study of Northern Flickers found that they had fewer offspring in their second nesting attempt. Despite the level of harassment by starlings, most evicted species have not declined significantly in abundance.

Starlings contribute to the spread of invasive plants that have fleshy fruits, such as English Ivy and blackberry. They are also considered an agricultural pest upon fruit and grain crops. Large flocks of European Starlings can interfere with airplanes at airports. Their feces damage buildings and transmit diseases, such as histoplasmosis, a fungal disease that infects the lungs causing flu-like or occasionally more severe symptoms.

Control Methods and Management

Large-scale attempts to reduce European Starling populations are ineffective. Poisoning, as well as scare tactics such as loud noises and moving objects, have moderate, short-term success in preventing crop damage, but over the long term birds habituate. Broadcasting recordings of their own distress calls is effective in moving roosts but not in eliminating them. Providing nest boxes with small entry holes can prevent European Starlings from displacing smaller native birds. Nest boxes with entry holes large enough to accommodate woodpeckers, however, allow entry to starlings, which may result in an increase in their nesting. European Starlings, like many invasive species, are closely linked to human development. Minimizing disturbance to natural areas may reduce their spread into those areas.

Life History and Species Overview

European Starlings are cavity nesters, using either natural cavities made by woodpeckers or man-made cavities such as openings in buildings or nest boxes. The female lays 4–6 pale greenish-blue eggs and both parents incubate the eggs and feed the young. Young birds leave the nest after 21 days, at which time the female may start a second brood. European Starlings start breeding at age 1 and can live up to 21 years. Their diet is varied and changes seasonally. It consists largely of invertebrates such as beetles, grasshoppers, earthworms, and earwigs during the spring and summer. Fleshy fruits, which increased with domestic fruit production, the use of ornamental plants, and the spread of invasive plants such as blackberries, provide important food sources in colder months. Starlings also eat grain from feedlots. Human-modified habitats often provide the open foraging areas, such as lawns and agricultural fields, that European Starlings prefer, when flanked by buildings or patches of woodlands where they find nesting sites.

European Starlings are native to Eurasia, from the British Isles to Lake Baikal in western Russia, with wintering populations extending into North Africa, northern India, and parts of the Middle East and China. European Starlings are partially migratory, which means that some individuals within a population migrate while others do not. They are also highly social, foraging communally and roosting in large numbers. Winter flocks can be as large as 100,000 birds and a roost can have over a million birds. Starlings have an impressive range of vocalizations. Ten different calls have been identified, and their warbling song is composed of 35–67 individual song elements. They often mimic the songs of other birds and can even mimic the human voice.

History of Invasiveness

Many efforts were made to establish the European Starling in North America, from introductions in Cincinnati, OH, and QC, Canada, in the 1870s, to Portland, OR, in 1889 and 1892. These populations all died out within a decade despite initial increases. Releases were mainly sponsored by local

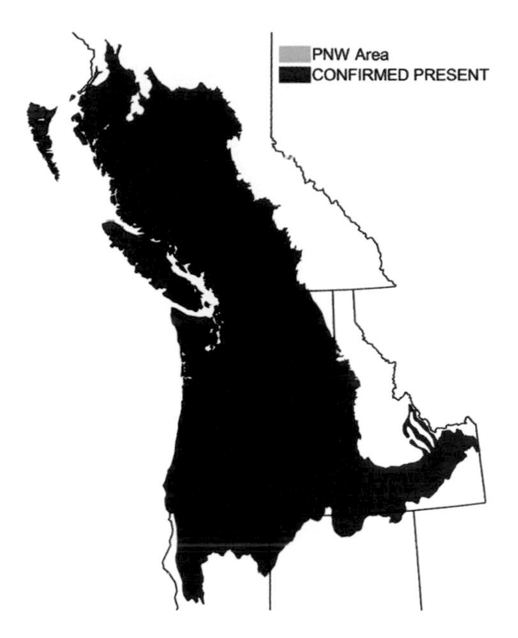

acclimatization societies whose mission was to deliberately introduce exotic plants and animals for use as food, sport, or for aesthetic reasons. An introduction of approximately 100 birds in 1890 and 1891 into Central Park in New York City was successful. These birds thrived but dispersed little within the first decade. Subsequently their range expanded rapidly, with winter migrants traveling to more distant places and breeding populations establishing several years later. In this manner the European Starling migrated in waves across the continent, taking only 50 years to spread from coast to coast. Genetic testing supports the claim that the North American population of European Starlings all stem from the Central Park birds.

The European Starling arrived in the PNW in 1942, in Siskiyou County, CA, and was seen in OR and WA by 1943. By the early 1950s, it had spread throughout the PNW. The increase in human-modified habitat over the last century, which increased food resources and nesting sites for starlings, may help explain why early introductions failed, while the Central Park introduction was so successful.

Other Sources of Information: 59, 60
References: 47, 108, 189, 197
Author: J. Katie Barndt

Mountain Goat *Oreamnos americanus*

Species Description and Current Range

Mountain Goats have a white coat with a few brown hairs, black horns, and black scent glands behind each horn used for territorial marking. Total length is 1.2–1.8 m, tail length 10–20 cm, horn length 20–30 cm in both sexes, weight 50–80 kg, with females about 15% smaller than males. Annual layers on the horns are used to age animals. Mountain Goats prefer the rockiest, steepest, and highest parts of alpine or subalpine areas where terrain is naturally unstable. Their large oval hoofs, hard on the outside with a soft convex middle, provide traction and adhesion on steep slopes. Goats are never far from rocky outcrops, important habitat features used to escape predators.

Invasive populations of goats are found on the Olympic Peninsula in WA. Although native to the Cascade Range, goats were not present in prehistoric times in the Olympic Mountains. The Olympic Peninsula has long been isolated by saltwater on three sides and lowlands on the fourth, making it unlikely that goats reached these mountains on their own. Consistent with this is the absence of goat fossil remains or native oral histories and art on the Olympic Peninsula. Blood analysis of 71 goats in the Olympics showed virtually no genetic variation, suggesting a founder effect from 12 goats released in the 1920s. In the mid-1970s, approximately 1,200 goats were on the Olympic Peninsula, with the largest subpopulation of over 100 animals at Klahhane Ridge. Goat control reduced the total population to its current level of about 300, mostly concentrated in the central mountains (e.g., the head of Quinault River, Bailey Range, and Mount Olympus).

Impact on Communities and Native Species

Mountain Goats create wallows, trample and eat vegetation, and make trails, all activities that can harm plants and reduce their habitat. Goats are generalists, feeding on grasses and herbaceous plants. They eat endemic plants but their digging and trampling are probably more damaging than their feeding. Goats may not significantly damage rare plants at the population level and because they are at home at or above timberline, there is little if any competition with native herbivores. In the 1980s, when goat populations peaked, lichen and moss cover decreased and soil erosion increased. Non-native herbivores like goats lower plant biomass, reduce primary production, lower species diversity, and increase potential extinctions, all reasons to remove them from national parks.

Control Methods and Management

Invasive species are only one of the problems facing national parks. When the invasive is as charismatic and popular as the Mountain Goat, control becomes controversial. After years of study and reams of reports, the Olympic National Park concluded that Mountain Goats met the criteria for mitigation: non-native status, threat to park resources, and feasible and prudent control. Park personnel removed 509 goats between 1981 and 1989, mainly by live capture and relocation. Goats were also shot, 99 by hunters outside the park and 19 for research. The National Park Service also tried control by contraception. Control is expensive, difficult, and potentially dangerous. Tens of thousands of dollars were spent on control but more money was spent on environmental impact statements, public hearings, and reports. Sportsmen also objected to goat control, and there was a goat hunting season on adjacent state land, now closed.

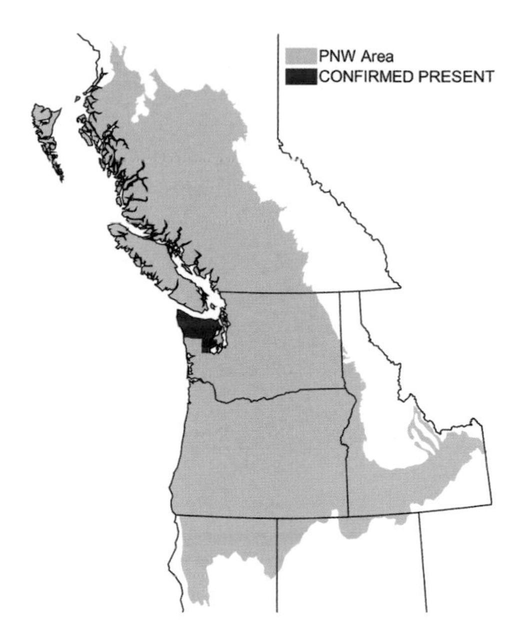

Life History and Species Overview

The Mountain Goat's native range is from southeastern AK to WA, western MT, and central ID. Mountain Goat fossils are found as far south as Mexico. The goat is one of 12 species of subalpine and alpine mammals—including Canadian Lynx, American Pika, Hoary Marmot, and Bighorn Sheep—native to the Cascades but absent from the Olympics. About 100,000 Mountain Goats are found in North America, high in the mountains where climate is often extreme and plant populations are in a continuous state of flux from gravity, freezing, and thawing. In winter, goats migrate to lower elevations, retreating to forests to eat foliage of lesser quality than their preferred alpine plants. Grasses, woody plants, mosses, lichens, and herbaceous plants make up their diet. Like cows, Mountain Goats are ruminants that can survive on relatively low-quality food because of their efficient digestion. Their four-chambered stomach acts as a storage space for unchewed food that can be regurgitated, chewed, and digested later. They crave salt, particularly in rainy climates where salts are leached from the soil. In spring, goats will travel miles to find salt. Vehicles often hit goats licking salt off the roadways in Glacier National Park.

The breeding season is from late November to early January. Males fight for access to mates by slashing their horns at the sides of rivals. The brief courtship includes kicking of the front legs, chasing, mounting, and vocalizations (loud baas). Gestation lasts 5–6 months, with females giving birth to one or sometimes two kids in May or June. The young are precocious, able to walk and stand shortly after birth. Females aggressively protect their kids with their sharp horns. Kids feed on vegetation within 2 weeks and are largely nutritionally independent in a month. They are weaned at 3 or 4 months but remain with the mother until she gives birth in the following year. If she does not give birth or loses her kid, offspring may stay with the mother until they are 2 years old. Both males and females reach sexual maturity when they are older than 2 years. During the warm summer months, goats are most active early in the morning and late in the afternoon, using the middle of the day for rest and digestion. Bedding depressions and wallows 25–50 cm deep are dug where they rest or wallow.

History of Invasiveness

In 1925, 4 adult goats from the Selkirk Mountains, BC, were released near Lake Crescent, WA, and in 1927 and 1929, local people released 8 more from AK. By 1979 the population had grown to 700. Mountain Goats were using most of the alpine and subalpine habitat available. Trapping and removal, plus hunting, reduced the population to fewer than 400 by the early 1990s. The Mountain Goat has been successfully introduced into several other western states, including OR, NV, UT, CO, WY, and SD.

References: 163, 282
Author: P. Dee Boersma

Black Rat *Rattus rattus*
Norway Rat *Rattus norvegicus*

Species Description and Current Range

The Norway Rat and Black (or Roof) Rat are members of the biggest family of rodents, the Muridae, which also includes mice and voles. Their genus, *Rattus*, contains over 50 other species, most of which are not invasive and live only in native habitats. Both the Norway and the Black Rat originated in Asia and are now on every continent except Antarctica. Rats are intelligent, adaptable, learn quickly, and are easily trained to complete complex tasks. They are trained as "bio-robots" by the Pentagon, to find land mines. As adept as dogs at locating mines and small enough not to trigger them, they may be even better for the job. Both species are commonly kept as pets, and the Norway Rat has been bred to produce close to 100 forms of fancy rat that compete in shows. The well-known white lab rat is descended from the Norway Rat.

The Norway Rat, the larger of the two species, weighs 200–500 g, the Black Rat only 77–108 g. Norway Rats are generally bigger, lighter in color, have smaller ears, and have a shorter tail than Black Rats. Both species are commensal with humans, but Black Rats are more agile, living in trees and other elevated structures and are usually the stowaways on ships, colonizing new areas and islands. Norway Rats are not arboreal and often dig holes and live in a system of burrows. They generally do better in colder climates and are the more common species in the PNW, where they live primarily in urban areas. Black Rats live primarily in rural areas. Because they are both dependent on human food sources, they are less likely to be found in native habitats.

R. rattus

Impact on Communities and Native Species

Rats are omnivores and eat nearly every edible object they can find. Not surprisingly, their economic impact on agriculture is huge, estimated at $100 million to $1 billion/yr in the US, and billions globally. Rats eat many native species, including small mammals, reptiles, amphibians, invertebrates, plants, and especially birds. On islands in the tropics Black Rats have decimated bird populations. In 1996, when they were eradicated from Bird I. in the Seychelles, they had reduced the population of tropic birds to just six breeding pairs. In northern regions Norway Rats tend to be more destructive. On Campbell I., NZ, they (along with introduced cats) caused the local extinction of at least six bird species within 200 years. Before being eradicated in 2000, Norway Rats caused massive declines in Ancient Murrelet populations on Langara I., BC. In 1981, 200,000 pairs of Ancient Murrelets bred on the 3,000-hectare island. Seven years later only 24,000 pairs remained but the population is poised for a rebound as rat removal increased the breeding area available to the birds. Rats carry diseases, including bubonic plague, typhoid, salmonella, leptospirosis, trichinosis, and tularemia. In the last 10 centuries, rat-borne diseases have killed more humans than all the wars ever fought. As disease carriers, they can affect many mammal species, especially other rodents. Two species of rat found only on Christmas I. in the Indian Ocean were extinct by 1908, most likely because of the arrival of Black Rats and the diseases they carried.

Control Methods and Management

Rat control and eradication methods include anticoagulant poisons, traps, and removal of habitat. Although rats are difficult to eradicate once established, some efforts have succeeded and resulted in bird populations rebounding. New Zealand successfully eradicated rats from Campbell I., which previously had the world's highest density of Norway Rats and is 11,300 hectares in size. In mainland areas where eradication is impossible, control is mainly by property owners. A Seattle city ordinance (#74182) enacted in 1945 requires property owners to contain food and control rats on their property. The best way to protect off-shore islands and wilderness areas is to prevent rats from being introduced.

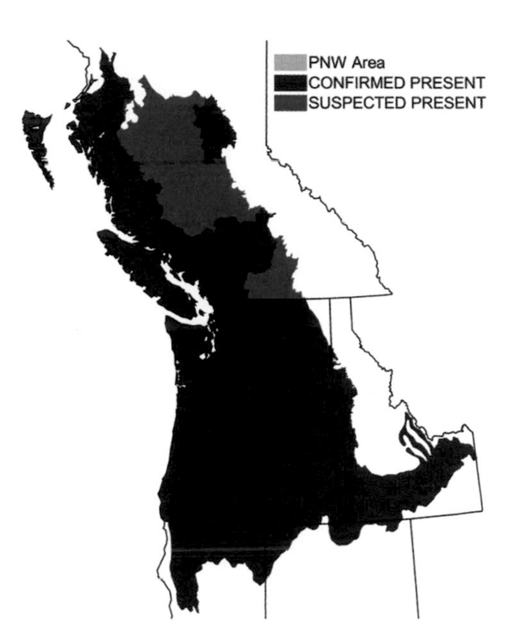

Life History and Species Overview

Prolific breeders, rats reproduce year-round if food is available, averaging 3–6 litters/yr. Females have 2–13 offspring/litter, averaging about 7, after a gestation period of 21–26 days. They are fertile and can breed again within 18 hours after giving birth. Theoretically a female could produce 80 offspring/yr. The young, born hairless and blind, mature quickly and wean within 3–4 weeks. Individuals become sexually mature at about 3 months. Both species are highly social, living in groups, raising young communally, and defending territories against other groups. Males defend an area containing several females and groups may have several dozen males. Norway Rat groups can have 200 rats, but Black Rat group sizes are smaller, up to 60 individuals. Usually only a single dominant male has access to communally nesting females, although sometimes there is a linear male hierarchy.

In Hindu religion, rats are seen as companions to the god Ganesha and are treated with great respect; Hindu temples often house and feed thousands of them.

History of Invasiveness

Norway Rats spread throughout Europe both overland (they can travel up to 7 mi/day) and aboard trading vessels during the 17th and 18th centuries. The species arrived in North America in 1775, perhaps in boxes of grain brought by German mercenaries hired by the British to fight the American colonists. As people moved westward so did the Norway Rats, assisted along the way by habitat modification, removal of predators, and poor human sanitation practices. They continued to spread to islands, by swimming (they can swim for up to 72 hours straight), aboard boats, and in grain shipments. The Black Rat arrived in Europe long before the Norway Rat, perhaps as early as the 4th century, and is the creature credited with carrying fleas that spread the Black Death (bubonic plague) during the 14th century. From Europe, they stowed away on explorer and pirate ships and made it to islands and continents all over the world. Black Rats first appeared in North America with the Jamestown, MA, colonists in the early 17th century and, like the Norway Rat, spread westward following human settlements. Black Rats were originally more common than Norway Rats but decline when Norway Rats are present, perhaps due to size asymmetries. In Europe, Black Rats are now rare (due to the presence of Norway Rats) and are listed as one of the top 10 endangered rodents. In VA, they have been listed as endangered.

Other Sources of Information: 50
References: 13, 99, 127, 235, 339
Author: Sacha N. Vignieri

Ring-necked Pheasant *Phasianus colchicus*

NOT LISTED

Species Description and Current Range

The Ring-necked Pheasant is native to Eurasia. The male's breast is a coppery chestnut with iridescent purple and copper hues. The head and neck are iridescent purplish or greenish, usually with a white neck ring; a red bare patch surrounds yellow eyes. Including its long, barred tail, males are 60–70 cm and weigh 0.9–3.0 kg. The female is smaller, 50–60 cm and 0.5–1.1 kg, and mottled brown with small brown spots on its back and a chestnut wash on its neck.

Taxonomy for the Ring-necked Pheasant is complex, including a division into two major groups considered by some to be separate pheasant species: Ring-necked (31 subspecies) and Green (3 subspecies). Multiple introductions of numerous subspecies from both major groups and subsequent hybridizations cloud taxonomic status.

The Ring-necked Pheasant has been introduced into a larger area of the world than almost any other bird species by individuals and governmental agencies. Such intentional introductions continue throughout much of this game bird's current range and in new areas throughout the world. Most populations are associated primarily with mid-altitude agricultural lands. Although its distribution is discontinuous, the Ring-necked Pheasant is found throughout the entire PNW, except for the most northern region. Other well-established game birds introduced to the PNW include Chukar (*Alectoris chukar*) and Gray or Hungarian Partridge (*Perdix perdix*). The California Quail (*Callipepla californica*) and Mountain Quail (*Oreortyx pictus*), though native to some PNW regions, were introduced into most of their current PNW range.

Impact on Communities and Native Species

In the Midwest, the Ring-necked Pheasant negatively impacts remnant populations of the Greater Prairie Chicken (*Tympanuchus cupido*) and Lesser Prairie Chicken (*T. dicinctus*) through nest parasitism, harassment, and direct competition for resources. Direct resource competition is likely in the PNW, where ranges of introduced game birds overlap with native grouse species such as the Greater Sage Grouse (*Centrocercus urophasianus*) and Sharp-tailed Grouse (*T. phasianellus*), which are in serious decline. At present, however, no evidence exists that introduced game birds have negative impacts on communities or native species in the PNW.

Control Methods and Management

Where attempted, removal of Ring-necked Pheasants has not benefited declining Prairie Chicken populations. Ironically, nearly all management for introduced game birds is focused on activities to increase populations, not control them. No efforts are under way to control any introduced game-bird species in the PNW.

Life History and Species Overview

The Ring-necked Pheasant is usually nonmigratory, requiring summer cover of grasses high enough to conceal nests but still allow easy travel and winter cover of shrubs and dense grasses. Foods include primarily seeds, but also fruits, leaves, and insects. In spring, the males' crowing is a common sound.

Vocalizations differ between sexes and are variable, with up to 24 different calls. While capable of strong short flights, Pheasants often prefer to run. The mating system is polygynous, and females raise only one brood per year, laying 7–15 eggs, which they incubate for 3–4 weeks. Young are precocial, fledge at 70–80 days, and are sexually mature at one year.

History of Invasiveness

Benjamin Franklin's son-in-law, Richard Bache, made the first documented attempt to introduce the Ring-necked Pheasant into the US in 1790. As with most game-bird introductions, his efforts were unsuccessful. The first successful introduction was in the PNW. Judge O. N. Denny, American consul-general in Shanghai, made the first introductions of Pheasants in the PNW in 1881 and 1882 into the Willamette Valley, OR. Two years later, the Ring-necked Pheasant was introduced on Protection I. in WA. These PNW introductions, made with Chinese-trapped birds, were highly successful, and many US and Canadian populations of Ring-necked Pheasants are descendants of these early PNW introductions. Innumerable efforts to introduce dozens of other game birds met with almost universal failure.

Introduced game birds remain immensely popular. The Ring-necked Pheasant is the state bird of SD, one of three states with non-native state birds (the other two states, RI and DE, selected varieties of the domestic chicken [*Gallus gallus*]). In WA, both Ring-necked Pheasants and *A. chukar* are included on the state's list of "Priority Habitats and Species," a list of priorities for conservation and management. The popularity of Ring-necked Pheasants and other game birds, and their intentional introduction, continues in many states and provinces (including the entire PNW). These introductions are not only authorized and encouraged by most states but are often undertaken by the states themselves. Throughout much of the late 19th century and the entire 20th century, vast sums of money were spent on game-bird introductions, most of which were unsuccessful. In the PNW, all introduced game birds are in decline, despite tremendous efforts to increase

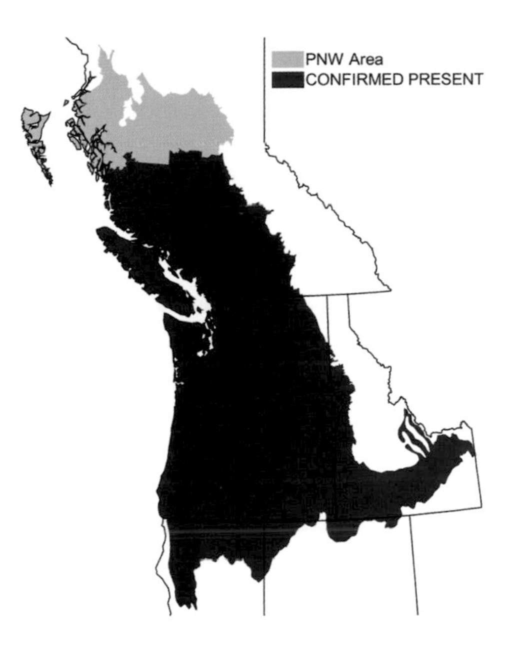

populations. This decline is associated with changes in agricultural practices that leave less uncultivated habitat. Huge investment in reintroducing exotic game birds and managing their populations continues at present. The substantial efforts and financial investment to enhance populations of introduced game birds mean fewer resources for native game birds, a group already in serious decline.

Other Sources of Information: 77, 130
References: 41, 121, 129, 133, 326
Author: J. Alan Clark

Eastern Gray Squirrel *Sciurus carolinensis*
Fox Squirrel *Sciurus niger*

Species Description and Current Range

The Eastern Gray and the Fox Squirrel are two species of tree squirrels found in the PNW. Due to their coloration and similar size, Foxes are easily mistaken for Grays. Foxes are slightly larger, have a yellow-orange belly, and tend to have more color (reds, grays, and blacks) than their gray cousins. Although predominantly reddish throughout their range, black-colored Grays dominate in the Vancouver, BC, area. Despite occasional overlap in the range of these species, as in Marion County, OR, usually only one of the species is introduced in any given area. Adult Grays vary in length from 385–500 mm, while Foxes can reach 474–565 mm. Both squirrels were introduced to the West Coast around 1925, primarily as pets and as charismatic lawn ornaments on estates and campuses. They are still found primarily in urban areas because of their taste for a wide variety of foods, from apples to human refuse. No surveys have been conducted, so any spread in their range may have been missed. Releasing nuisance squirrels, often trapped in house attics, into new areas increases their range. Grays replace native squirrels in new subdivisions within 5–10 years of development. Foxes in the Southern Okanagan Valley, BC, are believed to come from the Okanogan County, WA, population, suggesting that these squirrels are expanding their range.

Impact on Communities and Native Species

Gray and Fox Squirrels occupy the tree squirrel's niche in metropolitan areas of the PNW, perhaps through competition or disease. Despite a lack of evidence for their expansion into nonurban areas, they remain a potential invasive threat. In Europe, introduced Grays caused considerable damage to native plant populations by eating the nuts, flowers,

S. niger

seedlings, and buds and stripping the bark off of, most often, nut trees. They prefer oaks, walnuts, hickory, and other trees regularly found in their native habitats. Their habitat versatility and propensity for bark stripping made them a serious pest in Britain, where they are sometimes called "tree rats." Gray Squirrels may carry parapoxvirus, a disease deadly to Red Squirrels, but Grays who are exposed remain unaffected. These species have a potential impact on native species that live within city limits, as they prey on small birds, eggs, frogs, carrion, and even other young squirrels. Their food-hiding behavior may also have an effect by introducing seeds from non-native plants into the local environment. Why these two potentially destructive invaders have yet to expand their range significantly should be studied.

Control Methods and Management

Currently there is no official management procedure outlined for either invasive squirrel in the PNW. In the eastern US, traditional squirrel hunting remains popular and brings in at least $12.5 million/yr. In areas where squirrels have a serious invasive impact, as in Britain, squirrels are trapped and killed. In South Africa they were classified as vermin in the early 1900s, and bounties were paid for their tails. In some areas of the US (e.g., PA), bounties have been paid for furs and tails since the mid to late 1700s. The rewards began at just a few cents. Bounties may still be paid in areas where these animals are severe nuisances (there are no bounties in the PNW).

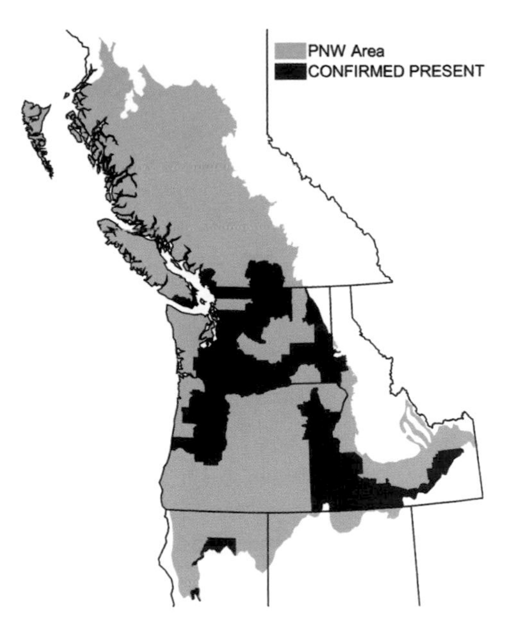

Life History and Species Overview

Grays breed in winter and summer. Foxes can mate year-round, but typically have only one litter. A single female will be chased by and mate with several males. Both species have a gestation period of 44 days, with up to 7–8 young/litter. Each female is reproductive for several years and can have 16 young/yr which gives this species potential for rapid expansion. Gray females can breed at 5 mo, but rarely breed before their first birthday (Fox females will breed at 8 mo.) Females are the sole caregivers. Lactating Grays are aggressive, especially when disturbed nestlings give their alarm squeal. Intruders handling nestlings are often leaped on from the branches above and bitten. Rabies has been reported in squirrels, but transmission to humans is not documented. Male-dominated hierarchies are established, and are more heavily enforced during the breeding season. Young males kept near a dominant adult male may show delayed maturation. Fox Squirrel home territories are 0.4–16.4 hectares. Gray Squirrels generally have smaller home territories, about 0.5–1.8 hectares. In both species, males hold larger territories than females. For Grays, territory size fluctuates depending on the time of year; when females are caring for young, their home ranges shrink dramatically, sometimes decreasing by half.

Both squirrels are active during the day (diurnal) and visible. They make nests, often a large ball of leaves high off the ground, in oaks and walnuts, and other nut-bearing trees, but can reside in conifer stands. Unlike native tree squirrels, these animals are not picky about their food, but prefer nuts and seeds, which they may hide in trees or bury for winter. Cached food stores are later found by smell, but seeds buried deeper than 7 cm are undetectable to squirrels and later become trees, which bear fruit for future generations. Up to 85% of cached seeds may be reclaimed after burial. These squirrels do not hibernate, and can use stored fat to survive only about a day of adverse weather, after which they must eat or they become too weak to forage. They are exceptionally good scavengers and will feast on hamburgers or fries thrown into a trash can. Eastern Gray Squirrels generally live no longer than 7 years, but can live 15–20 years in captivity. Mass migrations of squirrels have been reported and are more likely to occur when nut-bearing trees are less productive.

History of Invasiveness

By 1830, Grays were introduced over much of the world as pets. In Africa, they were introduced to tree plantations. Britain, the primary importer of Grays, found that the squirrels stripped the bark from trees because they had few predators and, most likely, were finding alternative food sources during lean and high population times. In the early 1900s Gray and Fox Squirrels were imported into South Africa and the western US, mostly for novelty purposes. In Europe and South Africa, the two species expanded rapidly and were declared vermin. The IUCN ranked the Eastern Gray Squirrel, one of only 14 mammals listed, on their list of 100 of the Worst Invasive Alien Species (see Appendix 1.).

References: 12, 78, 82, 188, 310
Author: Mari Aiko Tokuda

Dogwood Anthracnose *Discula destructiva*

Species Description and Current Range

Discula destructiva is the fungus that causes the disease Dogwood Anthracnose. This fungus is host-specific, meaning it only infects certain plant species. The hosts affected by the fungus are Pacific Dogwood (native to the PNW), Flowering Dogwood (native to the eastern US), and the ornamental Japanese Dogwood. The fungus appears on the leaf as an irregular-shaped brown to charcoal spot that can range in size from a pinhead to 0.7 cm. A distinctive purplish rim surrounds each spot. The name Anthracnose is derived from the Latin word for charcoal, because the fungus produces charcoal-colored leaf spots. The fungus can only infect if a host is available. The natural range of Pacific Dogwood begins in southern BC and extends to central CA. Pacific Dogwood is mostly found along forest edges, streams, and riverbanks. A disjunct population exists along the Selway, Lochsa, and Clearwater rivers in ID. The fungus favors stressed dogwoods found in moist, shaded environments with poor air circulation. Moisture is a requirement, as the fungus is spread primarily by water-splashed spores. *Discula destructiva* finds many stressed dogwoods in the urban environment and is considered a predominantly urban problem in the PNW. However, the disease is present and is having an effect, slowing the growth and eventually causing the death of Pacific Dogwoods within their natural environment.

Impact on Communities and Native Species

Symptoms of infection first occur on the leaves as brown spots in late spring. Pacific Dogwood may prematurely drop its leaves to fight the infection. The premature leaf drop hinders the annual growth of the tree, by reducing its photosynthesis area. Severely infected leaves turn brown and die but remain hanging on the tree. These remaining leaves are referred to as "mummies," because they

stay on the tree undisturbed throughout the fall, housing the dormant fungus. The infection will eventually spread to the branches and trunk causing cankers (swollen areas of dead tissue under the bark) that can strangle the tree slowly by constricting the vascular tissue, impeding water and nutrient flow. An infected tree may die 1–3 years after infection. *Discula destructiva* devastates native populations of Flowering Dogwood in the eastern US. The US Forest Service reported up to 88% mortality in stands throughout the Appalachian Mts. Such widespread destruction alters plant communities and changes forest structure.

More than 40 species of birds, such as Wild Turkey and Northern Mockingbirds, as well as many species of mammals feed on the fall fruits of Flowering Dogwood. Whether other plant species colonizing the gaps left by Flowering Dogwood attract the same variety of birds and mammals is unknown. The fungus apparently does not affect the Pacific Dogwood in the western US as severely as it does the eastern Flowering Dogwood. Unfortunately, little documentation exists detailing the level of impact this fungus has had on native dogwood populations. However, Dogwood Anthracnose kills or severely weakens native Pacific Dogwoods, increasing susceptibility to infection from root diseases and insects. The highest mortality rates in Pacific Dogwood have been recorded in the disjunct, higher-density populations in ID. Higher-density populations favor the water-splashed dispersal of the fungus, increasing the likelihood of infection.

Control Methods and Management

Current management techniques include chemical and cultural control, and breeding disease-resistant dogwoods. Cultural control means keeping the dogwood healthy by watering and mulching, especially during dry periods, pruning out infected twigs and foliage, and removing infected leaf litter from the site to reduce the level of fungal spores. Healthy dogwoods prevent the potential spread of the fungus to natural areas. Chemical control is not widely used in natural areas, as the application may cause the eradication of beneficial species; it can also be expensive and hard to apply chemicals accurately over large areas. Breeding for fungal resistance is under study in Flowering Dogwoods that did not become infected during an epidemic at Catoctin Mountain Park, MD, in 1977.

Life History and Species Overview

Fruiting bodies overwinter in the leaf litter or as mummies in infected trees, producing thousands of spores in the spring. The spores are primarily spread by water splash and wind and infect the newly emerging foliage. Cool, moist spring and fall conditions increase the likelihood of infection. If conditions are favorable, secondary infection from asexual spores is common in the late spring or summer, further weakening the tree. Little is known of the sexual stages of the fungus.

History of Invasiveness

Discula destructiva was first noticed on Pacific Dogwood in Clark County, WA, in 1976. At the same time, effects of the disease were observed on Flowering Dogwood in NY. The origin of the fungus is unknown, but shipments of ornamental Japanese Dogwood into Seattle and New York are thought to be responsible for the invasion. How the disease spread into natural areas is not known. Organisms that consume dogwood fruit and subsequently disperse the seed are possible vectors for the disease, as current research has found evidence of

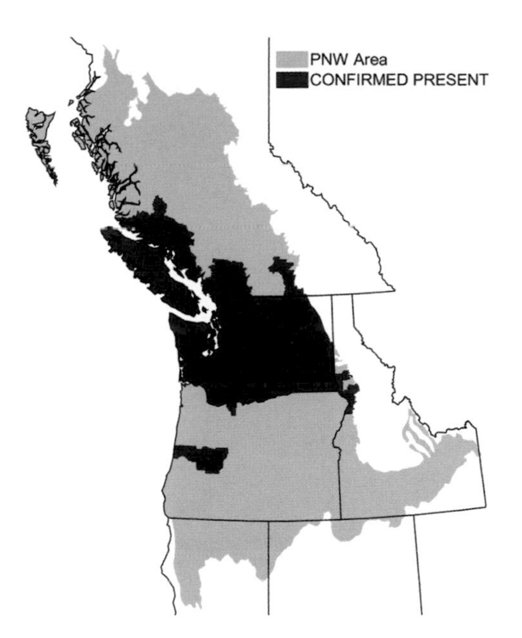

Anthracnose infection inside dogwood seeds. Furthermore, fungal spores have been found on the bodies of ladybugs, suggesting insects as possible vectors of infection. The fungus spread more rapidly, causing more destruction on the East Coast than the West Coast of the US. Several theories account for the differences in spread and severity of infection. First, Flowering Dogwood in the eastern US tends to grow in denser stands, favoring the windborne and water-splashed spore dispersal of the fungus. Pacific Dogwood rarely grow in stands except in the undisturbed, disjunct population of ID, where the dogwoods are heavily infected with the fungus. Second, host populations of Flowering Dogwood are more abundant and tend to be understory species, which grow in moist shaded environments favorable to the fungus. Pacific Dogwood is more commonly found along rivers and stream banks in open, well-ventilated conditions. Third, current research suggests that the eastern fungus is a more aggressive and genetically varied strain than the PNW fungus, making it a more effective invader.

Other Sources of Information: 68, 106, 117
References: 54, 83, 98, 220, 277
Author: Rachel Price-Rayner

Newcastle Disease *Rubula virus*

Species Description and Current Range

Newcastle Disease (ND) is a highly contagious avian virus with virulent and non-virulent strains. The most virulent ND strains lead to 100% mortality in poultry populations, causing bleeding lesions in the digestive tract or damage to the respiratory or neurological tissues. Although some birds die without overt symptoms of an ND infection, other birds display behaviors that are characteristic of the infection, such as sneezing, gasping for air, nasal discharge, coughing, greenish and watery diarrhea, musculature tremors, drooping wings, twisting of the head and neck, circling, complete paralysis, partial to complete drop in egg production, and swelling of the tissues around the eyes and in the neck. Though ND has a global distribution, virulent strains of ND are rare in the US and mostly come into the country with exotic pet birds, mainly parrots. ND killed Double-crested Cormorants at the Columbia River in 1997, and a nonlethal form of the virus is present in Cormorants in BC. Eggs from colonies in western Canada, including BC, test positive for antibodies to the disease. Adelie Penguins in the Antarctic test positive for ND antibodies, and may have been infected by chicken parts discarded by researchers.

Impact on Communities and Native Species

More than 230 species of birds are known to contract ND. Domestic fowl, turkeys, pigeons, ducks, geese, canaries, and parrots are highly susceptible to ND infection. ND outbreaks throughout North America have killed or partially paralyzed thousands of Double-crested Cormorants. Mortality from ND occurs during the breeding season, with the disease affecting young birds, not adults. Deaths of shag, gannet, White Pelican, Ring-billed Gull, California Gull, and turkey have been associated with ND outbreaks in Cormorants in Canada, though ND is often not the direct cause of death in these species. ND can be devastating to the poultry business. During an outbreak in the 1970s, almost 12 million domestic birds in CA were killed and $56 million spent to stop the spread of the virus. Humans and other mammals are rarely infected with ND, and infections are isolated to those in close proximity to infected poultry. Infection in humans leads to conjunctivitis and mild flu-like symptoms.

Control Methods and Management

The largest threat of ND is to poultry farmers. Because ND is passed from bird to bird via contact with bird droppings and secretions from the nose, mouth, and eyes, it is rapidly transmitted within a single farm. To prevent its spread to other poultry farms and wild animals, farmers destroy their entire flock when the disease is discovered in just one bird. The virus can remain infectious in water and on many surfaces, from cloth to plastic, so anything traveling between poultry farms, including clothing, gear, and vehicles, must be disinfected to prevent infecting other colonies. Poultry owners should take precautions to prevent the spread of ND. Exotic pets can be a source of ND. For example, Amazon Parrots from Central America can carry and shed the virus for up to 400 days without showing signs of an infection, giving ample time for one infected individual to infect many others. Many parrots, such as Amazons, do not breed easily in captivity, so be sure not to buy illegally acquired birds. Request certification papers from bird suppliers, and quarantine all new birds for at least 30 days. The Wild Bird Conservation Act of 1992 makes it illegal to purchase wild birds from foreign countries if the species' populations are not strictly monitored. Penalties for violating this law range from fines up to $25,000 to jail time of up to 2 years. Vaccination for ND is used in some situations, though is not 100% effective. Disease outbreaks in wild birds are not treated or contained.

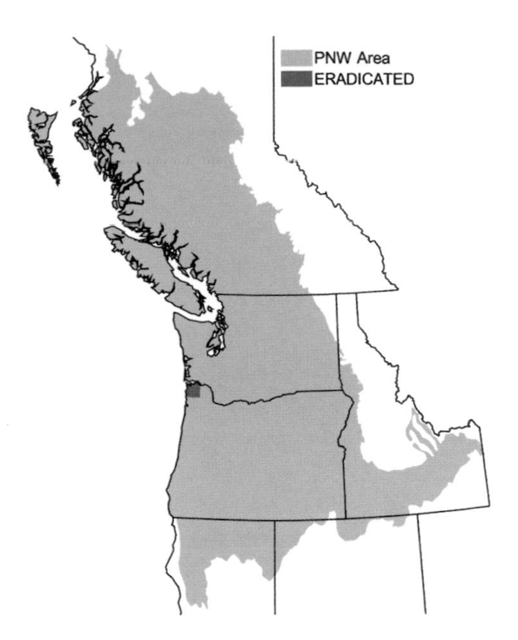

Life History and Species Overview

Most of the strains of ND from wild birds do not produce disease symptoms. However, the strength of a strain varies among different bird species, making wild birds resistant to strains lethal in poultry and vice versa. Furthermore, benign strains of the disease in wild waterfowl can transform and become lethal to poultry over time. Wild birds, especially aquatic species, are generally more resistant to ND and may act as reservoirs for the virus.

ND virus can survive in an infectious state for several weeks in warm, humid environments such as bird feathers and manure. The virus has remained infectious for 255 days in a henhouse that ranged in temperature from –11 to 36°C. It can survive for at least 3 hours at 56°C and in environments with pH 2–12. In frozen material, the virus can survive indefinitely. ND is killed by dehydration and exposure to the sun's ultraviolet rays. Usual transmission occurs by inhalation or ingestion of infected matter. The incubation period for the virus in infected animals is 2–15 days, but is usually 5–6 days.

History of Invasiveness

ND was first discovered in 1926. Since then, three major, global outbreaks of virulent strains have occurred. The first outbreak was probably due to transportation of poultry, other domestic birds, and their products. The second outbreak, which began in the late 1960s and continued into the early 1970s, was linked to the movement of caged parrots. The third outbreak, which started in the late 1970s, most likely initiated at pigeon races and bird shows and through international trade in these pigeons and doves. Outbreaks of ND in wild Cormorants in North America began in 1990, and since then over 10 outbreaks have occurred in Cormorant populations. In the 1992 Great Lakes/Midwest outbreak, more than 20,000 Cormorants died. Mortality in the other outbreaks was not as severe. In 2002–3, ND outbreaks occurred in poultry in AR, CA, NV, and TX, and over 3 million domestic birds were culled in CA alone. These recent ND outbreaks might be part of a fourth global outbreak.

Trade in exotic birds is the primary source for introduction of new and/or lethal strains of ND. Between October 1973 and September 1981, e.g., US authorities refused 6.5% of imported bird lots because some birds were infected with a form of ND lethal to poultry. However, Cormorants infected with ND might acquire the disease on their wintering grounds (Gulf of Mexico to Cuba and the West Indies) prior to migration into the US.

Other Sources of Information: 101, 102
References: 92, 246
Author: D. Shallin Busch

DISEASES

Port Orford Cedar Root Rot
Phytophthora lateralis

H

Species Description and Current Range

Port Orford Cedar Root Rot is caused by the fungus *Phytophthora lateralis*. The chance that a Port Orford Cedar (POC) (*Chamaecyparis lawsoniana*) will be attacked depends on its probability of coming into contact with the fungus. In 1942 the disease spread into nurseries and plantings so rapidly that it nearly destroyed the industry in southern OR within 10 years. It can live over 7 years in cool, moist conditions, but suffers quicker mortality in warmer and dryer conditions.

Mycelia, the underground vegetative parts of the fungus, contact root structures to spread and eventually girdle the tree's primary root collar. If a host rootlet is not found, other surfaces are contacted, agitation occurs, and a zoospore forms a cyst. In contact with a host, the cyst can germinate and form a mycelium that infects the host, or it can form a sporangium and release more zoospores. Infected foliage turns a lighter hue, withers, and eventually turns bronze. Finally, the inner bark dries up and darkens, and the foliage turns a light brown. Seedlings die within a few weeks, but old trees may live 2–4 years.

The disease is limited to southwestern OR and northwestern CA, the range of POC. The disease occurs most commonly along roadsides, near construction sites, or along riparian zones that have some contact with infected areas upstream. Except in the northern parts of its range, most POCs are found along riparian zones, but in zones of higher precipitation the tree is found on mountain slopes up to 6,400 feet. It is also present in some garden plantings in WA.

Impact on Communities and Native Species

POC is the only species damaged by the fungus, but Pacific Yew (*Taxus brevifolia*) can get this disease when it grows with POC.

The disease can cause major alterations in ecosystems, as POC is replaced with Sitka Spruce, Western Hemlock, White Fir, Redwood, Red Fir, mixed-pine, and mixed-evergreen forests. POC grows in a wide variety of soil types, but must have a consistent supply of groundwater.

Litter and soil under POC stands are less acidic than those under the conifer forests that commonly replace them. Snags and logs are an important habitat component for both terrestrial and aquatic wildlife habitat, and POC decomposes much more slowly than many of its associates. POC also often provide the only large-tree components of forest stands on ultramafic (high mineral) soils. Loss of POC in these regions has affected habitat and nutrient systems. When shade from these trees is lost along streams, water temperatures rise, reducing survival of desirable fish species.

Mortality of young trees at the lower size limits of marketability causes financial losses.

Other costs come from reduced productivity, replacement of dead ornamental trees, and costs of conversion to other crops. POC plays a significant role in the cultural, medicinal, and religious life of the Hoopa Tribe as well as other tribes in POC's region.

Life History and Species Overview

Once a root system is infected, the roots look water-soaked and subsequently darken, and fine roots quickly disintegrate. The inner bark and cambium of larger roots discolor progressively. Infected bark turns a deep cinnamon brown color, contrasting with the light cream color of healthy bark. Spread of the disease up the trunk is limited to the distance of about twice the stem diameter as the crown dies and tissues dry.

The fungus forms a tangle of very small filaments, forming the mycelium. Sporangia containing swimming spores grow at the filament tips and thick-walled, spherical resting spores form on the sides of the mycelium. When water saturates the soil the spores burst forth and travel with surface water. Resting spores spread the fungus as the soil that contains them is transplanted by vehicles, shoes, cattle, and wildlife. The fungus spreads uphill via interconnecting root systems between infected and uninfected trees. Surface-water movement is the main vector spreading the disease downslope, but root systems can also spread the disease to uninfected trees through contact.

Control Methods and Management

The disease is mostly spread by humans moving infected soil into uninfected areas on the heels of their shoes or the tires of their vehicles. In some locales, vehicles are banned from sensitive roads. Other areas enforce road closures during high-risk periods, usually October 1–June 1, or during particularly moist months when spores are likely to cause infection. Washing vehicles also helps remove fungus spores.

Management includes the removal of POC from riparian zones that run alongside roads to reduce the risk of vehicles picking up the fungus. Minimizing the use of outside soil sources in road building reduces spread, and bleach in water can kill spores and sterilize equipment, also reducing spread. Breeding programs for selecting the most resistant strains of POC are under way and may reduce the impacts of the fungus on POC.

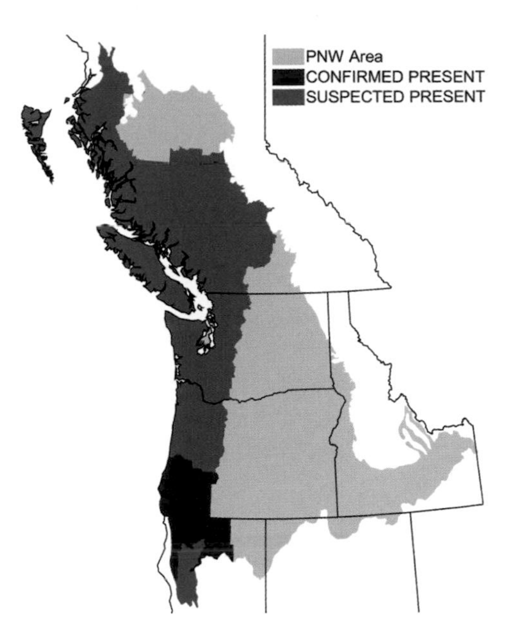

History of Invasiveness

The disease was first reported near Seattle in 1923 in nursery stock imported from France, the only other country to have the fungus. The disease is believed to have been introduced from either Asia or Europe, though neither of these regions has the disease now. Asiatic species of *Chamaecyparis* are resistant to this fungus, suggesting a long association with this pathogen.

Other Sources of Information: 97, 98, 103, 106

References: 138, 139, 270, 271, 296

Author: Joshua C. Misenar

DISEASES

Sudden Oak Death *Phytophthora ramorum*

HA

Species Description and Current Range

First observed in 1995 and identified in 2000, *Phytophthora ramorum* is the plant pathogen that causes Sudden Oak Death (SOD). SOD is related to the potato blight that brought about the great Irish Potato Famine of the 1840s, but despite its name is not sudden, not limited to oaks, and is not entirely fatal. Nearly 40 species of trees and shrubs occurring in CA and PNW ecosystems, including Redwood, Douglas-fir, and rhododendron, are susceptible to SOD, and more will be discovered. Trees most severely affected by SOD are tanoaks and the red oaks. White oaks have not been infected in the field. SOD infection manifests itself in two forms: (1) stem cankers that appear as brown, water-soaked lesions of dead bark, which may be seeping black to amber ooze; and (2) foliar lesions and/or twig infections that may result in branch-tip dieback in some hosts. Stem cankers often lead to crown death by girdling trees and/or predisposing the tree to secondary pests such as decay fungi and bark beetles.

Currently, SOD exists in mixed evergreen and redwood forests from Monterey County, CA, to southern OR. While the disease is most prevalent in the southern portion of its range near the ocean, there are isolated infestations in Mendocino and Humboldt counties in CA and in Curry County, OR. Based on climate and host species distribution, this northern portion of the range may be the most susceptible to widespread infestations.

A nationwide distributor of ornamental plants discovered nursery stocks of Camellia infected with SOD in 2004. More than 100 nurseries in 13 states received infected plants from this single source, greatly increasing the potential for spread of the disease into unaffected areas.

Impact on Communities and Native Species

Because SOD was only recently described, little is known about its ecology and potential effects on natural ecosystems. SOD killed tanoak and Coast Live Oak over a large swath of the CA coast. Over 300 species of wildlife, some of which are at risk for extinction, rely on these forests for food and shelter, and the loss of dominant trees could have dramatic implications for the diversity and function of these ecosystems. While no one knows exactly what the impacts of the disease will be in the PNW, many commercially and ecologically important species could be at risk.

Control Methods and Management

Resource management agencies are taking the threat of SOD very seriously. In 2004, the USDA allocated over $15 million toward identifying and combating the spread of the disease nationwide. Special task forces in CA and OR are coordinating regional research, monitoring, management, and education efforts toward minimizing the impacts of this disease. Recently approved application of a fungicide directly to the lower trunk of oaks and tanoaks can prevent or slow the growth of the pathogen. However, no practical treatment in forested situations exists.

Measures are in place to quarantine the disease to existing locations and many areas of the PNW are actively monitored for SOD. In OR, eradication efforts are under way to prevent the establishment of the pathogen. Strict protocols govern the transport of both dead wood and horticultural material to and from areas known to be infected. There are also a number of precautions individuals should take to limit the spread of SOD.

1. Do not transport plant material or soil out of infested areas. Regulations in CA and OR prohibit the removal of material from susceptible species out of regulated areas. The same applies for materials from Europe.

2. Clean shoes *before* leaving natural areas infected with SOD. Avoid muddy areas and remove all soil from your shoes. Disinfect with a common household disinfectant or mild bleach solution. Also thoroughly clean mountain bikes and car wheels if they have been off-road.

3. Report plants with symptoms of SOD. Symptoms include stem cankers with red-brown to black discoloration and isolated bleeding through intact or fractured bark, lesions on branch tips or leaves, and sudden branch dieback. Symptoms of SOD are similar to other, less lethal plant pathogens and visual identifications should be confirmed through tissue culture or molecular techniques. If you suspect SOD infection of a known susceptible species in or near infected areas, contact your state or local agriculture departments.

Life History and Species Overview

SOD is most closely related to water molds but is visually very similar to fungus with filamentous growth forms and reproductive spores. The pathogen's current geographic range includes areas that are relatively temperate and moist, conditions typical of the western side of the Cascade Range. SOD spreads its spores aerially by the nonlethal leaf and twig infections. Thus, foliar hosts may serve as reservoirs for the spread of the pathogen. In CA, mortality of oaks is strongly associated with the presence of Bay Laurel, which has high rates of leaf infection. It is currently unclear if SOD will spread throughout the PNW, since the range of Bay Laurel is limited to northern CA and southern OR, or if other suitable hosts that exist throughout the region (e.g., rhododendron, Bigleaf Maple, and Douglas-fir) can also act as a host and spread the disease.

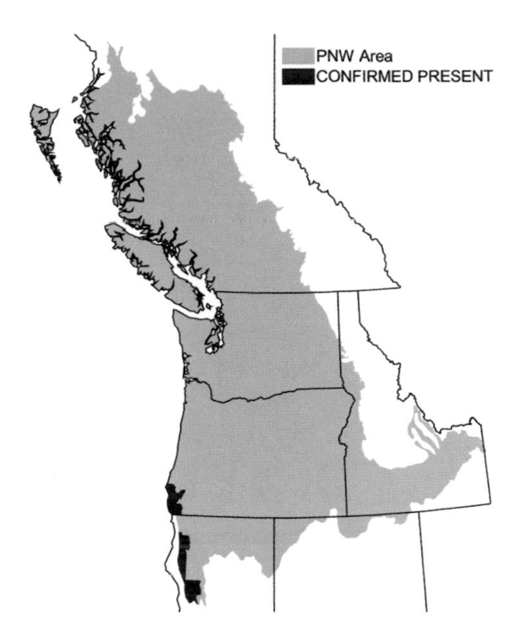

History of Invasiveness

The origin of SOD in North America is not known. Since it has only recently been discovered, is highly virulent, and has a small geographic range within the wider range of its hosts, this suggests that it is a recent arrival on this continent. Genetic analyses have shown the European strain of SOD to be distinct, raising the possibility of both strains being imported from an unknown third location. Over time the strain is expected to become less virulent — but that may be too late for many species.

Other Sources of Information: 19, 21, 72, 120
References: 84, 115, 265
Author: Gordon W. Holtgrieve

DISEASES

West Nile Virus

H

Species Description and Current Range

In 1999, West Nile Virus (WNV) appeared for the first time in North America, isolated from a dead crow. This was followed by reports of encephalitis in humans and horses. WNV is transmitted by mosquitoes and spread by infected migrating birds. It is a member of the flavivirus family that causes inflammation of the spinal cord and brain, and results in fever and death. A dead raven near Newport, WA, tested positive for WNV in 2002. That year two more birds in WA, one in Pend Oreille County and one in Snohomish County, had the virus. Two horses also tested positive for the virus, one in Island County and one in Whatcom County. In 2004, 740 humans got WNV on the West Coast, 737 in CA, 2 in ID, and 3 in OR. In WA, HI, and BC, human WNV has not been reported yet. WNV spreads erratically; it moves as a wave, hitting hardest at its perimeter and leaving some level of resistance in its wake. More than 20 birds died at the Bronx Zoo in 1999. WNV in zoos has the potential to eliminate entire groups of exotic species and to concentrate in rodents and other animals that are difficult to monitor, creating an epidemic in highly confined environments. CA had 20 human deaths in 2004.

Impact on Communities and Native Species

When infected mosquitoes bite, they transmit West Nile fever, causing flu and fever-like symptoms that last a few days but may not cause long-term health effects. Severe infections of the virus can lead to West Nile encephalitis, meningitis, or meningoencephalitis. Encephalitis is an inflammation of the brain, meningitis an inflammation of the membrane around the brain and the spinal cord, and meningoencephalitis an inflammation of the brain and the surrounding membrane. People 50 years old and over have weaker immune systems and are more susceptible to WNV.

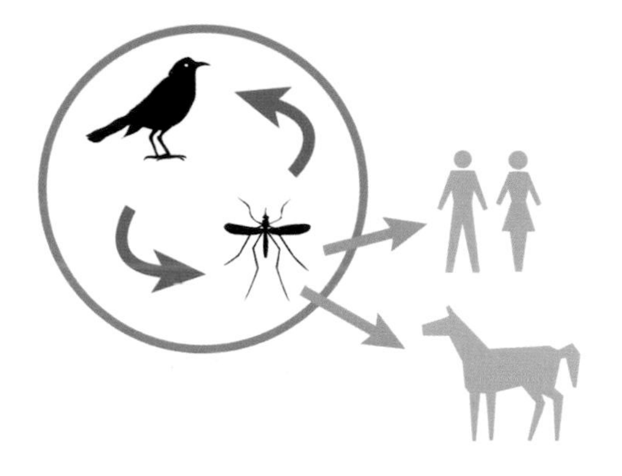

The presence of WNV and increased numbers of infected mosquitoes increase virus transmission to people and birds. WNV is known to be in 138 species of birds. Native corvids, such as crows, blue jays, and ravens, are more susceptible to the virus; research into the cause of bird susceptibility is ongoing. In horses, early virus detection and treatment usually save the animals. Increased mosquito control and the collection of mosquito distribution data allow mapping of WNV vector distributions and foster better control practices.

In WA nine mosquito species that are known to carry West Nile and/or West Nile viral ribonucleic acid (RNA) have been identified: *Aedes vexans, Aedes cinereus, Anopheles punctipenis, Coquillettidia perturbans, Culiseta inornata, Culex pipiens, Culex tarsalis,* and *Ochlerotatus japonicus* (an Asian species identified in WA, January 2001). Of these, *C. pipiens*, the common house mosquito, is the most widely distributed and of greatest concern. Mosquitoes are virus vectors; they take the virus from organism to organism and are not affected by it. Only two of the nine mentioned species are non-native, but they can all transmit WNV.

Control Methods and Management

Managers apply *Bacillus thuringiensis israelinsis* (Bti) to mosquito breeding sites for mosquito control. A naturally occurring soil bacterial larvicide that inhibits mosquito larval development, Bti is usually applied once every 2–3 weeks. Application is dependent on temperature, which affects larval production. Granulated Bti costs only $4–9/kg, but the cost of employing stormwater maintenance personnel for multiple applications is high. In addition to costs, the harmful effect of Bti on other aquatic organisms is little understood. In 1998 a University of Minnesota study showed that repeated applications of Bti drastically reduced the diversity of non-target insects and other invertebrates such as chironomids, Tipulids, and plecopterans (midges, crane flies, and stone flies). Alternative biological controls mosquito larvae are being explored, such as copepods (small crustaceans), mosquito fish, and frogs. Mosquito egg traps may prove effective at controlling mosquito populations.

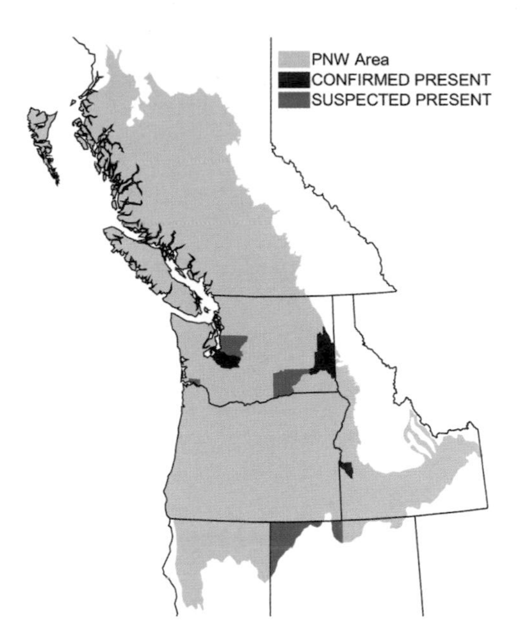

Life History and Species Overview

WNV is carried by mosquitoes and is transmitted when a mosquito bites an infected bird and then bites another animal or human. Mosquitoes transmit WNV 10–14 days after initial infection. A dead or sick bird, or flu-like symptoms appearing after being bitten by a mosquito, can be indications of the disease. Symptoms in humans take 3–15 days to appear. WNV affects the central and peripheral nervous system and can cause the brain to swell until death. WNV is a member of the flavivirus family (flaviviridae), which includes the viruses that cause Japanese encephalitis, Bovine Virus Diarrhea, St. Louis encephalitis, pestivirus diseases such as Hog Cholera and Border Disease, Hepatitis C, Dengue fever, and Yellow fever. The virus is not transmitted from person to person and handling a live or dead infected bird doesn't cause infection. The virus's primary carrier is the mosquito. Water where mosquitoes breed is among the first places to look to detect the virus. Tires, birdbaths, puddles, roadside ditches, and stormwater retention ponds are breeding grounds for mosquitoes. Crows and ravens (mainly) are part of the virus transmission cycle, so the virus can be detected in birds. Spread of WNV can be slowed by eliminating water bodies and containers breeding mosquitoes in and around the home, and by reporting dead or sick birds to see if they have the virus.

History of Invasiveness

WNV was first isolated from a febrile woman in Uganda in 1937. It has been described in Africa, Europe, the Middle East, west and central Asia, Oceania, and, most recently, North America. Outbreaks of WNV encephalitis in humans occurred in Algeria in 1994, Romania in 1996–97, the Czech Republic in 1997, the Democratic Republic of the Congo in 1998, Russia in 1999, the US in 1999–2004, and Israel in 2000. Epizootics of disease in horses occurred in Morocco in 1996, Italy in 1998, the US from 1999 to the present, and France in 2000. Epizootics in birds were found in Israel in 1997–2001 and in the US from 1999 to the present.

References: 151, 205, 228, 332, 337
Author: Sergio Camacho

White Pine Blister Rust

Species Description and Current Range

White Pine Blister Rust is a disease caused by the fungus *Cronartium ribicola* that affects white pine species (pines with clusters of five needles). Visible evidence of infection on white pines includes branches and trunks lined with hundreds of blisterlike, thimble-sized storage sacs containing millions of orange spores, bunches of yellowing needles, dead (brown-needled) branches in the canopy, diamond-shaped cankers (swellings on the branches with rough, dead bark and yellow-green margins), and resin dripping from these cankers. White Pine Blister Rust, native to north-central and eastern Asia, is widespread throughout the PNW and is common from lowland to subalpine forests, through almost the entire range of Western White Pine (*Pinus monticola*), Whitebark Pine (*P. albicaulis*), and Sugar Pine (*P. lambertiana*). White Pine Blister Rust is also invasive in Europe.

Impact on Communities and Native Species

In Europe, Blister Rust destroyed forests of Eastern White Pine (*P. strobus*), introduced and native to the US, and has severely impacted native white pines. In the PNW, White Pine Blister Rust kills 142,000 m^3 of white pine trees/yr, and those not killed are highly susceptible to pests. Subalpine Whitebark Pine communities have been especially devastated: only 1% of Whitebark Pines are resistant, 70% of exposed trees become infected, and 90% of infected trees die. The near destruction of this community type has likely had substantial impacts on wildlife such as grizzly bears and squirrels that eat pine seeds, as well as on important ecosystem processes, such as stream flow (through changes in snowpack duration) and regeneration. Most research on the effects of the disease on PNW white pines uses the more commercially viable Western White Pine (typical infection rate, 75%; mortality rate, 70%). The Blister Rust–induced mortality of Western White Pine and Sugar Pine impacted a valuable timber economy, and today's high timber prices for these species reflect the costs of Blister Rust control. White Pine Blister Rust is the most destructive and costly conifer disease in all of North America.

Control Methods and Management

Initially managers tried to control White Pine Blister Rust by removing *Ribes* shrubs, the alternate hosts for *C. ribicola*. In 1968, after spending millions of dollars with little success, removal was abandoned. This method was mostly unsuccessful due to the abundance of *Ribes*, their ability to resprout, the longevity of their seeds, and the long-distance dispersal ability of the *C. ribicola* spores traveling from pines to Ribes. Chemical control techniques are not effective. Pruning is costly but may be appropriate when the potential for recovery is high (e.g., branch cankers are not close to reaching the stem). Control methods center on planting resistant strains of white pines; however, even these trees can become infected. Natural or controlled reproduction of resistant white pines over several generations may increase resistance and allow trees to be maintained despite high levels of disease. Today, Blister Rust–resistant Western White Pine seeds are available to the public and are about 66% resistant.

Life History and Species Overview

Cronartium ribicola produces five types of spores and has two hosts during its life cycle —white pines and shrubs of the genus *Ribes* (currants and gooseberries). During the aecial stage in the spring (named after the fruiting body of the rust fungus), the millions of aeciospores that line white pine branches are dispersed by the wind to *Ribes* plants (distances up to 500 km) where they produce very small, orange, horseshoe-shape colonies. As summer nights become cooler, telial columns, containing thick-walled teliospores, develop on Ribes, and when the weather holds at 100% relative humidity for 6–8 hours, each column produces around 6,000 smaller basidiospores. These spores travel by the wind to nearby white pines, where they infect the pine host, entering through the stomata (the openings in the needles that allow for gas exchange during photosynthesis). As the resulting Blister Rust fungus grows over the next 6 months, a red or yellow spot develops on the needle. Further development of the disease results in diamond-shaped cankers and pycniospores (another type of sexual spore) that fertilize compatible spores, returning the cycle to the aecial stage, during which the fruiting body produces aeciospores that infect *Ribes* plants. After 2 years, the aeciospores erupt from the surface of the canker. They keep erupting each year, so that eventually the branch is girdled, resulting in the death of branches and of the tree if the main trunk is infected.

History of Invasiveness

White Pine Blister Rust was discovered in Russia in 1854, and by 1900 it had spread throughout northern Europe. Much of its spread in Europe started from an initial introduction of infected Siberian Stone Pine (*Pinus sibirica*), but the extensive introduction of Eastern White Pine around 1705 and

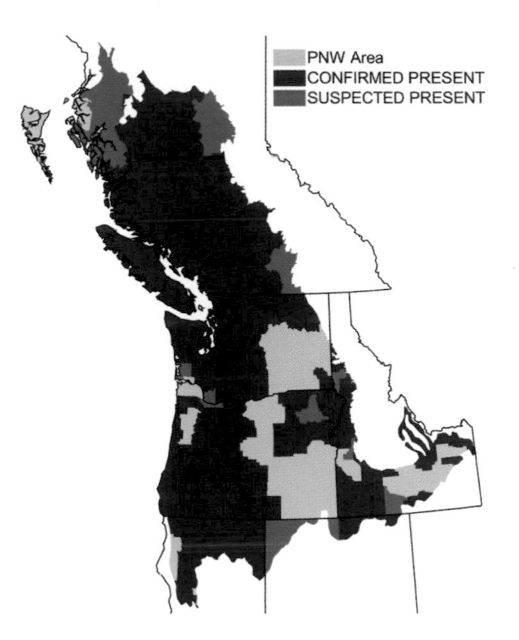

the abundance of cultivated European Black Currant (*Ribes nigrum*) in white pine forests helped its spread. Damage, and often mortality, from the Blister Rust fungus was rapid, and European foresters stopped managing white pine for timber because of the rust. White pine seedlings were exported to North America after inspection, but it was not known that the host could be carrying the disease without outward signs of infection (the "blister rust") for up to a year. Consequently, Blister Rust fungus was unintentionally shipped to North America on millions of white pine seedlings and distributed to hundreds of locations throughout the US. When Blister Rust was noticed on seedlings in the US, actions were taken to eradicate infected seedlings and *Ribes* shrubs, but the number of infected plants proved too numerous, and the disease continued to spread. In the PNW, the infection started at a tree nursery near Vancouver, BC, in 1910, and favorable climate conditions in following years allowed the disease to spread to southern WA and ID by 1923, southern OR and MT by 1929, mid CA by 1941, and southern CA and NM by 1970.

Other Sources of Information: 3, 125a
References: 204, 209
Author: Melisa L. Holman

African Waterweed *Lagarosiphon major*

Species Description and Current Range

The African Waterweed is a hardy and fast-growing aquatic plant from South Africa that has become problematic over the past few decades in both New Zealand and the UK. It takes root in the bottom sediment of freshwater bodies (e.g., lakes, ponds, and slow-moving streams). After it is firmly rooted, its stem then grows upward, filling the water column with vegetation. It spreads out into a surface mat up to 1 m thick, blocking sunlight from reaching anything below the mat. It also clogs ditches and drainage vents and impedes recreational swimming, fishing, and boating. Despite its domineering qualities, the Waterweed's attractive curved stems and its tiny (3 mm) pinkish-white springtime flowers make it a popular choice for aquaria and ponds in Europe. African Waterweed is not found in the US, but it nonetheless appears on the Federal Noxious Weeds list and a few lists of prohibited plants compiled by states, primarily because of its biological similarity to two other harmful invasive species, both of which are now well established in the PNW: Hydrilla (*Hydrilla verticillata*) and Brazilian Elodea (*Egeria densa*). If African Waterweed were to be accidentally released into Northwest waterways, it is likely to become established and cause problems similar to those observed in New Zealand and the UK.

Hydrilla, Brazilian Elodea, and African Waterweed are all members of the same taxonomic family, Hydrocharitaceae. They also all bear a strong resemblance to the American Waterweed (*Elodea canadensis*), which is native to the PNW. However, a few distinguishing characteristics make differentiating the species fairly easy. While the leaves of the African Waterweed occur alternately on the stem and recurve back toward it (it is also known as "Curly Water Thyme"), the leaves of the others grow in clusters or whorls. Hydrilla's leaves, which occur in

whorls of five, are about 1 cm long and feel rough around the edges. Hydrilla can also be distinguished by the small, potato-like tubers that grow near its roots. Brazilian Elodea leaves are 1–3 cm in length and cluster into whorls of four or eight. The American Waterweed's leaves grow in threes and are less than 1 cm long.

Impact on Communities and Native Species

Waterweed canopies can block up to 99% of sunlight from the lower water column. The plant also grows fast, easily outcompeting other plants. As the invader synthesizes sunlight into energy, calcium carbonate coats the upper surface of its leaves, increasing the pH of surrounding water and making it more basic (in small ponds the pH can increase to 10). This weed takes up carbon as bicarbonate from the water, and the calcium carbonate is a by-product. Its unusual physiology, along with its ability to elevate pH levels, is the key to the Waterweed's success in taking over waterways. Almost no other submerged macrophytes can photosynthesize effectively at such a high pH. After prolonged periods of high pHs, the African Waterweed typically competitively excludes all other macrophytes.

Control Methods and Management

No method of control is completely effective against African Waterweed. In the UK wetland managers cut the plant back in the spring, but if fragments break off and float away, cutting may actually facilitate its spread. Waterweed can be kept in check by using targeted applications of herbicide, but with limited success and negative impacts on other species. Wetland managers in New Zealand tried Chinese Grass Carp as a biological control, but though the Grass Carp had a slight preference for the Waterweed, they ate other plants and harmed native species. If the infestation is in a small, relatively contained area, e.g., around a dock, covering the area with a dark plastic cover for a month, depriving the Waterweed of sunlight, may kill the plants. Unfortunately, this shade treatment is not 100% effective and may have to be continued for longer than a month to achieve total success. Despite this species' invasiveness, it is still marketed in nurseries and aquarium supply stores, particularly in Europe. Even though there is often less oxygen in waters surrounding areas of dense Waterweed infestation, the British Broadcasting Company's (BBC) 2004 gardening Web site touts it as a terrific "oxygenator," with no mention of its negative impacts to natural communities.

Life History and Species Overview

African Waterweed is a hardy, semi-evergreen perennial. Its Latin name, *Lagarosiphon major*, means "large, thin, tubes," which likely refers to the skinny tubules that grow up from about 0.25 m below the water's surface and support its tiny flowers. It can reproduce sexually through fertilization of its flowers, but in most places only the female plant is present, so its primary means of reproduction is through fragmentation. The brittle stems break off, float downstream, put down roots, and form new plants. Worldwide there are 15 species of *Lagarosiphon*. Nine, including the African Waterweed, originate in Africa. The others come from Madagascar and India.

History of Invasiveness

African Waterweed is native to the high mountain streams and lakes of South Africa. It was first discovered outside its native range

growing in northern England in 1944. It is also an established and problematic invasive species in Ireland and New Zealand. It is unknown how the Waterweed came to those areas originally, but gardeners are a likely vector since the plant is a well-loved ornamental in ponds and aquaria. Once the species is established, boating and fishing help spread it as the stems break off, tangle on propellers and fishing gear, and then drop off elsewhere. The nearshore waters of Lake Wanaka in New Zealand are severely infested with Waterweed, and no efforts to control it have been very successful. In 1977 there were 45 separate patches of the species distributed primarily in the southern arms of the lake. By the 1980s the number of patches jumped to 79 and by the late 1990s its distribution was nearly continuous in the south and was creeping into the north.

The best method of control for this plant is to prevent establishment in the first place by banning its sale and use outside its native range.

Acknowledgment: Thanks to Kathy Hamel of the Washington State Department of Ecology.

Other Sources of Information: 24

References: 58, 103, 104

Author: Stacey Solie

THREATS Terrestrial Plant

Bamboo
Family: Poaceae

Species Description and Current Range

At first glance, one could be deceived by bamboo's treelike structure—but bamboos are actually members of the grass family (Poaceae). They are distinguishable from other grasses by their woody stems, which have a variety of colors, their branched growth, and in some species their very large size. Although dwarf bamboos are as small as 30 cm, the giant timber bamboos reach heights of 30 m. There are more than 1,200 known species of bamboo in the world, and over 300 of these have been introduced throughout the US. In contrast, the US has only 2 native species (neither is native to the PNW). Bamboo is common in urbanized areas, including gardens and unlandscaped yards. Hardier temperate bamboos persist in the coldest areas of New England and BC.

Impact on Communities and Native Species

Some bamboos are invasive. Such species are hearty, adaptable grasses and can survive, if not thrive, in many environments. Depending on the rhizome type, bamboo grows to full height in as little as 5–8 weeks, and the rhizomes of running bamboo grow as much as 1.5 m/yr. Many running bamboos can send out rhizomes up to 9 m away from the bamboo hedge, although they rarely grow that fast in the PNW. All it takes for a new stand to develop is the escape of one rhizome. This ability to propagate so readily allows a thicket or forest of bamboo to spread from one accidental dispersal into our forests. Bamboo can form thick groves with a dense mid-canopy leaf layer, diminishing the amount of light and water reaching the forest floor. By limiting the access of seedlings and other understory plants to these resources, the bamboo thicket can form a permanent monoculture that eventually eradicates other plants, including the native vegetation surrounding the grouping. Once a bamboo establishes, it is difficult to destroy. The PNW has no invasive bamboos yet but the threat is real. Its climate, latitudinal range, and similar understory floral structure are comparable to those of the forests of northern Japan. Moreover, there are 3 species in western Argentina, particularly *Chusquea culeous*, and 12 species of bamboo in Chile that are cold hardy, shade tolerant, or able to sprout following a fire, which are likely to thrive in the PNW.

Control Methods and Management

Controlling or exterminating bamboo is labor intensive and treatment must often be repeated. Bamboos may be either running or clumping types. If allowed to grow beyond its intended area, running (rhizomatous) bamboo may require much effort, persistence, and regular monitoring, depending on such factors as soil texture, compaction, and available nutrients. In the PNW many cultivated bamboos are species of the genus *Phyllostachys*, which is a hardy temperate runner. A recommended way to permanently remove a bamboo plant of either the running or clumping type is to dig up and remove every piece of the root system. This is a difficult task because of the distance rhizomes extend from the plant. You will need to dig up all the soil around the plant, but even then some piece of root may escape your attention and sprout. Or you can kill the rhizomes in the ground using mechanical methods similar to those used on blackberries. Other options include restricting the plant's expansion with sturdy underground barriers, planting in a bucket or pot, or applying a nonselective herbicide to the newly emerging leaves. Mowing close to the clump or spading around the perimeter for the clumping variety can restrict its growth. Even sturdy barriers will degrade over time, and the beak-shaped rhizomes will often find a way above, under, or around a barrier. Herbicides will have to be applied for 2–3 yrs, and mowing rarely works to kill the plant. If you plant bamboo, be aware of the varieties chosen and avoid those in the *Phyllostachys* and *Sasa* genera—all are types of running bamboo. Ask local nurseries for suitable alternatives. Ensure that disposed bamboo rhizomes are completely dead. Avoid planting any variety near running water as the rhizomes can survive for years, be carried away by the water, and propagate in a new area without your knowledge. Also, never purposely spread rhizomes into natural environments.

Life History and Species Overview

Bamboo is an ancient group of grasses with a large diversity in forms, including climbing vines, fern-like herbaceous plants, groundcovers, and shrubs. The earliest bamboos apparently originated in the Southern Hemisphere and slowly radiated into the temperate zones of the Northern Hemisphere. Most grasses are found in open habitats but bamboos have adapted to forest environments and therefore can often tolerate low light conditions. Running bamboo species are the

most commonly distributed and the most invasive. Clumping bamboos are generally limited to tropical species although *Chusquea* and *Fargesia* species grow in the PNW as well. Most species only flower, often in unison and over a large area or their entire range, once every 60–100 years. When it does set seed, one plant may produce more than a million tiny seeds.

History of Invasiveness

There are no recorded observations of invasive bamboo in the PNW, but if we follow the trend of other unlikely invasives it is only a matter of time. There are quite a few problems with the introduction of bamboo in HI and other coastal islands throughout the Pacific Ocean. Bamboo thrives in many places, including potentially the PNW, replacing the current vegetation and expanding at a rapid rate. Please use care when growing bamboo species and report any wild populations to the appropriate state agencies.

Other Sources of Information: 1
References: 201, 218
Author: Allisyn Hudson

Chinese Mitten Crab *Eriocheir sinensis*

Species Description and Current Range

The Chinese Mitten Crab is easily identified by the brown hairy patches (resembling mittens) on its white-tipped, equal-sized claws. It is brown to green in color, has a deep notch between the eyes, and a carapace up to 80 mm wide with four prominent spines on either side. The legs are typically twice as long as the carapace is wide. Mitten Crabs are efficient travelers, capable of traveling hundreds of kilometers in a lifetime. During migration, they meet very few natural barriers to their passage; they are known to leave the water and walk on land to bypass obstacles. They have been found on airport runways and city streets, in swimming pools, and even inside houses. More commonly they inhabit marine areas, coastal streams, and estuaries in Asia, Europe, and the western US. The Mitten Crab, although not established in the PNW, has already arrived several times and will likely come again through intentional release or ballast water.

Impact on Communities and Native Species

Mitten Crabs are a threat because they alter aquatic communities through predation, competition, habitat alteration, and food web disturbance. They compete with and prey on economically valuable species such as clams and mussels and eat the eggs of endangered salmon and sturgeon. Mitten Crabs harm native vertebrate and invertebrate species by consuming them or their prey. Crabs can be a nuisance for commercial and recreational fishers because they steal bait off hooks, damage fishing nets, and directly damage catches when their spiny carapaces puncture the fish. Mitten Crabs live in sediments with high levels of contaminants and often consume filter-feeding invertebrates such as clams and mussels, bioaccumulating toxins and heavy metals and passing them up the food chain.

Control Methods and Management

Management methods are preventative, focusing primarily on blocking introduction and limiting dispersal to new areas. Although education may help in delaying their introduction, Mitten Crabs are likely to reach the PNW by deliberate release to establish a fishery or accidental release via ballast water. Current ballast-water exchange regulations protect WA from introduction via foreign sources but not via other US ports such as San Francisco, CA.

In 1989, the US Fish and Wildlife Service listed the Mitten Crab as an "injurious species." The federal Lacey Act makes it illegal to import eggs or live specimens of any species of Mitten Crab to the US. It is illegal to import or transport live Mitten Crabs in West Coast states and BC. Mitten Crabs are listed as a "prohibited species" in CA, where they may be fished recreationally but must be killed immediately upon capture. In an effort to control already established crab populations, researchers are studying the possibility of using a lethal fungi as a biological control agent.

Life History and Species Overview

Mitten Crabs are catadromous, meaning that they reproduce in saltwater and migrate to freshwater to grow to adult size. They typically mature at 1–2 years but may take as long as 5 years to become adults. They mate during the winter in brackish or salty water; females carry up to 1 million eggs under their abdominal flaps until spring when the larvae hatch. Adults reproduce only once and die several months after mating. Mitten Crabs start life as planktonic larvae living in saltwater for 1–2 months; they then move through several developmental stages, eventually sinking to the bottom to develop into juvenile crabs. Juveniles migrate into freshwater, where they may seek refuge in submerged aquatic vegetation or burrow in the mud (riverbanks, levees, etc.). Mitten Crabs burrow for protection from predators and to escape desiccation during low tides. Burrows may be 20–30 cm deep and extremely close together. Researchers in CA observed three burrowing Mitten Crabs/30 cm^2 in a freshwater tributary of San Francisco Bay. Mitten Crab juveniles are predominantly vegetarian, but adults will eat almost anything. Adults typically eat small invertebrates but may also feed opportunistically on plants or animals or may scavenge dead animals.

History of Invasiveness

Native to coastal areas and rivers of the Yellow Sea in China and Korea, Mitten Crabs were accidentally introduced into Germany in the 1930s. Populations grew rapidly, and burrowing caused tremendous structural damage by accelerating erosion rates of levees and riverbanks. Within a few decades, their distribution expanded throughout many northern European rivers and estuaries (e.g., Netherlands, Belgium, and England).

In North America, Mitten Crabs have been found in the Detroit River, MI, the Great Lakes, OH, and WI, CA, and HI (one crab). Mitten Crabs were first caught in CA in 1992 when shrimp trawlers in San Francisco Bay hauled them in with their catch. Since then, they have invaded the Sacramento and San Joaquin rivers, CA, upstream of the delta. Mitten Crabs can clog pumps and fish screens,

causing problems in CA's water-diversion and fish-passage facilities. One individual of another invasive species, the Japanese Mitten Crab (*E. japonica*), was found in OR in 1997 near the mouth of the Columbia River. The CA and OR incidences likely were intentional releases to establish a fishery. Mitten Crabs are a desired food product in Asian countries, and there is an increasing demand for Mitten Crab on the black market where crabs frequently sell for $20 each. Shipments of live, illegally imported Mitten Crabs have been confiscated in Los Angeles, San Francisco, Seattle, and New York.

Acknowledgments: I wish to thank Cindy Messer and Tanya Veldhuizen, CA Dept. of Water Resources; Kathy Hieb, CA Dept. of Fish and Game; and David Bergendorf, USFWS Aquatic Nuisance Species Program

Other Sources of Information: 17, 85, 90, 108

References: 274, 302, 328

Author: Casey L. Ralston

THREATS

Dead Man's Fingers
Codium fragile ssp. *tomentosoides*

Species Description and Current Range

Dead Man's Fingers, a green marine algae, is a member of the *Codium fragile* species complex. Both the introduced subspecies, *tomentosoides*, and the native West Coast subspecies, *mucronatum*, share a variety of common names—Dead Man's Fingers, Felty Fingers, Green Sea Fingers, Sea Staghorn, and Sputnik Weed—and differentiating the introduced from the native subspecies can be tricky. The species occasionally varies in appearance depending on environmental conditions. The pale to dark green algae grow as a series of spongy, cylindrical, fingerlike branches, 3–10 mm in diameter, that branch repeatedly in a dichotomous, or Y-shaped, pattern, to form a bushlike structure up to 1 m in length and weighing up to 3.5 kg. Tips are broadly rounded. Occasionally, when plants are stressed, the surfaces form numerous long hairs that give the plant a fuzzy appearance. The hairs can be shed at any time. Its surface may also sometimes be covered with epiphytic algae. A perennial, broad, spongy basal holdfast attaches it to hard substratum. It is sometimes mistaken for a sponge, and juveniles produced from female gametes sometimes appear more like a fuzzy algal mat. Dead Man's Fingers is coenocytic, that is, the entire body is one large cell having many nuclei and chloroplasts.

The *tomentosoides* subspecies has colonized around the world, including both North American coasts. On the West Coast, the only positive identification of this species was in 1977, from San Francisco Bay, CA, and nearby Tomales Bay to the north. Sightings in OR and Prince William Sound, AK, turned out to be false positives, but the plants were so close to *tomentosoides* in appearance that identification required DNA analysis.

Impact on Communities and Native Species

Dead Man's Fingers settles and grows on any surface providing a firm substrate for attachment, including docks, rocks, pilings, reefs, marine farming equipment, mussels, and shellfish. Attachment to shellfish can hinder shellfish movement and feeding, and in heavy waves large plants may be swept away, carrying the oysters or mussels with them. Its impact on aquaculture led to another common name, the "oyster thief." Dead Man's Fingers has also been implicated in the loss of kelp communities. When kelp is abundant, Dead Man's Fingers will not move in to establish, but where sea urchin predation removes kelp, exposing bare substrate, Dead Man's Fingers may invade, forming large meadows and excluding kelp reestablishment. Unlike kelp and some other macrophytes, which hold much of their biomass on stalks above the substrate, a meadow of bushy Dead Man's Fingers provides little understory for fish or large invertebrate species to hide in or move through easily. It inhibits seabed-foraging species, which, in turn, alters the subtidal community structure.

Control Methods and Management

As eradication of large populations is not very successful, prevention is the best control, particularly through management of ship fouling. Native herbivores appear to place little grazing pressure on Dead Man's Fingers. Although sea slugs may control some local populations, using biocontrols is risky. The sea slug, *Aplysia*, has an interesting relationship with Dead Man's Fingers, sucking out and storing its chloroplasts, which continue to photosynthesize within the slug, producing a slime the slug uses to lubricate its movement.

Life History and Species Overview

The invasive subspecies, *tomentosoides*, is found in marine and estuarine rocky intertidal, subtidal, and tide-pool habitats, and generally likes more sheltered habitats than the native subspecies, *mucronatum*. It is somewhat sensitive to wave action and is generally found in more sheltered shallow areas. In CA, the native subspecies is not likely to be found on docks and pilings because it is restricted to the outer coast. It does not enter San Francisco Bay, so it is not distributed in the same areas as *tomentosoides*, which is restricted to quieter waters. *Tomentosoides* was found as deep as 22 m, but there is a question as to whether the subspecies was correctly identified.

Dead Man's Fingers reproduces both sexually and asexually, and reproduction is triggered by exposure to light and tides. It has motile reproductive cells, but can also reproduce parthenogenetically, releasing female biflagellate gametes that settle in a day and germinate, developing a mat-forming stage, which can persist for months to years. It also reproduces asexually from plant parts and vegetatively from buds that attach to substrate and form new plant bodies (thalli). Whole plants may also drift to new locations and establish new populations. Optimal growth temperature is 21°C, but it needs 12–15°C at least part of the year for successful reproduction. Adults can survive temperatures as low as –2°C.

History of Invasiveness

Assumed to be native to Japan, the subspecies *tomentosoides* is considered one of the most invasive seaweeds in the world. It is now reported from Europe, New Zealand, Australia, the Mediterranean, and both coasts of North America. It probably originally entered the US as a fouling organism on ship hulls and was first reported from Long Island, NY, in 1957. It is now found along the coast from NS to SC. Plants on the East Coast vary widely in characteristics, so it is not known whether the species was introduced several times or whether it is simply very adaptable to differing environments. Dead Man's Fingers withstands wide variations in temperature, salinity, light exposure, and nutrient condition. It uses a variety of nitrogen sources, grows and matures rapidly, and reproduces in a variety of ways. This plant is a successful invader because it can spread by the drift of detached plants, ship hull or fishnet fouling, through oyster introductions, or in seafood packing material. Gametes are short-lived, so ballast water transport may not be an important factor in its spread.

References: 43, 288, 311
Author: Joan Cabreza

Fishhook Water Flea *Cercopagis pengoi*

Species Description and Current Range

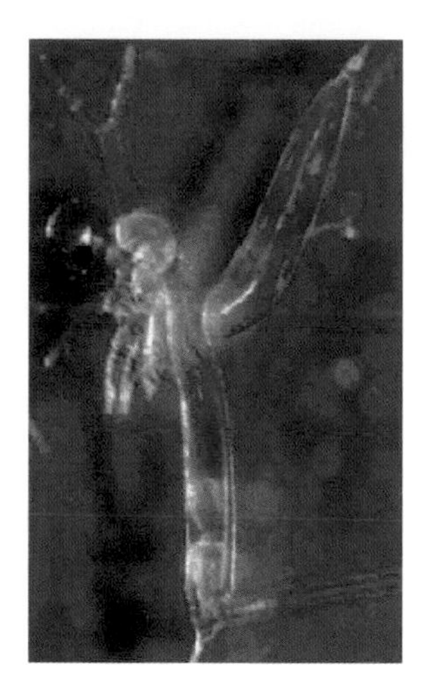

The Fishhook Water Flea is a small predatory crustacean that lives in warm saltwater and freshwater lakes and has a remarkably troublesome tail. While its milky-white body is only about 2 mm long, its tail is five times that length with a loop, or "hook," on the end. Its hook catches on the tails of other water fleas, causing them to collect in massive sheets with densities of up to 600 individuals/m^3. The water fleas clog fishing nets and lines and impede recreational activities, such as swimming and boating. When clumped, they look like wet cotton batting. Water flea heads are dominated by a large, single, black eye, visible without the aid of a microscope.

The Fishhook Water Flea was first found in the US in Lake Ontario. Since then, it has spread rapidly throughout the Great Lakes region, along with its closely related cousin, the Spiny Water Flea (*Bythotrephes cederstroemi*), another non-native introduced in the 1980s. Both species are from the Ponto-Caspian region of northern Europe. Researchers suspect that the water fleas arrived in the US in the ballast water of transoceanic ships coming from the Gulf of Riga, a bay off the Baltic Sea located between Latvia and Estonia. The Fishhook Water Flea population has grown exponentially for the last 10 years in the Gulf of Riga. Though it hasn't yet been found in the waters of the PNW, it has been identified in the ballast water of ships harbored in Vancouver, BC. The prevalence of ports frequented by transoceanic ships in the Northwest indicates the region's vulnerability to invasion by this and a host of other species that are carried inadvertently in ballast water.

Impact on Communities and Native Species

The Fishhook Water Flea likely has some impact on the food-web interactions among many species in the Great Lakes. Large fish such as adult Yellow Perch, Alewife, walleye, and salmon feed on the Water Flea during at least some parts of the year, but its long tail makes it too big for the mouths of juvenile and larval fish, thus excluding it from their diet. Both the Fishhook Water Flea and Spiny Water Flea are predatory species: they feed primarily on other zooplankton. The invasive water fleas consequently compete directly with juvenile fish for food. Also, after the introduction of the Spiny Water Flea to the Great Lakes in the 1980s, native populations of the slightly smaller native zooplankton, Daphnia, decreased. Three Daphnia species declined, leaving only the *Daphnia galeata mendotae* to dominate in offshore areas. Another native water flea, *Leptodora kintii*, declined when the Spiny Water Flea was introduced, though scientists do not know the exact mechanisms behind this shift. Once the invasive water fleas are introduced, general population changes suggest that large-bodied zooplankton are favored over smaller species.

Control Methods and Management

Stopping the water fleas' spread to other lakes is the main concern. In the Great Lakes region, regulations stipulate that fishing gear is to be thoroughly cleaned. Large cargo ships are also required to exchange their ballast water when they are out at sea, instead of near shore. Ships carry ballast water to help maintain balance when their cargo loads are light, and the practice of dumping contaminated ballast waters has been common and unregulated until recent years. However, even "empty" ballast tanks contain up to 150 tons of residual mud and water, which can harbor hundreds of millions of invertebrate eggs, many of which are known invaders.

To curtail further spreading of invasives, researchers are experimenting with different filtration and water-treatment methods. Using physical filters in combination with a secondary treatment is effective. Water-treatment methods include heating water before it is discharged or subjecting it to intense ultraviolet (UV) radiation, to ozone, or to a biocide like vitamin K, which kills most microorganisms within a 24–hour period before it breaks down. Another treatment method adds nitrogen to ballast water. Unfortunately, there currently are no regulations that require these treatment methods, and the shipping industry has not widely embraced these potentially important control mechanisms because they can be costly.

Life History and Species Overview

All water fleas belong to the taxonomic order Cladocera, and most are active swimmers, moving with a characteristic flea-like jump. However, both the Fishhook and the Spiny Water Flea drift rather than swim. They use one of two pairs of antennae to achieve some motility, and in the Gulf of Riga they migrate up and down in the water column, sinking 50–60 m during the day to cooler waters and resurfacing in the evening. Fishhook and Spiny Water Fleas can live and thrive in freshwater lakes, reservoirs, and river mouths, as well as in brackish water with up to 8% salinity.

The Fishhook Water Flea is a prolific reproducer, capable of bearing broods of up to 13 larvae every 2 weeks all summer (June–September, sometimes longer if the water is warm enough). It reproduces parthenogenically (by producing clones that develop from unfertilized eggs) during the

summer and into fall, but as temperatures drop, the female begins to produce male offspring. The males then fertilize her eggs, which are protected by a thick shell. This allows them to survive the low temperatures as well as to pass through a predator's digestive tract unharmed. Water fleas reproduce so fast that a single individual could potentially populate an entire lake.

History of Invasiveness

Fishhook and Spiny Water Fleas are native to the Ponto-Caspian area of northern Europe, including the Caspian, Aral, and Azov seas. Fishhook populations also reach up into surrounding rivers like the Danube, Don, and Volga in Germany and Russia. By 1992 the species had spread to the Baltic Sea, into the Gulf of Bothnia in Finland, and also to coastal lakes and reservoirs in Bulgaria. It was found in Lake Ontario in 1998 and had spread to half a dozen of the Finger Lakes in NY and into both ends of Lake Michigan a year later. It has since been found in Lake Erie, OH, and likely will cross into the Mississippi River drainage basin.

Other Sources of Information: 32, 62, 70
References: 20, 224, 261
Author: Stacey Solie

THREATS Freshwater Vertebrate

Northern Snakehead *Channa argus*

Species Description and Current Range

Channidae is a family of air-breathing, predatory fishes ranging in size from 17 cm to 1.5 m and weighing up to 7 kg. Snakeheads get their name from their elongated cylindrical shape, enlarged scales on the head, and eyes located on each side near the top front of the head. A large mouth with a protruding lower jaw and canine-like teeth are also characteristic. Dorsal and anal fins are elongated. Pelvic fins are close to pectoral fins and gills. Coloration and size vary between species and individuals, but the Northern Snakehead is generally tan with dark brown mottling. The only similar-appearing species is the native Bowfin. The snakehead family is noted for its ability to survive in freshwater lakes, swamps, ponds, streams, and rivers under adverse conditions, but it cannot live in salt water. Northern Snakeheads tolerate temperatures from 0°C to greater than 30°C. Snakeheads are not currently in the PNW, though the Northern Snakehead could easily establish in this region. Six adults and over 1,300 juvenile Northern Snakeheads were discovered in Crofton, MD, in 2002. The pond was treated with rotenone and all fish removed, but 19 more fish of the same species have been caught in the Potomac River basin since April 2004, indicating an established population. The Northern Snakehead is popularly called "Frankenfish."

Impact on Communities and Native Species

Snakeheads have a voracious appetite, and at all stages of growth they compete with other fish species for food. Adult snakeheads prey on fish, crustaceans, frogs, and small reptiles, birds, and mammals, including native, endangered, and threatened species such as shad. At risk are 16 species of amphibians, 115 species of fish, and 5 species of crustaceans. Snakeheads' predilection for other fish drastically reduces valuable populations of native, recreational, and food fish species. Rearrangement of both food chains and the ecological balance of freshwater habitats is a worrisome consequence of established snakehead populations. Their high reproductive rate, protection of young, survival capabilities, and maneuverability make snakeheads a definite threat to native species of fish. Snakeheads also carry epizootic ulcerative syndrome and parasites, such as the helminth parasite *Gnathostoma*, which affect humans as well as fish.

Control Methods and Management

Once a population of snakeheads has been established, options are limited. Physical removal by electroshock is not guaranteed to remove all the fish. Chemical removal by rotenone is effective but kills everything in the treatment area. Rotenone must be neutralized by potassium permanganate, but it can persist in surface water for 7 weeks and in cold water for longer than 9 months. Rotenone treatment is also expensive. Treatment of 5,000 acre-feet requires 3,333 gallons of liquid rotenone at a cost of $200,000. Monitoring, chemical application, fish removal, and restocking in an area of that size cost nearly a million dollars. The best way of preventing the spread of snakeheads is not to release overgrown aquarium fish into the wild. If you catch a snakehead, do not release it. Instead, kill it and put it on ice, and notify your state authorities immediately.

Life History and Species Overview

Other fish make up 64–70% of the juvenile diet and 90% of the adult snakehead diet. The snakehead feeds on species up to 33% of its body length. When food is scarce, snakeheads can cannibalize their own young. Snakeheads become capable of aerial breathing at the juvenile stage. They are capable of overland migration and can survive for a considerable length of time out of water under moist conditions; the fish secrete mucus to prevent drying and to facilitate breathing. Northern Snakeheads are less adept due to their rounded bodies and can only move when water is present. Age is not easily determined but is roughly proportional to length. The longest Northern Snakehead reported, 1.5 m, indicates a long-lived species. Most snakeheads become sexually mature within 2 years, and after this point fecundity increases with size. Fecundity varies, but Northern Snakeheads can produce 50,000 eggs/spawning. Spawning occurs mainly in June–August and individuals may spawn up to 5 times/yr. Hatching time depends on water temperature but is within the range of 28–120/hr. Parental care of offspring is characteristic; snakeheads attack and in some cases are reported to have killed humans who ap-proached the young. Adults guard the fry until the early juvenile stage; once juveniles reach 4 cm in length, they begin to feed on fishes.

History of Invasiveness

Snakeheads are native to southeastern Asia, ranging from southeastern Iran and Afghanistan to Malaysia, Korea, China, and Siberia. First introduced to the US in CA in 1997, 5 species of snakehead have been caught in the open waters of 9 states in the US. Three species established reproducing populations, one was cultured in AR until prohibition in 2002, and another is cultured in HI. The live-food fish trade was a common mode of introduction; snakeheads are favored food fish in India and Asia and were imported to the US for sale in live-fish markets, increasing the risk of the fish escaping and becoming established. The aquarium fish trade of snakeheads is especially important in Japan and Europe and was to a lesser extent in the US until its prohibition in 2002. Because snakeheads require expensive live food and outgrow their aquariums, they are frequently released into nearby ponds where they consume native populations of fish.

Snakeheads were listed by the US Fish and Wildlife Service as injurious wildlife under the Lacey Act in October 2002, and importation and interstate transport of live snakeheads, either for food markets or aquaria, is now prohibited in the US. Currently, at least 23 states in the US prohibit possession of live snakeheads but many may remain in aquaria around the country, and there have been several violations of the regulations. Several species have been marketed in Canada, and the importation and sale of snakeheads is still legal in BC. An April 2001 shipment of live ling cod from a Canadian fish wholesaler, destined for a Seattle seafood distributor, included three open boxes of Northern Snakeheads from China that had been shipped to Canada without water. Upon inspection in Blaine, WA, most of the snakeheads were alive and capable of movement. Despite the shipment driver's attempts to kill them with a board, the fish remained alive. Finally, they were seized by the Department of Fish and Wildlife and placed in a freezer. Even after 30 minutes of freezing, most were still alive.

Other Sources of Information: 53, 112
References: 24, 48, 73, 93
Author: Diana Thayer

Wood Wasp *Sirex noctilio*

Species Description and Current Range

Adult Wood Wasps are metallic blue with four transparent yellow wings and can be 9–36 mm long. Females have reddish-brown legs, and a spikelike projection that points backward from under their abdomen. Males have a wide orange band on their abdomen and thickened hind legs that are dark brown to black in color. Wood Wasps don't look like most wasps because they lack the characteristic wasp waist (constricted abdomen). In the larval stage, the Wood Wasp looks like a cylindrical yellowish-white grub with a round head, no legs, and a black spine at the rear of the abdomen. Native to the pine forests of Europe, Asia, and northern Africa, Wood Wasps reach their greatest density in Mediterranean areas. No established populations are known within the US.

Impact on Communities and Native Species

Although considered to be a minor pest in its native habitat, this wasp is highly destructive in new environments. In Australia, the Wood Wasp killed more than 5 million pine trees in commercial forests in 3 years. As US markets open to more international imports, the potential for an accidental introduction of a wood-boring insect, like the Wood Wasp, increases. Raw logs are one pathway of introduction. Larvae can even survive in logs that have been air-dried or kiln-dried at low temperatures. Green untreated lumber can also be a pathway. Wood chips are unlikely to carry wood-borers because the chipping process kills the developing larvae and makes the wood unsuitable for further habitation. Because the PNW does not import raw logs, the most likely way a Wood Wasp would enter this region is in wooden packing material. Approximately 52% of the maritime cargo and 9% of air cargo that enter the US include some sort of wood-packing material (e.g., pallets, crates, bracing). For example, Asian Longhorned Beetles entered the US in or on crates used to ship heavy machinery from Asia. To minimize costs, shippers use any wood available, including wood salvaged from stands of unhealthy or dead trees, which may be infested with wood-boring insects. Since port inspectors are able to look at only a small percentage of the cargo in a shipment, many insects escape detection at the port of entry. In addition, inspections are difficult because wooden packing materials are rarely identified on shipping documents, and cargo containers, which are increasingly being used, hide wood packing materials from view. Even manufactured products imported into the US, such as artificial Christmas trees made with real wood poles, have been infested with Longhorned Beetles.

A pest risk assessment, conducted to evaluate the potential economic impact of an infestation of Wood Wasps on Atlanta, GA, Minneapolis, MN, and San Francisco, CA, predicted Atlanta would suffer the greatest damage, ranging from $48 to $607 million. The monetary loss in timber production around San Francisco over 30 years was estimated to be between $7 and $77 million, depending on the insects' rate of spread.

A tree may remain healthy after a Wood Wasp attack by flooding the holes containing wasps' eggs with resin, or by producing substances in the stem that prevent growth of a fungus that the wasp introduces when it lays its eggs. Even if an attack is unsuccessful, the timber is degraded because of the accumulation of resin and dead wood around the entry hole. Sometimes only one section of the trunk is affected, resulting in a strip of dead wood in an otherwise healthy tree.

Control Methods and Management

Several forms of biological control are effective. The most common uses microscopic worms called nematodes. These parasitic worms sterilize the adult female wasp by entering her reproductive organs and penetrating her eggs. These sterile eggs are dispersed when she lays them in the surrounding trees. But nematodes don't stop the fungus that the wasp introduces to feed its larvae. Another form of biological control uses parasitic wasps, which drill through the wood until they encounter a Wood Wasp egg or larva. The larva is paralyzed with a sting, and then an egg is laid on it. When the egg of the parasitic wasp hatches, it then feeds on the Wood Wasp larva.

Life History and Species Overview

Wood Wasps are solitary. When an adult emerges from a tree, its fully developed reproductive system is ready to deposit eggs. It is attracted to the volatile substances produced in the stems of weak and dying trees. It prefers to lay its eggs in pines, but it may also attack spruce, larch, and firs. Once it locates a tree, the wasp drills a hole and deposits a creamy-white, sausage-shaped egg, about 1.35–1.56 mm long. Typically one egg is laid per hole. Females deposit 20–500 eggs, depending on their size. Only mated females produce female offspring, so males outnumber females 20:1. When depositing an egg into a tree, the wasp also injects a toxic mucus and spores of a specific fungus that it carries in pocket-like organs on each side of its body. The mucus causes the tree's needles to wilt and turn from light green to yellow. This weakens the tree, creating a better environment for the fungus to spread. As the fungus grows, it disrupts the movement of water in the xylem, drying out the wood and creating a favorable environment for the eggs to hatch. The fungus and mucus act together to kill the tree. Once the eggs hatch, the larvae tunnel through the wood, feeding on the fungus.

Their life cycle ranges from 10 months to 2 years in cooler climates. In its adult stage, the Wood Wasp can live up to 12 days, although a female who has deposited all her eggs may live as little as 3–4 days. Adults are capable of flying several km in search of suitable trees for breeding sites.

History of Invasiveness

The Wood Wasp first became established outside its native environment in the 1950s when it was introduced to New Zealand. By the 1960s, it had infected pine trees in Australia, in spite of strict quarantine measures. It was first reported in South America in the 1980s, in Uruguay and Argentina. By the late 1980s, it had migrated to southern Brazil, where it is now widely distributed. By 1994, Wood Wasps were detected in South Africa, where they were impacting pine forests, spreading at a rate of 48 km/yr.

References: 144, 294, 318, 335, 362
Author: Darren A. Linker

Zebra Mussel *Dreissena polymorpha*

Species Description and Current Range

Zebra Mussels are freshwater bivalves in the family Dreissenidae. At present, Zebra Mussels have invaded North America only east of the Continental Divide. Once they were introduced into the Laurentian Great Lakes in the mid-1980s, they spread throughout all of the Great Lakes and connected waterways including the St. Lawrence, Hudson, and Mississippi rivers. They have also spread to inland lakes through the activity of recreational boaters; however, this expansion proceeded much more slowly than through connected waters. Zebra Mussels are aggressive invaders worldwide and are continuing to spread throughout Europe and North America.

Zebra Mussels are the only freshwater animal of any size to attach to hard substrates. Like marine mussels, they have a hard two-part shell and attach to hard substrates, including each other, with strong threads (byssus). This tendency allows them to form clumps, or druses, that are able to live on sandy and muddy areas of lake bottoms. The coloring of their shells is quite variable, but often has stripes or zigzag patterns. Zebra Mussels live in lakes, reservoirs, canals, and rivers. They are much less common in streams. The Zebra Mussels that have invaded North America are intolerant of salt water, and therefore their distribution within estuaries is limited to freshwater reaches. Zebra Mussels are the only freshwater mollusks that have a dispersal phase in plankton.

Impact on Communities and Native Species

Zebra Mussels are ecosystem engineers; they cause dramatic changes to communities and ecosystem functions. They impact other species by attaching to surfaces (including the shells of native mussels and clams), creating three-dimensional spaces, and filtering water. Like most mussels, they feed by filtering fine particles, including plankton, from the water

column. They are one of the most efficient filterers, and can remove almost all of the particles from water they filter. This can have large impacts on the plankton. Zebra Mussels can provide habitat for some species that are not normally part of the community, changing the composition of the community of bottom-dwelling organisms.

Zebra Mussels are estimated to cost the US economy over $100 million/yr. They clog municipal water systems, including turbines and cooling systems for power plants. They attach to boat hulls, locks, docks, and all other human-made structures.

Control Methods and Management

Once Zebra Mussels invade, they are virtually impossible to control. Humans are the transporters of Zebra Mussels, introducing Zebra Mussels to freshwater ports via shipping and dumping of ballast water. Because Zebra Mussels invade lakes through recreational boating, their spread can be controlled by requiring boats to be washed before allowing access to lakes. This removes any attached Zebra Mussels, as well as Waterweeds, which often have attached mussels.

Life History and Species Overview

Zebra Mussels are relatively small (0.3–3 cm) and have thin shells. They have separate sexes, which are genetically determined, and free spawn; both males and females shed their gametes into the water column. Spawning occurs once water temperatures warm to 12°C, and a single female can release up to 10^6 eggs and males $>10^8$ sperm. Larvae are in the plankton for 5–30 days, depending on water temperature, and can be carried by water currents. Then they move to the bottom and attach to hard substrates and generally settle in very high densities that can cover all available space in a lake. During warm weather Zebra Mussels can grow very rapidly. Mussels that settle in early July are reproductive by September. In many North American lakes Zebra Mussels probably live 3–5 years.

Zebra Mussels feed by filtering particles from the water column. Organic particles, such as microalgae, are consumed. Other particles that are not ingested are deposited on the bottom. Zebra Mussels filter large volumes of water and, because of their high densities, they can have large impacts on the water column, clearing the water and allowing more light to penetrate. They also create habitat for other species of benthic animals.

Zebra Mussels live in many different types of lakes, rivers, and canals but are less abundant in streams. They do well in hard-water lakes in most temperate regions with sufficient calcium and mild pH, but are not tolerant of high levels of pollution, acidification, low oxygen, or very hot temperatures. Although Zebra Mussels are edible, they are small and concentrate toxins such as metals and PCBs, so are usually not eaten by humans. They are eaten by diving ducks, crabs, crayfish, and some fish that can eat hard-shelled benthic prey. In North America, 15 species of fish have been found to feed on Zebra Mussels, including sturgeon, drum, carp, and another invader, the Round Goby. Another 15 are predicted to become predators of Zebra Mussels based on their natural diets. However, even in Europe, predators do not limit Zebra Mussel populations. Zebra Mussels settle in very high densities, up to 50,000/m², and due to their rapid growth can clog an intake pipe or overgrow the shells of other animals, such as crayfish, in a matter of weeks to months.

History of Invasiveness

Zebra Mussels are native to the freshwater and brackish regions of the Caspian and Black seas. Over 200 years ago they spread across Europe through canals built to connect the Caspian and Black seas with the Baltic. Since that time they have spread to connected and isolated waters across eastern and western Europe, and continue to spread to this day. In the mid-1980s Zebra Mussels invaded North America. They were first introduced to the shipping ports in the Laurentian Great Lakes and rapidly spread to all water bodies connected to the Great Lakes. They have continued to spread to inland and isolated lakes, mainly through transport on recreational boats and trailers, especially on Waterweeds that become attached to trailers.

Other Sources of Information: 58, 85
References: 40, 175, 176, 177, 259
Author: Dianna K. Padilla

Appendixes

Appendix 1. IUCN's 100 of the World's Worst Invasive Alien Species

This list includes 100 of the world's worst invasive species, as identified by the World Conservation Union's Invasive Species Specialist Group. These species are known to have serious impacts on biological diversity or human activities. Highlighted species are those found in the PNW.

Appendix 2. Habitat Associations of PNW Invasive Species

As part of the ranking process, authors identified the major and minor habitat types in which a particular invasive species is found. The habitat types we used derive from a variety of sources and are intended to reflect the full diversity of terrestrial, freshwater, and marine habitats found in the PNW. Items in boldface are major habitat types; minor habitat types are listed following each major habitat type.

Appendix 3. Vectors of Spread for PNW Invasive Species

As part of the ranking process, authors identified the primary vectors that facilitate the spread of a particular invasive species. Although the vectors listed derive from information about the individual invasive species included in the book, they are representative of vectors described for invasive species in general.

Appendix 4. The Questions

Each species considered for inclusion in this book was evaluated using a set of invasion threat/impact criteria in a series of 47 questions. Using a combination of published information, unpublished data, and expert opinion, we answered each of the questions for each species. Because of the taxonomic breadth of our species list, answers represent conditions or characteristics relevant to the particular taxon being evaluated.

Appendix 5. Answers to the Questions and Invasive Scores

For each of the invasive species, answers to the questions were used to determine an invasive score that reflects how big of a problem the invasive species is or would likely become. The highest score (A–D, U) within a section (1–3) was used as the overall score for that section. The three section scores were then entered into the scoring matrix to determine a species' invasive threat category (High, High Alert, Medium, etc.).

Appendix 6. Code of Conduct

Appendix 1. IUCN's 100 of the World's Worst Invasive Alien Species

Aquatic Plants

Caulerpa Seaweed	*Caulerpa taxifolia*
* Common Cordgrass	*Spartina angelica*
Wakame Seaweed	*Undaria pinnatifida*
Water Hyacinth	*Eichhornia crassipes*

Aquatic Invertebrates

Chinese Mitten Crab	*Eriocheir sinensis*
Comb Jelly	*Mnemiopsis leidyi*
European Green Crab	*Carcinus maenas*
Fish Hook Water Flea	*Cercopagis pengoi*
Golden Apple Snail	*Pamacea canaliculata*
Marine Clam	*Potamocorbula amurensis*
Mediterranean Mussel	*Mytilus galloprovincialis*
Northern Pacific Seastar	*Asterias amurensis*
Zebra Mussel	*Dreissena polymorpha*

Fish

Brown Trout	*Salmo trutta*
Common Carp	*Cyprinus carpio*
Largemouth Bass	*Micropterus salmoides*
Mozambique Tilapia	*Oreochromis mossambicus*
Nile Perch	*Lates niloticus*
Rainbow Trout	*Oncorhynchus mykiss*
Walking Catfish	*Clarias batrachus*
Western Mosquito Fish	*Gambusia affinis*

Amphibians and Reptiles

American Bullfrog	*Rana catesbeiana*
Brown Tree Snake	*Boiga irregularis*
Cane Toad	*Bufo marinus*
Caribbean Tree Frog	*Eleutherodactylus coqui*
Red-eared Slider	*Trachemys scripta*

Land Plants

African Tulip Tree	*Spathodea campanulata*
Black Wattle	*Acacia mearnsii*
Brazilian Pepper Tree	*Schinus terebinthifolius*
Cogon Grass	*Imperata cylindrica*
Cluster Pine	*Pinus pinaster*
Erect Pricklypear	*Opuntia stricta*
Fire Tree	*Myrica faya*
Giant Reed	*Arundo donax*
Gorse	*Ulex europaeus*
Hiptage	*Hiptage benghalensis*
Japanese Knotweed	*Fallopia japonica*
Kahili Ginger	*Hedychium gardnerianum*
Koster's Curse	*Clidemia hirta*
Kudzu	*Pueraria montana* var. *lobata*
Lantana	*Lantana camara*
Leafy Spurge	*Euphorbia esula*
Leucaena	*Leucaena leucocephala*
Melaleuca	*Melaleuca quinquenervia*
Mesquite	*Prosopis glandulosa*
Miconia	*Miconia calvescens*
Mile-a-minute Weed	*Mikania micrantha*
Mimosa	*Mimosa pigra*
Privet	*Ligustrum robustum*
Pumpwood	*Cecropia peltata*
Purple Loosestrife	*Lythrum salicaria*
Quinine Tree	*Cinchona pubescens*
Saltcedar (Tamarisk)	*Tamarisk ramosissima*
Shoebutton Ardisia	*Ardisia elliptica*
Siam Weed	*Chromolaena odorata*
Strawberry Guava	*Psidium cattleianum*
Wedelia	*Sphagneticola trilobata*
Yellow Himalayan Raspberry	*Rubus ellipticus*

Birds	
European Starling	*Sturnus vulgaris*
Indian Myna Bird	*Acridotheres tristis*
Red-vented Bulbul	*Pycnonotus cafer*

Mammals	
Black Rat	*Rattus rattus*
Brushtail Possum	*Trichosurus vulpecula*
Domestic Cat	*Felis catus*
Domestic Pig	*Sus scrofa*
Eastern Gray Squirrel	*Sciurus carolinensis*
European Rabbit	*Oryctolagus cuniculus*
Goat	*Capra hircus*
Macaque Monkey	*Macaca fascicularis*
Mouse	*Mus musculus*
Nutria	*Myocaster coypus*
Red Deer	*Cervus elaphus*
Red Fox	*Vulpes vulpes*
Small Indian Mongoose	*Herpestes javanicus*
Stoat	*Mustela erminea*

Microorganisms	
Avian Malaria	*Plasmodium relictum*
Banana Bunchy Top Virus	*Banana bunchy top virus*
Rinderpest Virus	*Rinderpest virus*

Land Invertebrates	
Argentine Ant	*Linepithema humile*
Asian Longhorned Beetle	*Anoplophora glabripennis*
Asian Tiger Mosquito	*Aedes albopictus*
Big-headed Ant	*Pheidole megacephala*
Common Malaria Mosquito	*Anopheles quadrimaculatus*
Common Wasp	*Vespula vulgaris*
Crazy Ant	*Anoplolepis gracilipes*
Cypress Aphid	*Cinara cupressi*
Flatworm	*Platydemus manokwari*
Formosan Subterranean Termite	*Coptotermes formosanus shiraki*
Giant African Snail	*Achatina fulica*
Gypsy Moth	*Lymantria dispar*
Khapra Beetle	*Trogoderma granarium*
Little Fire Ant	*Wasmannia auropunctata*
Red Imported Fire Ant	*Solenopsis invicta*
Rosy Wolf Snail	*Euglandina rosea*
Sweet Potato Whitefly	*Bemisia tabaci*

Macrofungi	
Chestnut Blight	*Cryphonectria parasitica*
Crayfish Plague	*Aphanomyces astaci*
Dutch Elm Disease	*Ophiostoma ulmi*
Frog Chytrid Fungus	*Batrachochytrium dendrobatidis*
Phytophthora Root Rot	*Phytophthora cinnamomi*

* Highlighted species are in the PNW.

Modified from: Lowe et al. 2004. *Global Invasive Species Database:* www.issg.org/database
Lowe et al. (2004) selected species by two criteria: their serious impacts on biological diversity, and on human activities.

Appendix 2. Habitat Associations of PNW Invasive Species

	Stream or River	Alpine	Low velocity	High velocity	Low order (1 – 3)	Medium order (4 – 6)	High order (6 and above)	Lowland	Montane	**Wetland**	Bog	Fen	Floodplain	Lowland	Marsh	Slough	Swamp	**Lake**	Alpine
African Waterweed	**x**																	**x**	
American Bullfrog	**x**		x					x		**x**				x	x	x	x	**x**	
American Shad	**x**																		
Ascidians																			
Asian Freshwater Clam	**x**																	**x**	
Asian/European Gypsy Moth																			
Atlantic Salmon	**x**																	**x**	
Atlantic Slipper Limpet																			
Balsam Wooly Adelgid																			
Bamboo																			
Bass-sunfish	**x**																	**x**	
Blackberries	**x**									**x**								**x**	
Brazilian Elodea	**x**		x					x										**x**	
Broom & Gorse																			
Butterfly Bush																			
Carp, Black	**x**				x														
Carp, Common	**x**																	**x**	
Carp, Bighead	**x**				x														
Carp, Grass																		**x**	
Carp, Silver	**x**				x														
Catfish	**x**																	**x**	
Cheatgrass																			
Chinese Mitten Crab																			
Cirsium Thistles																			
Common Reed										**x**									
Common Velvet-grass										**x**									
Cordgrasses																			
Crayfish										**x**				x				**x**	
Dalmatian Toadflax																			
Dead Man's Fingers																			
Dogwood Anthracnose	**x**																		
Domestic Cat																			
Domestic Pig																			
Earthworms																			
English Holly																			
English Ivy										**x**				x					
Eurasian Watermilfoil	**x**																	**x**	
European Beachgrass																			

Freshwater	Lowland	Montane	Saline	**Canal**	**Pond**	Alpine	Freshwater	Lowland	Montane	Permanent	Seasonal	**Reservoir**	**Forests**	Coastal Rainforest	Dry Montane Forests	Lowland Moist Forest	Mixed-conifer forests	Subalpine Forests	**Grasslands (lowland)**	Coastal Grasslands	Glacial Outwash Prairies	Palouse grassland	Willamette Valley Prairies	**Woodlands**	Juniper	Oak
					x																					
x	x				**x**		x	x		x																
												x														
				x	**x**							**x**														
													x				x									x
					x																					
													x	x	x	x		x								
													x											**x**		
												x														
					x														**x**					**x**		
x	x			**x**	**x**		x	x				**x**														
																			x							
	x				**x**																					
	x			**x**	**x**		x	x																		
												x														
																			x							
																			x							
																			x							
													x						**x**				x			
	x				**x**			x																		
													x		x				**x**			x		**x**		
													x													
													x						**x**					**x**		
													x						**x**					**x**		
													x						**x**					**x**		
													x	x		x	x									
													x			x										
					x							**x**														

	Ponderosa Pine	**Riparian**	Alder Flats	Alpine Riparian	Forested Riparian	Grassland Riparian	Shrub-Steppe Riparian	**Semi-Natural Areas**	Rural / managed pasture	Rural Disturbed Areas	Urb./Suburb. Natural Areas	**Shrublands**	Bitterbrush Shrub-Steppe	Chaparral	Desert Shrub	Sagebrush Shrub-Steppe	**Special Minor Types**	Coastal Dunes	Interior Saline
African Waterweed																			
American Bullfrog																			
American Shad																			
Ascidians																			
Asian Freshwater Clam																			
Asian/European Gypsy Moth																			
Atlantic Salmon																			
Atlantic Slipper Limpet																			
Balsam Wooly Adelgid																			
Bamboo		x						x											
Bass-sunfish																			
Blackberries		x						x											
Brazilian Elodea																			
Broom & Gorse								x		x	x								
Butterfly Bush								x		x									
Carp, Black																			
Carp, Common																			
Carp, Bighead																			
Carp, Grass																			
Carp, Silver																			
Catfish																			
Cheatgrass								x		x		x				x			
Chinese Mitten Crab																			
Cirsium Thistles								x	x	x	x								
Common Reed		x						x											
Common Velvet-grass		x																	
Cordgrasses																			
Crayfish																			
Dalmatian Toadflax	x	x			x	x	x	x	x	x		x	x			x			
Dead Man's Fingers																			
Dogwood Anthracnose		x						x											
Domestic Cat		x						x	x	x	x	x							
Domestic Pig		x						x				x					x		
Earthworms		x						x	x	x	x	x					x		
English Holly		x			x			x			x								
English Ivy		x			x			x			x								
Eurasian Watermilfoil																			
European Beachgrass																		x	

Vernal Pools	**Non-Forested Alpine/Subalpine**	Lush Meadows	Talus Slopes	**Estuarine**	Artificial structures	Intertidal rock/ hard substrate	Pebbles, sand and mud	Lagoons	Saline marshes	Reefs	Subtidal rock/ hard substrate	Subtidal pebbles, sand, mud	**Intertidal**	Artificial structures	Intertidal mud	Intertidal Reef	Intertidal sand/gravel	Wave-exposed rocks	Wave-protected rocks	**Subtidal**	Docks, floats, artificial substrate	Subtidal reef	Subtidal rock/boulders	**Pelagic**
				x																				
													x	x					x	x	x			
				x					x															
				x																x				x
				x																				
	x																							
				x									x							x				
				x					x															
	x	x																						
				x			x	x	x				x		x		x							
				x		x					x		x						x	x			x	
	x																							
	x																							

	Stream or River	Alpine	Low velocity	High velocity	Low order (1 – 3)	Medium order (4 – 6)	High order (6 and above)	Lowland	Montane	**Wetland**	Bog	Fen	Floodplain	Lowland	Marsh	Slough	Swamp	**Lake**	Alpine
European Fruit Fly																			
European Green Crab																			
European House Sparrow																			
European Rabbit																			
European Starling																			
European Yellowjacket																			
Fanwort	**x**		x															**x**	
Fennel																			
Fishhook Water Flea																		**x**	
Garlic Mustard										**x**									
Giant Hogweed																			
Hedgehog Dogtail																			
Herb Robert																			
Houndstongue																			
Hydrilla	**x**																	**x**	
Japanese Eelgrass																			
Japanese Mussel																			
Knapweeds																			
Knapweed, Spotted																			
Knotweeds																			
Kudzu																			
Ladybeetles																			
Leafy Spurge																			
Longhorned Beetles																			
Marine Clams																			
Mediterranean Mussel																			
Medusahead																			
Mosquitoes										**x**	x	x	x		x	x	x	**x**	
Mountain Goat																			
Mudsnails																			
New Zealand Isopod																			
New Zealand Mudsnail	**x**																	**x**	
Newcastle Disease	**x**																	**x**	
Northern Snakehead	**x**				x	x				**x**			x		x	x	x	**x**	
Nutria	**x**									**x**								**x**	
Orchard Grass																			
Oyster Drills																			
Pacific Oyster																			

Freshwater	Lowland	Montane	Saline	**Canal**	**Pond**	Alpine	Freshwater	Lowland	Montane	Permanent	Seasonal	**Reservoir**	**Forests**	Coastal Rainforest	Dry Montane Forests	Lowland Moist Forest	Mixed-conifer forests	Subalpine Forests	**Grasslands (lowland)**	Coastal Grasslands	Glacial Outwash Prairies	Palouse grassland	Willamette Valley Prairies	**Woodlands**	Juniper	Oak
													x											**x**		
																								x		
																			x	x				**x**		x
																			x							
													x											**x**		
x					**x**		x																			
x			x																							
													x													
													x	x		x										
																			x					**x**		
													x			x										
				x	**x**							**x**														
													x		x		x		**x**					**x**		
													x		x		x		**x**					**x**		
													x													
																								x		
																								x		
													x											**x**		
																			x			x		**x**		x
x					**x**		x			x	x		**x**	x												
					x							**x**														
x				**x**	**x**		x			x	x	**x**														
					x																					
													x	x	x				**x**					**x**	x	x

	Ponderosa Pine	**Riparian**	Alder Flats	Alpine Riparian	Forested Riparian	Grassland Riparian	Shrub-Steppe Riparian	**Semi-Natural Areas**	Rural / managed pasture	Rural Disturbed Areas	Urb./Suburb. Natural Areas	**Shrublands**	Bitterbrush Shrub-Steppe	Chaparral	Desert Shrub	Sagebrush Shrub-Steppe	**Special Minor Types**	Coastal Dunes	Interior Saline
European Fruit Fly		**x**						**x**											
European Green Crab																			
European House Sparrow								**x**	x	x	x								
European Rabbit								**x**			x								
European Starling								**x**											
European Yellowjacket								**x**	x	x	x	**x**	x						
Fanwort																			
Fennel								**x**											
Fishhook Water Flea																			
Garlic Mustard								**x**											
Giant Hogweed		**x**			x			**x**			x								
Hedgehog Dogtail												**x**							
Herb Robert		**x**			x			**x**		x	x								
Houndstongue										x									
Hydrilla																			
Japanese Eelgrass																			
Japanese Mussel																			
Knapweeds		**x**						**x**	x	x		**x**					**x**	x	
Knapweed, Spotted		**x**				x		**x**	x	x		**x**					**x**	x	
Knotweeds		**x**						**x**		x	x								
Kudzu								**x**	x	x									
Ladybeetles								**x**	x	x	x								
Leafy Spurge								**x**		x									
Longhorned Beetles								**x**											
Marine Clams																			
Mediterranean Mussel																			
Medusahead	x											**x**	x	x		x			
Mosquitoes								**x**		x	x								
Mountain Goat																			
Mudsnails																			
New Zealand Isopod																			
New Zealand Mudsnail																			
Newcastle Disease																			
Northern Snakehead																			
Nutria																			
Orchard Grass	x											**x**				x			
Oyster Drills																			
Pacific Oyster																			

Vernal Pools	**Non-Forested Alpine/Subalpine**	Lush Meadows	Talus Slopes	**Estuarine**	Artificial structures	Intertidal rock/ hard substrate	Pebbles, sand and mud	Lagoons	Saline marshes	Reefs	Subtidal rock/ hard substrate	Subtidal pebbles, sand, mud	**Intertidal**	Artificial structures	Intertidal mud	Intertidal Reef	Intertidal sand/gravel	Wave-exposed rocks	Wave-protected rocks	**Subtidal**	Docks, floats, artificial substrate	Subtidal reef	Subtidal rock/boulders	**Pelagic**
				x	x	x		x	x				**x**	x					x					
				x									**x**											
				x	x		x	x	x			x												
													x		x		x		x					
				x	x			x					**x**	x				x	x					
	x		x																					
				x	x		x	x	x				**x**	x	x		x							
				x	x	x	x		x				**x**	x	x			x	x	**x**	x			
									x															
				x					x															
				x			x	x		x			**x**		x	x	x		x					
				x	x	x	x			x			**x**	x	x	x			x					

	Stream or River	Alpine	Low velocity	High velocity	Low order (1 – 3)	Medium order (4 – 6)	High order (6 and above)	Lowland	Montane	**Wetland**	Bog	Fen	Floodplain	Lowland	Marsh	Slough	Swamp	**Lake**	Alpine
Perennial Pepperweed										**x**									
Pike	**x**				x	x	x			**x**					x			**x**	
Port Orford Cedar Root Rot																			
Purple Loosestrife										**x**	x	x	x	x	x	x			
Rats																			
Reed Canarygrass	**x**						x	x	x	**x**			x	x		x		**x**	
Ring-necked Pheasant																			
Russian Olive	**x**							x		**x**			x						
Saltcedar	**x**									**x**								**x**	
Sargassum																			
Siberian Prawn	**x**		x		x			x										**x**	
Slender False-brome																			
Spiders																			
Spurge Laurel																			
Squirrel, Eastern Gray																			
Squirrel, Fox										**x**							x		
St. John's Wort																			
Sudden Oak Death																			
Sunfish	**x**			x														**x**	
Swollen Bladderwort																		**x**	
Tansy Ragwort																			
Temperate Bass, Striped	**x**																	**x**	
Temperate Bass, White	**x**																	**x**	
Terrestrial Mollusks																			
Thistles																			
Traveler's Joy																			
Tree-of-heaven																			
Trout	**x**	x						x	x									**x**	x
Turtles	**x**									**x**								**x**	
Water Primrose																		**x**	
West Nile Virus	**x**									**x**								**x**	
White Pine Blister Rust																			
Whitetop																			
Wood Wasp																			
Yellow Flag Iris										**x**								**x**	
Yellow Floatingheart																		**x**	
Yellow Perch	**x**																	**x**	
Zebra Mussel	**x**																	**x**	

Freshwater	Lowland	Montane	Saline	**Canal**	**Pond**	Alpine	Freshwater	Lowland	Montane	Permanent	Seasonal	**Reservoir**	**Forests**	Coastal Rainforest	Dry Montane Forests	Lowland Moist Forest	Mixed-conifer fcrests	Subalpine Forests	**Grasslands (lowland)**	Coastal Grasslands	Glacial Outwash Prairies	Palouse grassland	Willamette Valley Prairies	**Woodlands**	Juniper	Oak
																			x							
x																										
													x											x		
x	x			x	x		x	x				x							x				x			
																			x	x	x	x	x			
				x									x													
				x	x							x														
x	x				x		x	x		x																
													x											x		
													x			x								x		x
													x			x								x		x
													x											x		x
													x											x		x
																			x							
													x				x							x		x
					x							x														
x					x		x																			
													x	x	x	x	x		x	x	x		x	x		x
												x														
												x														
													x											x		
																			x							
													x		x											
													x													
	x	x			x	x		x	x	x																
					x																					
x					x		x																			
				x	x							x														
													x		x		x	x								
													x											x		
					x																					
					x							x														
					x																					
				x	x																					

	Ponderosa Pine	**Riparian**	Alder Flats	Alpine Riparian	Forested Riparian	Grassland Riparian	Shrub-Steppe Riparian	**Semi-Natural Areas**	Rural / managed pasture	Rural Disturbed Areas	Urb./Suburb. Natural Areas	**Shrublands**	Bitterbrush Shrub-Steppe	Chaparral	Desert Shrub	Sagebrush Shrub-Steppe	**Special Minor Types**	Coastal Dunes	Interior Saline
Perennial Pepperweed		**x**						**x**				**x**							
Pike																			
Port Orford Cedar Root Rot								**x**			x								
Purple Loosestrife																			
Rats								**x**		x	x								
Reed Canarygrass		**x**	x		x	x		**x**	x	x	x								
Ring-necked Pheasant		**x**	x	x		x	x	**x**	x	x	x	**x**	x	x	x	x			
Russian Olive		**x**			x			**x**	x	x	x								
Saltcedar		**x**															**x**		x
Sargassum																			
Siberian Prawn																			
Slender False-brome		**x**						**x**											
Spiders								**x**	x		x								
Spurge Laurel		**x**			x			**x**		x	x								
Squirrel, Eastern Gray																			
Squirrel, Fox																			
St. John's Wort		**x**						**x**				**x**							
Sudden Oak Death																			
Sunfish																			
Swollen Bladderwort																			
Tansy Ragwort	x	**x**	x		x	x		**x**	x	x	x						**x**	x	
Temperate Bass, Striped																			
Temperate Bass, White																			
Terrestrial Mollusks		**x**						**x**											
Thistles		**x**			x	x	x	**x**	x	x	x	**x**							
Traveler's Joy		**x**						**x**											
Tree-of-heaven		**x**						**x**	x	x	x								
Trout																			
Turtles																			
Water Primrose																			
West Nile Virus																			
White Pine Blister Rust																			
Whitetop		**x**				x	x	**x**	x	x									
Wood Wasp																			
Yellow Flag Iris		**x**																	
Yellow Floatingheart																			
Yellow Perch																			
Zebra Mussel																			

Vernal Pools	**Non-Forested Alpine/Subalpine**	Lush Meadows	Talus Slopes	**Estuarine**	Artificial structures	Intertidal rock/ hard substrate	Pebbles, sand and mud	Lagoons	Saline marshes	Reefs	Subtidal rock/ hard substrate	Subtidal pebbles, sand, mud	**Intertidal**	Artificial structures	Intertidal mud	Intertidal Reef	Intertidal sand/gravel	Wave-exposed rocks	Wave-protected rocks	**Subtidal**	Docks, floats, artificial substrate	Subtidal reef	Subtidal rock/boulders	**Pelagic**
				x	x	x				x	x		**x**	x		x		x	x	**x**	x	x	x	
				x					x															
				x																				

Appendix 3. Vectors of Spread for PNW Invasive Species

	Deliberately stocked or planted	Aquaculture	Brought in with oyster aquaculture	Biological control of other invasive species	Ship ballast (water and soil)	Hitchhiking in oversea imports	Hitchhiking on ship hulls and rotors	Hitchhiking on humans and vehicles	Hitchhiking in and on wildlife and livestock	Hitchhiking in wood, soil, and agricultural products	Hitchhiking in an infected host	Natural dispersal once arrived	Gardening and horticulture	Released pets and aquaria	Discarded bait buckets
African Waterweed													x	x	
American Bullfrog	x	x										x			
American Shad	x											x			
Ascidians			x		x		x								
Asian Freshwater Clam	x													x	x
Asian/European Gypsy Moth						x		x		x					
Atlantic Salmon	x	x										x			
Atlantic Slipper Limpet			x												
Balsam Wooly Adelgid								x	x	x		x	x		
Bamboo													x		
Bass-sunfish	x											x			
Blackberries									x			x			
Brazilian Elodea	x						x					x		x	
Broom / Gorse								x	x			x			
Butterfly Bush													x		
Carp	x	x		x								x			
Catfish	x											x			
Cheatgrass	x							x	x	x					
Chinese Mitten Crab	x				x										
Cirsium Thistles								x		x		x			
Common Reed					x			x				x			
Common Velvet-grass								x	x	x		x			
Cordgrasses	x				x				x			x			
Crayfish	x	x										x		x	x
Dalmation Toadflax								x	x	x					
Dead Man's Fingers							x					x			
Dogwood Anthracnose									x		x		x		

	Deliberately stocked or planted	Aquaculture	Brought in with oyster aquaculture	Biological control of other invasive species	Ship ballast (water and soil)	Hitchhiking in oversea imports	Hitchhiking on ship hulls and rotors	Hitchhiking on humans and vehicles	Hitchhiking in and on wildlife and livestock	Hitchhiking in wood, soil, and agricultural products	Hitchhiking in an infected host	Natural dispersal once arrived	Gardening and horticulture	Released pets and aquaria	Discarded bait buckets
Domestic Cat												X		X	
Domestic Pig	X													X	
Earthworms								X		X			X		X
English Holly									X						
English Ivy									X				X		
Eurasian Watermilfoil							X		X						
European Beachgrass	X											X	X		
European Fruit Fly						X		X		X					
European Green Crab					X	X	X					X			
European House Sparrow	X											X			
European Rabbit	X											X			
European Starling	X											X			
European Yellowjacket	X														
Fanwort							X							X	
Fennel	X								X			X	X		
Fishhook Water Flea					X		X								
Garlic Mustard								X	X						
Giant Hogweed								X				X	X		
Hedgehog Dogtail								X	X						
Herb Robert								X							
Houndstongue								X	X						
Hydrilla							X	X	X					X	
Japanese Eelgrass			X												
Japanese Mussel			X		X		X					X			
Knapweed, Spotted								X	X	X					
Knapweeds								X	X	X					
Knotweeds								X		X		X			

	Deliberately stocked or planted	Aquaculture	Brought in with oyster aquaculture	Biological control of other invasive species	Ship ballast (water and soil)	Hitchhiking in oversea imports	Hitchhiking on ship hulls and rotors	Hitchhiking on humans and vehicles	Hitchhiking in and on wildlife and livestock	Hitchhiking in wood, soil, and agricultural products	Hitchhiking in an infected host	Natural dispersal once arrived	Gardening and horticulture	Released pets and aquaria	Discarded bait buckets
Kudzu	x									x					
Ladybeetles	x														
Leafy Spurge									x			x			
Longhorned Beetles						x				x		x			
Marine Clams		x	x		x										
Mediterranean Mussel	x	x	x		x		x					x			
Medusahead								x	x	x					
Mosquitoes						x									
Mountain Goat	x														
Mudsnails			x									x			
New Zealand Isopod							x			x					
New Zealand Mudsnail							x	x	x						
Newcastle Disease						x					x				
Northern Snakehead														x	
Nutria	x			x								x			
Orchard Grass		x								x					
Oyster Drills				x											
Pacific Oyster		x	x	x									x		
Perennial Pepperweed									x		x		x		
Pike		x			x										
Port Orford Cedar Root Rot									x	x	x				
Purple Loosestrife									x	x				x	
Rats							x		x		x		x		
Reed Canarygrass									x	x			x		
Ring-necked Pheasant		x													
Russian Olive		x								x					
Saltcedar		x											x	x	

	Deliberately stocked or planted	Aquaculture	Brought in with oyster aquaculture	Biological control of other invasive species	Ship ballast (water and soil)	Hitchhiking in oversea imports	Hitchhiking on ship hulls and rotors	Hitchhiking on humans and vehicles	Hitchhiking in and on wildlife and livestock	Hitchhiking in wood, soil, and agricultural products	Hitchhiking in an infected host	Natural dispersal once arrived	Gardening and horticulture	Released pets and aquaria	Discarded bait buckets
Sargassum				X				X					X		
Siberian Prawn						X	X						X		
Slender False-brome									X	X			X		
Spiders							X						X	X	
Spurge Laurel										X				X	
Squirrels															X
St. John's Wort									X	X			X		
Sudden Oak Death											X				
Sunfish		X													
Swollen Bladderwort										X					X
Tansy Ragwort									X		X				
Temperate Bass	X											X			
Terrestrial Mollusks								X		X			X		
Thistles								X	X	X					
Traveler's Joy								X				X			
Tree-of-heaven	X											X			
Trout	X											X			
Turtles														X	
Water Primrose									X			X	X		
West Nile Virus									X		X				
White Pine Blister Rust										X	X	X			
Whitetop								X	X	X					
Wood Wasp						X				X					
Yellow Flag Iris													X		
Yellow Floatingheart													X		
Yellow Perch	X														X
Zebra Mussel					X		X					X			

Appendix 4: The Questions

Section 1. Ecological Impact

This section assesses the cumulative impact (e.g., over a period of several decades) of the species on the natural systems in the PNW or in other places with similar environmental conditions. The assessment applies to impacts within the area currently occupied by the species within the PNW (to the extent that this area is known).

1. How severely does the species alter the disturbance regime?
2. How severely does the species change geomorphological processes, such as increasing or decreasing erosion or sedimentation rates?
3. How severely does the species influence bioturbation, such as through changes in soil, bottom structure, or flow regimes?
4. How severely does the species change hydrological regimes by modifying soil or water tables?
5. How severely does the species modify biogeochemical cycling, such as nutrient and mineral dynamics, salinity, alkalinity, pH, DO, or oxygen in sediments?
6. How severely does the species change light availability for other species, such as by covering the water surface or shading out other species?
7. How severely does the species cause impacts to native species through changes in community composition?
8. How severely does the species cause change in the physical structure of communities?
9. How severely does the species change biomass, productivity, or standing stock?
10. How severely does the species change the strength of interactions among species?
11. How severely does the species change the outcome of species interactions?
12. How severely does the species increase fragmentation of the community?
13. How severely does or can the species cause extirpation or endangerment of existing native species or populations?
14. How severely does or can the species cause elimination or a significant reduction in native species habitat, including nesting or foraging sites, cover, or other critical resources?
15. Characterize the severity of native species displacement.
16. Characterize the severity of change to the connectivity or reduction of migratory corridors.
17. How severely does or can the species interfere with or damage native pollinators?
18. How severely does or can the species injure native species by anti-digestive or toxic chemicals that can poison, reduce fecundity, or harm native species?
19. Characterize the species' importance as a reservoir or vector of diseases, pests, or pathogens that can or do impact native species.
20. Characterize the importance of hybridization or non-native genes for individuals within the native species.
21. Characterize the importance of hybridization or non-native genes for the native species population.

22. Characterize this species' ability to produce fertile or sterile hybrids that outcompete or breed with the native species.

23. Characterize this species' ability to negatively impact one or more rare species, habitats, or communities.

Section 2. Invasive Potential

The questions in this section rate a species' potential to establish itself, spread, and increase in abundance in wildlands. If the species has already invaded 75% or more of the habitat in the PNW in which it is likely to be successful, this section receives an overall score of A.

24. What is the potential for this species to invade in the absence of disturbance? Disturbance = a major or biologically significant change in biomass/species composition or the abiotic habitat that occurs unpredictably (does NOT occur on a predictable/annual/seasonal basis); evaluate on a spatial scale appropriate for the system and invasive organism in question. Examples of disturbances or disturbance regimes: roads, dams, landslides, cultivation, nutrient loading, grazing, rooting by domestic livestock, trails, construction, or windthrow.

25. What is the potential for long-term dormancy?

26. How high is the potential for this species to be released from its natural predators or pathogens?

27. How likely is the species to tolerate low food conditions or to withstand other severe perturbations?

28. Characterize the ability of individuals of this species to survive at high densities.

29. What is the reproductive potential (fertility) of the invader compared to other species (fertility)?

30. How successful are alternative methods of this species for spreading, such as fragmentation, vegetative reproduction, or asexual reproduction?

31. What is the species' potential for rapid sexual maturity?

32. What is the species' potential to breed regularly and often?

33. In the absence of management, how fast does this species spread from small, local infestations?

34. Characterize the importance of commercial sales in agriculture, ornamental horticulture, aquariums, pet trade, or aquaculture to the potential spread of this species.

35. Characterize the importance of intentional introductions to the wild, such as game animals, stocked fish, discarded baitfish, or biocontrol agents, to the potential spread of this species.

36. Characterize the importance of forage, erosion control, or revegetation for the species spread.

37. Characterize the importance of contaminants (seeds/propagules/disease) in seed, hay, feed, soil, or packing material to the potential spread of this species.

38. Characterize the importance of spread along transportation corridors such as highways, railroads, trails, canals, power lines, or drainage ditches to the potential spread of this species.

39. Characterize the importance of transport by humans, such as on or in boats, railroad cars, airplanes, cargo shipments, boat trailers, buses, cars, trucks, bikes, backpacks, or clothing, to the potential spread of this species.

40. Characterize the importance of transport by shipping discharge or ballast-water discharge.

41. What is the potential for dispersal of this species by wind, birds, or other animals that travel long distances?

42. Characterize the effectiveness of the dispersal traits of the species. Good dispersers or survivors are likely to be broadcast spawners and have long-lived larval stages, resistant eggs, dormant stages, and the ability to withstand harsh conditions or move to more suitable places.

43. How often does the species disperse long distances? (e.g., can disperse/jump from one habitat patch to another).

44. Characterize the species rate of increase in the PNW.

45. How many different ecological types or host organisms has this species invaded in other regions that are analogous to types in the PNW?

Section 3. Ecological Amplitude and Distribution

This section rates the number and proportion of different ecological types invaded. The "ecological amplitude" of the species indicates the diversity of ecological types invaded. The "distribution" addresses the extent of infestation in any given ecological type.

46. Ecological Amplitude: How many different ecological types or host organisms has this species already invaded in the PNW?

47. Distribution: For the ecological types or host organisms that have already invaded the PNW, what proportion of the total area has been invaded?

Possible Answers:

Questions 1 – 44

A. Severe, high, rapidly, frequently, or many
B. Moderate, medium, less rapidly, occasionally, or some
C. Minor, low, stable, rarely, or few
D. No effect, negligible, declining, never, or none
U. Unknown, no data or no expert opinion

Questions 45 and 46

A. At least three major ecological types or at least five minor ecological types invaded
B. Two major ecological types or four minor ecological types invaded
C. Only one major ecological type or two to three minor ecological types invaded
D. No major ecological types or only one minor ecological type invaded
U. Unknown, no data or no expert opinion

Question 47

A. More than 50% of the total area
B. 20–50% of the total area
C. 5–20% of the total area
D. Less than 5% of the total area
U. Unknown, no data or no expert opinion

Appendix 5: Answers to the Questions and Invasive Scores

QUESTION	1	2	3	4	5	6	7	8	9	10	11	12	13	14	15	16	17	18	19	20	21
African Waterweed	D	D	C	D	A	A	A	—	A	U	U	U	A	A	A	D	D	D	D	D	D
American Bullfrog	U	U	U	D	U	U	A	C	B	A	A	A	B	A	B	B	D	C	C	D	D
American Shad	D	D	D	D	D	D	U	D	A/B	U	U	D	U	D	U	B	D	D	U	D	D
Ascidians	D	D	D	D	D	D	B	B	C	C	C	C	B	B	B	D	D	D	D	D	D
Asian Freshwater Clam	C	C	B	C	B	C	A	A	A	D	D	C	B	A	A	D	D	D	D	D	D
Asian/European Gypsy Moth	B	B	D	D	D	D	B	D	A	A	A	D	B	B	B	D	D	D	D	D	D
Atlantic Salmon	U	B	B	D	B	U	U	U	U	U	U	U	U	U	C	D	D	D	B	U	D
Atlantic Slipper Limpet	—	A	A	B	A	D	A	A	A	A	B	B	A	B	B	C	—	U	U	—	—
Balsam Wooly Adelgid	B	C	C	C	C	B	B	A	A	C	C	C	C	C	C	C	D	D	D	D	D
Bamboo	U	U	U	U	U	U	U	U	U	U	U	U	U	U	U	U	U	U	U	U	U
Bass-sunfish	D	D	D	D	D	D	D	D	D	B	B	D	A/B	D	D	D	D	D	D	D	D
Blackberries	D	A	D	D	D	A	A	B	D	D	C	B	A	B	A	D	D	D	B	NA	NA
Brazilian Elodea	U	A	A	A	B	A	A	A	A	U	U	B	A/B	A	A	B/C	U	B	U	D	D
Broom & Gorse: Gorse	A	B	D	B	A	A	A	A	A	U	A	D	A	B	A	D	D	D	D	D	D
Broom & Gorse: Scotch Broom	C	B	D	B	A	A	A	A	A	U	A	D	A	B	A	D	D	B	B	D	D
Butterfly Bush	U	B	U	U	U	A	A	B	U	U	U	B	U	A	A	U	B	D	U	D	D
Carp: Black, Bighead, & Silver	D	D	C	D	B	D	A	A	A	A	A	B	A	B	A	D	D	D	B	D	D
Carp: Common & Grass	D	D	B	D	B	A	A	A	A	B	B	B	A	A	A	D	D	D	C	D	D
Catfish	D	D	D	D	D	D	B	D	C	B	C	D	A	D	C	D	D	D	D	D	D
Cheatgrass	A	B	D	A	B	C	A	A	A	A	A	D	A	B	A	D	D	D	D	D	D
Chinese Mitten Crab	D	D	D	D	D	D	A	B	B	B	B	D	A	B	B	D	D	D	D	D	D
Cirsium Thistles	C	C	U	C	B	B	C	C	C	U	U	U	B	B	B	U	B	B	C	U	B
Common Reed	B	A	B	A	U	B	A	B	B	U	U	A	A	A	A	D	D	D	D	D	D
Common Velvet-grass	B	A	C	B	A	C	A	C	A	B	B	C	A	A	A	U	B	A	U	C	U
Cordgrasses	A	A	A	A	A	A	A	A	A	A	A	A	B	A	A	A	U	U	U	D	D
Crayfish	D	B	B	D	D	B	A	B	A	B	B	D	A	B	B	D	D	D	A	C	C
Dalmatian Toadflax	B	B	B	C	U	A	A	B	A	B	B	B	A	A	A	B	U	D	D	D	D
Dead Man's Fingers	D	D	D	D	D	A	A	A	A	A	A	C	B	A	A	D	D	D	U	U	U
Dogwood Anthracnose	A	U	U	U	C	A	C	C	B	C	C	C	B	B	C	D	U	C	U	D	D
Domestic Cat	D	D	D	D	D	D	U	D	C	U	U	D	A	D	U	D	D	D	B	D	D
Domestic Pig	A	A	A	A	A	D	A	A	A	A	A	A	A	A	A	A	C	D	B	D	D
Earthworms	B	A	A	B	A	D	B	B	A	B	C	D	B	B	B	D	D	D	A	U	U
English Holly	B	B	A	C	C	B	A	A	A	B	B	A	B	B	A	A	C	B	C	D	D
English Ivy	C	B	B	D	U	A	A	A	B	U	U	U	B	A	A	U	U	U	B	D	D
Eurasian Watermilfoil	A	A	A	A	B	A	A	A	D	A	D	D	A	B	A	C	D	D	D	D	D
European Beachgrass	A	A	B	A	B	A	A	A	B	A	A	A	A	A	A	B	C	D	C	D	D
European Fruit Fly	D	D	D	D	D	D	A	D	A	U	U	U	A	A	A	U	D	D	D	U	U

SCORING MATRIX

Sec. 1	A	A	A	B	B	B
Sec. 2	A,B	A,B	C,D	A,B	A,B	C,D
Sec. 3	A,B	C,D,U	A-D	A,B	C,D,U	A-D
Invasive Threat	High	High Alert	Medium	Medium	Medium Alert	Low

22	23	Sec. 1	24	25	26	27	28	29	30	31	32	33	34	35	36	37	38	39	40	41	42	43	44	45	Sec. 2	46	47	Sec. 3
D	A	**A**	—	U	A	U	A	A	A	A	A	A	A	D	D	D	D	D	D	D	B	—	NA	A	**A**	A	NA	**A**
D	A	**A**	B	D	B	U	B	B	D	C	A	B	B	A	U	D	B	C	D	D	A	B	A	A	**A**	A	B	**A**
D	D	**B**	—	—	—	—	—	—	—	—	—	—	—	—	—	—	—	—	—	—	—	—	—	—	**A**	A	A	**A**
D	B	**B**	A	D	B	D	A	A	A	A	A	B	D	D	D	D	D	B	A	C	B	C	B	C	**A**	C	B	**B**
D	B	**A**	A	D	A	A	A	A	D	A	A	A	A	A	D	A	A	A	B	C	A	A	D	A	**A**	A	C	**A**
D	B	**A**	A	D	A	C	C	A	D	D	D	A	B	D	D	D	A	A	D	D	C	D	D	A	**A**	D	D	**D**
U	U	**B**	C	D	B	U	A	D	D	D	C	D/U	A	A	D	D	D	A	D	D	A	A	C	D	**A**	A	D	**A**
—	A	**A**	A	—	B	B	A	A	D	A	A	A	A	A/D	D	B	A	A	A	C	C	C	C	A	**A**	A	B	**A**
D	B	**A**	A	A	A	A	A	A	A	A	A	A	A	D	D	D	C	B	D	A	A	A	A	C	**A**	C	B	**B**
U	U	**U**	B	D	B	B	B	B	D	B	B	B	B	C	C	C	C	C	D	C	C	C	U	B	**B**	U	U	**U**
D	B	**B**	—	—	—	—	—	—	—	—	—	—	—	—	—	—	—	—	—	—	—	—	—	—	**A**	A	A	**A**
NA	B	**A**	C	A	B	A	A	A	A	A	A	A	D	C	A	D	D	D	D	A	A	A	U	U	**A**	A	U	**A**
D	A/B	**A**	A	D	A	A/B	B	D	A	D	D	A	A	C	C	D	B	B	C	C	C	D	A	A	**A**	A	C/D	**A**
D	A	**A**	C	A	A	A	A	B	B	A	A	B	C	D	C	A	B	A	D	B	B	B	B	A	**A**	A	C	**A**
D	A	**A**	C	A	A	A	A	A	D	A	A	B	C	D	C	A	B	A	D	B	B	B	B	A	**A**	A	B	**A**
D	U	**A**	U	A	U	A	A	A	D	A	A	A	A	D	D	D	B	B	D	A	A	C	B	A	**A**	B	U	**B**
B	A	**A**	A	D	U	B	B	A	D	B	A	A	A	A	D	D	C	D	D	D	B	D	U	B	**A**	D	D	**D**
C	A	**A**	A	D	U	B	B	A	D	B	B	B	A	A	D	D	B	D	D	D	B	D	B	A	**A**	A	B	**A**
D	A	**A**	—	—	—	—	—	—	—	—	—	—	—	—	—	—	—	—	—	—	—	—	—	—	**A**	A	A	**A**
D	A	**A**	A	C	C	B	A	A	D	A	A	A	D	D	C	A	A	A	D	A	A	A	B	A	**A**	A	A	**A**
D	A	**A**	A	D	B	B	A	A	D	B	B	A	C	A	D	D	C	C	A	D	B	B	D	A	**A**	D	NA	**D**
B	B	**B**	D	D	C	C	C	A	D	A	A	C	A	D	D	A	B	C	D	A	B	B	B/C	A	**A**	A	U	**A**
D	A	**A**	B	U	A	B	A	A	A	NA	NA	A	D	D	A	C	A	A	D	B	U	B	A	A	**A**	A	U	**A**
U	A	**A**	—	—	—	—	—	—	—	—	—	—	—	—	—	—	—	—	—	—	—	—	—	—	**A**	A	A	**A**
D	A	**A**	A	U	A	A	A	A	A	A	A	A	C	C	C	C	A	A	A	A	A	U	A	A	**A**	A	C	**A**
D	A	**A**	A	B	C	C	B	B	D	C	D	C	B	B	D	D	C	D	D	D	C	D	A	A	**A**	B	C	**B**
D	A	**A**	A	A	A	A	A	B	A	A	A	B	C	D	D	A	A	C	D	C	B	C	B	U	**A**	A	A	**A**
U	B	**A**	C	D	U	B	B	B	A	B	A	B	D	D	D	B	D	B	C	A	A	D	NA	D	**A**	D	D	**D**
D	B	**A**	B	D	A	C	A	A	A	A	B	U	A	D	C	C	D	A	D	C	C	C	C	C	**A**	B	U	**B**
U	A	**A**	—	—	—	—	—	—	—	—	—	—	—	—	—	—	—	—	—	—	—	—	—	—	**A**	A	U	**A**
D	A	**A**	A	D	B	C	B	B	D	B	A	A	D	A	D	D	D	D	D	D	C	A	D	A	**A**	A	C	**A**
U	A	**A**	C	D	U	B	A	U	D	D	B	B	A	A	B	A	A	A	B	C	B	D	B	A	**A**	A	U	**A**
U	B	**A**	C	B	B	B	B	A	A	C	A	B	A	C	B	D	D	D	D	A	A	A	A	A	**A**	A	U	**A**
D	B	**A**	B	B	A	B	A	B	A	C	A	A	A	D	B	D	C	A	D	A	A	A	A	A	**A**	A	U	**A**
D	D	**A**	A	C	D	B	A	A	A	C	A	A	B	D	D	D	D	A	D	B	A	B	A	A	**A**	A	B	**A**
D	A	**A**	—	—	—	—	—	—	—	—	—	—	—	—	—	—	—	—	—	—	—	—	—	—	**A**	D	A	**A**
U	B	**A**	—	—	—	—	—	—	—	—	—	—	—	—	—	—	—	—	—	—	—	—	—	—	**A**	A	U	**A**

SCORING MATRIX

Sec. 1	C	C	C	C	C	D
Sec. 2	A	A	B	B	C	A-D
Sec. 3	A,B	C,D	A	B-D	A-D	A-D
Invasive Threat	Medium	Low	Medium	Low	Low	Not listed

QUESTION	1	2	3	4	5	6	7	8	9	10	11	12	13	14	15	16	17	18	19	20	21
European Green Crab	D	D	C	D	D	D	B	D	A	D	D	D	C	D	B	D	D	D	C	D	D
European House Sparrow	D	D	D	D	D	D	D	D	D	C	C	C	B	D	C	D	D	D	U	D	D
European Rabbit	B	B	D	D	D	D	A	B	B	C	D	D	A	B	A	D	D	D	B	D	D
European Starling	D	D	D	D	D	D	D	D	D	D	D	D	C	C	D	D	D	D	D	D	D
European Yellowjacket	D	D	D	D	D	D	D	D	D	C	C	C	A/B	A/B	U	D	D	D	U	U	U
Fanwort	D	B	B	D	B	B	A	A	B	B	B	B	A	A	A	B	D	D	D	D	D
Fennel	A	U	U	U	U	A	A	B	U	A	A	A	A	A	A	U	U	U	U	U	U
Fishhook Water Flea	D	D	B/U	D	B/U	D	B	B	A	B	B	U	B/U	D	D	D	D	D	D	D	D
Garlic Mustard	D	D	D	D	D	U	A	A	A	A	A	U	B	U	U	U	B	B	U	U	U
Giant Hogweed	C	B	U	D	D	A	B	C	B	U	U	D	B	B	B	C	U	D	D	D	D
Hedgehog Dogtail	U	U	U	B	U	U	A/B	U	U	U	U	D	A/B	—	A/B	D	D	D	D	D	D
Herb Robert	D	D	D	D	D	D	C	D	D	D	C	D	D	D	D	D	D	U	D	D	D
Houndstongue	B	C	U	U	U	B	B	B	C	U	U	B	U	C	U	U	U	A	D	U	U
Hydrilla	A	A	A	A	A	A	A	A	A	A	A	B	A	A	A	A	—	—	—	—	—
Japanese Eelgrass	B	A	B	C	B	B	B	A	A	U	U	B	C	B	C	D	U	U	D	D	D
Japanese Mussel	A	A	A	A	B	B	A	A	A	A	A	A	B	B	B	D	D	D	U	U	U
Knapweeds	C	B	U	D	C	B	A	A	A	A	A	A	A	A	A	B	B	A	D	D	D
Knotweeds	D	B	B	U	U	A	A	A	U	U	U	U	A	A	A	U	U	B	D	U	U
Kudzu	B	C	D	D	C	A	A	B	U	U	U	U	A	A	A	U	U	D	D	D	D
Ladybeetles	D	D	D	D	D	D	D	D	D	B	B	C	B	D	U	D	D	D	U	U	U
Leafy Spurge	U	U	U	U	U	B	A	A	A	U	U	U	A	B	A	U	D	A	D	D	D
Longhorned Beetles, Asian	C	B	U	D	D	A	B	C	B	U	U	D	B	B	B	C	U	D	D	D	D
Marine Clams	C	C	B	C	B	D	A	A	A	C	C	C	B	A	A	D	D	D	D	D	D
Mediterranean Mussel	U	U	U	D	U	U	U	U	U	U	U	U	U	U	U	U	D	U	U	U	U
Medusahead	A	B	B	U	A	B	A	A	A	B	A	U	B	A	A	U	U	U	U	D	D
Mosquitoes	D	D	D	D	D	D	A	D	D	D	D	D	A	D	A	D	A	A	A	A	A
Mountain Goat	B	A	B	D	D	D	C	B	C	U	U	D	B	B	B	U	U	A	U	D	D
Mudsnail, Asian	U	U	B	D	U	D	U	C	A	U	U	C	U	U	U	C	D	D	U	D	D
Mudsnail, Eastern	B	B	B	D	U	D	U	C	B	U	U	C	U	U	U	C	D	D	U	D	D
New Zealand Isopod	D	A	A	B	B	D	A	B	D	U	U	U	U	A	U	U	D	D	D	D	D
New Zealand Mudsnail	U	U	U	U	U	U	A	A	U	U	U	U	A	U	A	D	D	D	U	D	D
Newcastle Disease	D	D	D	D	D	D	C	D	D	D	D	D	C	D	D	D	D	C	D	D	D
Northern Snakehead	D	D	D	D	D	D	A	B	C/D	C	C	C	A	C/D	A	D	D	D	B	D	D
Nutria	C	C	C	C	C	C	C	C	D	D	D	C	C	D	C	D	D	D	C	D	D
Orchard Grass	B	B	C	D	B	C	B	C	C	D	D	D	B	C	B	D	D	D	D	D	D
Oyster Drills	D	D	D	D	D	D	C	C	C	C	C	D	B	D	B	D	D	D	D	D	D
Pacific Oyster	C	C	A	D	A	C	A	A	A	A	A	C	U	B	U	U	D	D	U	D	D

SCORING MATRIX

Sec. 1	A	A	A	B	B	B
Sec. 2	A,B	A,B	C,D	A,B	A,B	C,D
Sec. 3	A,B	C,D,U	A-D	A,B	C,D,U	A-D
Invasive Threat	High	High Alert	Medium	Medium	Medium Alert	Low

22	23	Sec. 1	24	25	26	27	28	29	30	31	32	33	34	35	36	37	38	39	40	41	42	43	44	45	Sec. 2	46	47	Sec. 3
D	C	A	A	D	A	A	A	B	D	A	A	A	A	D	D	D	D	A	A	D	A	A	B	A	A	C	D	C
D	D	B	—	—	—	—	—	—	—	—	—	—	—	—	—	—	—	—	—	—	—	—	—	—	A	B	A	A
D	A	A	—	—	—	—	—	—	—	—	—	—	—	—	—	—	—	—	—	—	—	—	—	—	A	B	C	B
D	D	C	—	—	—	—	—	—	—	—	—	—	—	—	—	—	—	—	—	—	—	—	—	—	A	B	A	A
U	D	A/B	—	—	—	—	—	—	—	—	—	—	—	—	—	—	—	—	—	—	—	—	—	—	A	A	A	A
D	A	A	B	U	A	C	A	B	A	U	U	D	A	D	D	D	D	A	D	D	D	U	D	C	A	C	D	C
U	A	A	B	A	A	A	A	A	A	A	A	A	A	A	A	U	A	U	D	A	U	A	U	A	A	B	U	B
D	U	A	—	C	A	A	A	A	A	—	A	A	A	D	D	D	U	A	A	C	B	A/C	NA	B	A	B	NA	B
U	U	A	A	B	A	B	A	A	A	A	A	A	B	D	U	U	A	A	D	D	C	B	B	A	A	A	U	A
D	B	A	B	A	U	A	A	B	C	D	C	B	B	D	C	D	B	C	D	D	B	B	B	A	A	A	D	A
D	A/B	A/B	C	D	D	U	B/C	D	D	A	A	B/C	D	D	D	U	U	U	D	D	C	C	C	U	A	A	U	A
D	C	C	D	C	B	D	C	C	D	D	C	B	C	D	D	C	A	A	D	A	A	A	B	A	A	A	C	A
D	C	A	C	B	A	A	A	A	D	C	B	A	A	B	C	B	A	A	C	A	A	A	A	A	A	A	C	A
—	A	A	A	—	—	B	A	A	A	A	A	A	A	A	A	A	A	A	C	C	A	C	A	A	A	C	A	A
D	D	A	A	C	B	B	A	B	B	B	B	B	B	B	A	C	C	B	C	C	B	B	B	C	A	C	B	B
U	A	A	A	D	U	B	A	A	D	A	C	B	C	B	C	D	C	B	A	A	A	B	U	A	A	A	U	A
D	A	A	B	B	B	A	A	A	B	A	A	A	D	D	A	A	A	A	D	A	B	A	A	A	A	A	C	A
A	A	A	U	U	A	A	A	U	A	A	U	A	A	D	B	A	A	A	D	U	A	D	A	A	A	A	U	A
D	U	A	B	C	B	B	A	C	A	B	C	B	D	D	A	B	B	D	D	D	C	C	C	B	A	D	D	D
U	D	B	—	—	—	—	—	—	—	—	—	—	—	—	—	—	—	—	—	—	—	—	—	—	A	B	A	A
D	C	A	C	A	A	A	A	A	A	A	A	A	D	C	B	B	B	B	B	A	A	B	C	A	A	A	C	A
D	B	A	B	A	U	A	A	B	C	D	C	B	B	D	C	D	B	C	D	D	B	B	B	A	A	A	D	A
D	C	A	A	D	A	D	A	A	D	A	A	A	A	A	D	A	D	A	B	A	A	A	A	C	A	C	A	A
U	U	D	U	D	U	U	A/U	U	D	U	U	U	A	D	D	D	D	U	U	U	U	U	U	C	A	D	U	D
U	B	A	B	B	A	B	A	A	D	A	A	B	C	D	C	B	C	B	D	C	A	U	B	A	A	A	B	A
A	A	A	A	A	A	A	A	A	D	A	A	B	C	D	D	D	A	C	D	D	A	C	C	A	A	D	D	D
D	A	A	B	D	B	B	B	B	D	D	C	B	D	C	D	D	D	D	—	D	D	C	B	C	B	C	A	A
D	U	A	A	D	B	B	A	B	D	B	B	U	C	C	D	C	D	C	D	D	D	D	U	C	A	C	U	C
D	U	B	A	D	U	U	A	B	D	B	B	U	C	C	D	C	D	C	C	D	A	U	U	C	A	C	U	C
D	D	A	—	D	U	B	A	—	D	D	D	C	C	D	D	D	D	C	D	B	C	C	U	A	A	A	U	A
D	A	A	A	D	A	U	A	A	A	A	A	A	D	D	D	B	B	A	B	B	B	A	A	A	A	A	A	A
D	D	C	D	B	D	D	C	C	B	B	B	B	B	D	D	C	C	B	D	B	B	C	C	A	A	D	D	D
D	A	A	A	A	A	A	A	A	D	D	A	B	A	D	D	D	C	A	—	D	A	B/C	D	U	A	U	U	U
D	C	C	C	D	C	D	B	B	D	A	A	C	C	C	C	D	D	D	D	D	D	D	C	A	A	A	B	A
D	A	A	B	D	C	C	A	A	D	D	D	B	C	D	A	D	B	D	D	B	B	U	B	A	A	A	A	A
D	B	B	B	D	C	B	B	B	D	B	B	C	B	D	D	D	C	B	D	D	C	D	B	D	B	B	B	B
D	U	A	A	D	U	B	A	A	D	B	B	B	A	A	D	D	D	C	C	D	A	A	B	A	A	A	C	A

SCORING MATRIX

Sec. 1	C	C	C	C	C	D
Sec. 2	A	A	B	B	C	A-D
Sec. 3	A,B	C,D	A	B-D	A-D	A-D
Invasive Threat	Medium	Low	Medium	Low	Low	Not listed

QUESTION	1	2	3	4	5	6	7	8	9	10	11	12	13	14	15	16	17	18	19	20	21
Perennial Pepperweed	B	C	C	C	A	A	A	A	A	C	C	B	B	B	A	C	B	D	U	U	U
Pike	D	D	D	D	D	D	A	A	A	A	A	U	A	D	A	D	D	D	D	D	D
Port Orford Cedar Root Rot	C	B	B	B	B	B	A	A	B	A	A	A	A	A	A	B	C	A	A	C	C
Purple Loosestrife	C	A	A	B	B	A	A	B	A	U	U	U	A	A	A	C	B	D	D	D	D
Rats	D	D	D	D	D	D	A	B	U	U	U	B	A	A	U	B	U	D	D	A	D
Reed Canarygrass	B	B	A	A	B	B	A	A	A	U	U	—	A	B	B	B	A	C	U	U	U
Ring-necked Pheasant	U	U	U	U	U	D	U	U	U	U	U	U	U	U	U	U	U	U	U	U	U
Russian Olive	A	A	B	B	A	A	A	A	B	U	A	U	A	A	A	A	U	A	U	D	D
Saltcedar	B	A	C	A	A	C	B	B	B	C	C	B	B	B	B	C	D	B	D	D	D
Sargassum	U	U	U	U	D	B	B	B	U	U	B	U	C	U	B	D	U	U	U	D	D
Siberian Prawn	D	D	U	D	D	D	B	D	A	D	U	D	A	D	U	D	D	D	D	D	D
Slender False-brome	C	U	U	U	U	A	A	B	B	A	A	B	A	A	A	D	A	C	D	U	U
Spiders	D	D	D	D	D	D	A	D	U	U	U	D	A	D	B	D	C	D	D	D	D
Spotted Knappweed	C	B	A	D	C	B	A	A	A	A	A	A	A	A	A	B	B	A	A	D	D
Spurge Laurel	D	C	C	U	C	B	B	B	B	U	U	U	B	C	B	U	U	A	U	D	D
Squirrels	D	D	D	D	D	D	B	C	D	C	U	U	A	B	C/D	D	U	D	B	U	U
St. John's Wort	D	D	D	D	D	D	B	C	D	C	U	U	A	B	D	D	U	D	B	U	U
Sudden Oak Death	D	D	D	D	D	D	U	U	B/C	—	U	U	B/C	A	D	U	D	D	A/B	D	D
Sunfish	C	C	C	C	C	C	A	C	B	C	B	C	B	B	B	C	U	U	U	B	B
Swollen Bladderwort	C	U	U	U	U	B	B	B	C	U	U	C	B	U	B	U	U	D	D	U	U
Tansy Ragwort	C	U	U	U	U	B	A	B	B	A	B	B	A	B	B	B	U	B	U	U	U
Temperate Bass	D	D	D	D	D	D	C	D	D	B	B	D	A/B	D	D	D	D	D	D	D	D
Thistle, Cotton	C	C	C	C	B	A	A	A	B	C	B	A	A	A	A	A	C	D	D	D	D
Thistle, Musk	C	C	C	C	B	A	A	A	B	B	C	B	B	A	A	A	D	A	D	D	D
Traveler's Joy	D	C	C	D	C	A	A	A	B	B	B	B	A	A	A	B	C	D	D	C	C
Tree-of-heaven	D	D	D	D	D	B	A	B	B	A	A	C	C	B	A	D	U	A	U	D	D
Trout	D	D	D	D	B	C	A	D	B	B	B	D	A	D	A	C	D	D	U	B	C
Turtles	U	D	D	D	D	D	B	B	U	U	U	D	B	B	B	D	D	D	A	D	D
Water Primrose	A	A	A	B	A	A	A	A	B	A	A	A	A	A	A	B	U	U	B	U	U
West Nile Virus	D	D	D	D	D	D	U	U	D	D	D	D	D	D	U	D	D	U	A	D	U
White Pine Blister Rust	A	C	B	C	C	D	A	B	A	B	B	B	A	B	A	B	D	D	A	D	D
Whitetop	C	C	D	B	U	C	B	C	U	U	U	U	B	C	U	D	C	B	U	D	D
Wood Wasp	A	D	D	D	D	D	D	A	A	U	U	U	A	A	D	U	D	A	A	D	D
Yellow Flag Iris	U	B	U	U	D	B	A	B	B	U	U	B	U	U	U	D	B	B	D	D	D
Yellow Floatingheart	A	A/B	B	B/C	A	A	A	B	A	A	A	B	B/U	C	B/C	B/C	U	U	U	D	D
Yellow Perch	D	D	D	D	D	D	A-D	U	B/C	B/C	B/C	U	A-D	D	A-D	U	U	U	D	D	D
Zebra Mussel	D	A	A	B	A	B	A	A	A	U	U	U	A	B	U	D	D	D	D	D	D

SCORING MATRIX

Sec. 1	A	A	A	B	B	B
Sec. 2	A,B	A,B	C,D	A,B	A,B	C,D
Sec. 3	A,B	C,D,U	A-D	A,B	C,D,U	A-D
Invasive Threat	High	High Alert	Medium	Medium	Medium Alert	Low

22	**23**	**Sec. 1**	**24**	**25**	**26**	**27**	**28**	**29**	**30**	**31**	**32**	**33**	**34**	**35**	**36**	**37**	**38**	**39**	**40**	**41**	**42**	**43**	**44**	**45**	**Sec. 2**	**46**	**47**	**Sec. 3**
U	B	**A**	B	C	B	A	A	A	A	—	A	B	A/B	D	A/B	A	A	A	C	B	A/B	B	B	A	**A**	A	B	**A**
D	A	**A**	D	D	C	C	C	C	D	D	D	C	C	A	U	U	C	U	—	U	C	U	C	C	**A**	C	U	**C**
D	A	**A**	C	B	D	B	B	U	B	B	B	B	A	B	B	A	A	A	U	B	B	B	B	C	**A**	B	B	**B**
D	A	**A**	A	B	A	B	A	A	C	C	A	A	A	D	D	B	A	A	A	B	B	B	B	C	**A**	C	B	**B**
D	A	**A**	C	D	C	B	A	A	D	A	A	A	B	D	D	C	A	A	D	D	B	B	C	C	**A**	B	A	**A**
U	A	**A**	B	C	U	B	B	A	A	A	B	B	B	D	A	C	A	B	C	A	B	C	B	A	**A**	A	B	**A**
U	U	**D**	—	—	—	—	—	—	—	—	—	—	—	—	—	—	—	—	—	—	—	—	—	—	**A**	A	U	**A**
D	A	**A**	A	A	A	A	A	A	A	A	A	A	A	A	A	U	A	C	D	A	A	A	A	A	**A**	A	C	**A**
D	A	**A**	C	C	A	B	B	A	B	C	C	A	C	D	C	C	A	C	D	A	C	A	B	U	**A**	A	C	**A**
D	D	**B**	C	C	B	U	A	A	B	B	A	U	D	D	D	D	D	B	C	D	B	B	U	D	**A**	A	B	**A**
D	A	**A**	A	D	A	U	A	B	D	A	B	A	C	C	D	D	D	D	A	D	A	D	B	A	**A**	B	D	**B**
U	A	**A**	B	C	B	U	A	B	C	C	B	B	U	D	D	C	B	A	U	B	B	B	A	U	**A**	A	D	**A**
D	A	**A**	B	D	B	A	B	B	D	B	A	B	A	D	C	D	B	B	D	C	B	C	B	U	**A**	A	B/C	**A**
D	A	**A**	A	B	B	B	A	A	D	B	A	A	C	C	A	A	A	A	D	A	B	B	A	A	**A**	A	A/B	**A**
D	B	**A**	B	B	A	A	B	B	D	B	A	B	B	D	D	D	B	C	D	A	A	A	A	A	**A**	A	D	**A**
U	B	**A**	B	U	U	U	B	C	D	B	A	U	A	A	D	D	D	D	D	D	D	D	U	B	**A**	C	U	**C**
U	B	**A**	B	U	U	U	B	C	D	B	A	U	A	A	D	D	D	D	D	D	D	D	U	B	**A**	A	C	**A**
D	A	**A**	C/U	U	D	U	U	U	A	U	U	B/U	A	D	D	D	D	A	D	D	A/U	B/U	C	A	**A**	C	D	**C**
A	B	**A**	D	D	C	C	B	C	D	C	C	D	A	A	D	D	D	D	D	D	D	D	D	A	**A**	A	A	**A**
U	B	**B**	A	U	U	U	A	B	B	B	B	B	B	B	D	D	D	B	D	A	B	D	B	A	**A**	B	C	**B**
U	A	**A**	C	B	C	B	A	A	B	A	B	B	A	C	U	A	A	B	D	C	C	U	B	A	**A**	A	C	**A**
D	A/B	**A/B**	—	—	—	—	—	—	—	—	—	—	—	—	—	—	—	—	—	—	—	—	—	—	**A**	A	A	**A**
D	A	**A**	B	A	C	A	A	A	A	C	B	B	B	D	A	B	A	A	D	B	A	B	B	A	**A**	A	A	**A**
B	A	**A**	B	A	C	B	A	A	A	C	B	B	B	D	A	A	A	A	D	B	A	B	C	A	**A**	A	B	**A**
C	A	**A**	C	B	B	A	A	A	A	A	A	A	A	D		D	A	B	D	A	A	B	B	B	**A**	A	C	**A**
D	B	**A**	B	D	A	A	A	A	A	B	B	B	B	D	D	D	A	C	D	A	A	B	U	A	**A**	B	D	**B**
C	A	**A**	—	—	—	—	—	—	—	—	—	—	—	—	—	—	—	—	—	—	—	—	—	—	**A**	A	A	**A**
D	A	**A**	A	C	D	C	B	C	D	D	U	D	A	A	D	D	B	C	D	D	B	C	U	B	**A**	B	U	**B**
U	A	**A**	B	U	U	A	A	A	A	B	A	C	B	A	C	C	B	B	B	B	B	C	C	A	**A**	B	D	**B**
U	A	**A**	U	B	A	A	A	A	A	A	A	B	U	A	D	D	A	A	A	A	A	A	B	A	**A**	A	U	**A**
D	A	**A**	—	—	—	—	—	—	—	—	—	—	—	—	—	—	—	—	—	—	—	—	—	—	**A**	B/C	A	**A**
D	A	**A**	C	D	D	B	A	B	A	B	B	B	D	D	D	A	A	A	C	C	B	B	B	A	**A**	A	U	**A**
U	A	**A**	B	B	U	U	A	A	D	A	A	A	A	U	D	A	A	A	B	C	A	U	U	B	**A**	U	U	**U**
D	B	**A**	A	B	U	B	A	U	A	U	U	A	A	C	D	D	A	C	D	A	A	U	B	A	**A**	A	U	**A**
D	B	**A**	U	U	U	U	A	A	A	B	A	C	A	A	D	D	C	A	D	A/B	A	D	C	B	**A**	C	D	**C**
D	D	**A**	B	D	A	B	A	A	U	B	B	B/C	D	B	U	U	C	B	C	U	U	U	B	B	**A**	A	B	**A**
D	B	**A**	A	D	C	B	A	A	D	B	B	A	C	C	D	D	B	A	A	C	A	C	NA	U	**A**	U	U	**U**

SCORING MATRIX

Sec. 1	C	C	C	C	C	D
Sec. 2	A	A	B	B	C	A-D
Sec. 3	A,B	C,D	A	B-D	A-D	A-D
Invasive Threat	Medium	Low	Medium	Low	Low	Not listed

Appendix 6. Code of Conduct

Twenty Rules to Help Reduce the Spread of Invasive Species

1. Never bring or release non-native species into native habitats, period.

2. Participate in early-warning systems by reporting new, unusual, or suspected invasive species you observe in your area to appropriate government agencies or nongovernmental organizations.

3. Be careful to look for possible invasive species when transporting raw wood, fruit, turf, nursery plants, mail-order plants, or pets. Invasives can hitch a ride.

4. Clean all plants and animals off boots, boats, or other equipment before moving to another water body.

5. Clean your car, bike, shoes, camping equipment, or off-road vehicle tires before entering wildlands.

6. Dump aquarium contents on your lawn or garden but not into lakes, rivers, bathtubs, or toilets.

7. Remove invasive plants from your land and replace them with noninvasive species suited to your site and needs.

8. Do not trade plants with other gardeners if you know they are species with invasive characteristics.

9. Ask for only noninvasive species when you acquire plants.

10. Never release pets into the environment.

11. Dump fishing bait buckets where contents can't survive.

12. Do not transport and release fish anywhere.

13. Discourage stocking or planting non-native species anywhere except in already degraded habitats such as ponds and yards.

14. Volunteer to assist ongoing efforts to diminish the threat of invasive plants in parks and natural areas.

15. Request that botanical gardens and nurseries promote, display, and sell only noninvasive species.

16. Urge florists and others to eliminate the use of invasive plant or animal material.

17. Help educate other local gardeners and your community through garden clubs and related groups.

18. Ask government officials and the media to emphasize the problem of invasive species and provide information.

19. Invite speakers who can educate the general public about invasive species.

20. Assist garden clubs and environmental organizations in creating policies for use of invasive species in horticulture activities, flower shows, parks, and plantings.

References

1. Abbott, R. T., and S. P. Dance. 1986. *Compendium of Seashells.* Melbourne, FL: American Malacologists, Inc.
2. Adams, M. J., C. A. Pearl, and R. B. Bury. 2003. Indirect facilitation of an anuran invasion by non-native fishes. *Ecology Letters* 6:343–351.
3. Adams, P. R. 2003. The impact of timing and duration of grass control on growth of a young *Eucalyptus globulus* labill plantation. *New Forests* 26 (2):147–165.
4. See entry 70–a.
5. American Pet Products Manufacturers Association. 2003. 2003–2004 APPMA National Pet Owners Survey.
6. Anderson, A. S., A. L. Bilodeau, M. R. Gilg, and T. J. Hilbish. 2002. Routes of introduction of the Mediterranean mussel (*Mytilus galloprovincialis*) to Puget Sound and Hood Canal. *Journal of Shellfish Research* 21(1):75–79.
7. Andreadis, T. G., J. F. Anderson, L. E. Munstermann, R. J. Wolfe, and D. A. Florin. 2001. Discovery, distribution and abundance of a newly introduced mosquito, *Ochlerotatus japonicus* (Diptera: Culicidae) in Connecticut, USA. *Journal of Medical Entomology* 38:774–779.
8. Anisko, T., and U. Im. 2001. Beware of butterfly bush. *American Nurseryman* 194:46–49.
9. Apfelbaum, S. I., and C. E. Sams. 1987. Ecology and control of reed canary grass (*Phalaris arundinacea* L.). *Natural Areas Journal* 7(2):69–74.

9a. Atkinson, I. 1989. Introduced animals and extinctions. In *Conservation for the twenty-first century*. D. Western and M. Pearl, eds. pp. 54–75. New York: Oxford University Press.

10. Aubry, K. B., and S. D. West. 1984. The status of native and introduced mammals on Destruction Island, Washington. *Murrelet* 65:80–83.
11. Bahls, P. F. 1992. The status of fish populations and management of high mountain lakes in the western United States. *Northwest Science* 66:183–193.
12. Barkalow, F. S., Jr., and M. Shorten. 1973. *The World of the Gray Squirrel.* Philadelphia, PA: J. B. Lippincott Co.
13. Barrett, S. 1975. *The Rat: A Study in Behavior*. Chicago: University of Chicago Press.
14. Baskin, C. C., and J. M. Baskin. 1998. *Seeds: Ecology, Biogeography, and Evolution of Dormancy and Germination*. San Diego, CA: Academic Press.
15. Baskin, Y. 2002. *A Plague of Rats and Rubbervines: The Growing Threat of Species Invasions.* Washington, DC: Island Press, Shearwater Books.
16. Baum, B. R. 1978. The genus Tamarix. The Israel Academy of Sciences and Humanities.
17. Beckenbach, A. T., and A. Prevosti. 1986. Colonization of North America by the European species *Drosophila subobscura* and *D. ambigua*. *American Midland Naturalist* 115:10–18.
18. Becker, G. C. 1983. *Fishes of Wisconsin*. Madison, WI: University of Wisconsin Press.
19. Beerling, D. J., J. P. Bailey, and A. P. Conolly. 1994. *Fallopia Japonica* (Houtt.) Ronse Decraene. *Journal of Ecology* 82(4):959–979.
20. Bielecka, L., et al. 2000. A new predatory cladoceran *Cercopagis pengoi* in the Gulf of Gdansk. *Oceanologia* 42(3):371–374.
21. Biesboer, D. D. 1998. Element stewardship abstract for *Euphorbia esula*. Arlington, VA: The Nature Conservancy, Wildland Weeds Management and Research Program.
22. Billington, H. L., et al. 1988. Divergence and genetic structure in adjacent grass populations. *Evolution* 42(6):1267–1277.

23. Binggeli, P. 1998. *An Overview of Invasive Woody Plants in the Tropics.* Bangor, UK: School for Agriculture and Forest Sciences. Publication No. 13, University of Wales, Bangor.
24. Böhme, M. 2004. Migration history of air-breathing fishes reveals Neogene atmospheric circulation patterns. *Geology* 32(5):393–396.
25. Borror, D. J., and R. E. White. 1970. *A Field Guide to the Insects of America North of Mexico.* Boston: Houghton Mifflin Company, pp. 31, 266.
26. Borror, D. J., C. A. Triplehorn, and N. F. Johnson. 1989. *Introduction to the Study of Insects.* 6th ed. Philadelphia: Saunders College Pub.
27. Bounds, D. L. 2002. Nutria: An invasive species of national concern. *Wetland Journal* 12(3):9–16.
28. Branson, B. A. 1980. Collections of gastropods from the Cascade Mountains of Washington. *The Veliger* 23(2):171–176.
29. Brayshaw, T. 1996. *Trees and Shrubs of British Columbia.* Vancouver, BC: University of British Columbia Press.
30. Brem, D. 2001. Epichloe grass endophytes increase herbivore resistance in the woodland grass *Brachypodium sylvaticum. Oecologia* 126:522–530.
31. Brenchley, G. A., and J. T. Carlton. 1983. Competitive displacement of native mud snails by introduced periwinkles in the New England intertidal zone. *Biological Bulletin* 165:543–558.
32. Brenton, R. K., and Klinger, R. C. 2002. Factors influencing the control of fennel (*Foeniculum vulgare* Miller) using Triclopyr on Santa Cruz Island, California, USA. *Natural Areas Journal* 22:135–147.
33. Brinkman, K. A. 1974. Rubus L.: Blackberry, raspberry. In *Seeds of Woody Plants in the United States.* C. S. Schopmeyer, ed. pp. 738–743. Agriculture Handbook No. 450. Washington, DC: U.S Department of Agriculture, Forest Service.
34. Britton, J. C., ed. 1986. Proceedings of the second international *Corbicula* symposium. *American Malacological Bulletin*, special ed. 2.
35. Britton-Simmons, K. H. 2004. Direct and indirect effects of the introduced alga, *Sargassum muticum*, in subtidal kelp communities of Washington State, USA. *Marine Ecology Progress Series* 277: 61–78.
36. Brock, J. H., M. Wade, and P. Pysek, eds. 1997. *Plant Invasions: Studies from North America and Europe.* Leiden, The Netherlands: Backhuys.
37. Brodeur, R. D., and M. S. Busby. 1998. Occurrence of an Atlantic salmon *Salmo salar* in the Bering Sea. *Alaska Fishery Research Bulletin* 5:64–66.
38. Brown, B. J., R. J. Mitchell, and S. A. Graham. 2002. Competition for pollination between an invasive species (purple loosestrife) and a native congener. *Ecology* 83:2328–2336.
39. Bruzzese, E. 1998. The biology of blackberry in south-eastern Australia. *Plant Protection Quarterly* 13(4):160–162.
40. Buchan, L. A. J., and D. K. Padilla. 1999. Estimating the probability of long-distance overland dispersal of invading aquatic species. *Ecological Applications* 9:254–265.
41. Bump, G., and C. S. Robbins. 1966. The newcomers. In *Birds in Our Lives.* A. Stefferud, ed. Washington, DC: U.S. Department of the Interior, Government Printing Office.
42. Burrill, L. C., R. H. Callihan, R. Parker, E. Coombs, and H. Radtke. 1994. *Tansy Ragwort,* Senecio jacobaea L. *Pacific Northwest Extension Publication* PNW 175.
43. Burrows, E. M. 1991. *Seaweeds of the British Isles.* Vol. 2, *Chlorophyta.* London: Natural History Museum.
44. Bury, R. B., and J. A. Whelan. 1984. Ecology and management of the bullfrog. USFWS Resource Publication 155, pp. 1–23.
45. Byers, J. E. 2002. Physical habitat attribute mediates biotic resistance to non-indigenous species invasion. *Oecologia* 130:146–156.
46. Byers, J. E., and L. Goldwasser. 2001. Exposing the mechanism and timing of impact of nonindigenous species on native species. *Ecology* 82:1330–1343.

47. Cabe, P. R. 1993. European Starling (*Sturnus vulgaris*). No. 48 in A. Poole and F. Gill, eds., *The Birds of North America*. Washington, DC: American Ornithologists' Union; Philadelphia: Academy of Natural Sciences.
48. Cailteux, R. L., L. DeMong, B. J. Finlayson, W. Horton, W. McClay, R. A. Schnick, and C. Thompson, eds. 2001. Rotenone in fisheries: Are the rewards worth the risks? Bethesda, MD: American Fisheries Society, *Trends in Fisheries Science and Management* 1.
49. Callaway, R. M., G. C. Thelen, S. Barth, P. W. Ramsey, and J. E. Gannon. 2004. Soil fungi alter interactions between the invader *Centaurea maculosa* and North American natives. *Ecology* 85(4):1062–1071.
50. Carlton, J. T. 1979. History, Biogeography, and Ecology of the Introduced Marine and Estuarine Invertebrates of the Pacific Coast of North America. PhD thesis, Department of Ecology, University of California, Davis. 904 pp.
51. Carlton, J. T., and J. B. Geller. 1993. Ecological roulette: The global transport of nonindigenous marine organisms. *Science* 261:78–82.
52. Carlton, J. T., and J. Hodder. 1995. Biogeography and dispersal of coastal marine organisms: Experimental studies on a replica of a sixteenth century sailing vessel. *Marine Biology* 121:721–730.
53. Carpenter, A. T., and T. A. Murray. 1998. Element stewardship abstract for *Linaria dalmatica*. Arlington, VA: The Nature Conservancy, Wildland Weeds Management and Research Program.
54. Carr, D. E., and L. E. Banas. 1998. Dogwood anthracnose (*Discula destructive*): Effects of and consequences for host (*Cornus florida*) demography. Charlottesville: University of Virginia, Department of Environmental Sciences and the Blandy Experimental Farm.
55. Carter, J., and B. P. Leonard. 2002. A review of the literature on the worldwide distribution, spread of, and efforts to eradicate the coypu (*Myocastor coypus*). *Wildlife Society Bulletin* 30(1):162–175.
56. Centre for Aquatic Plant Management. 2000. Annual Report. Reading, Berkshire, RG4 6TH, UK.
57. Chambers, K. L. 1966. Notes on some grasses of the Pacific Coast. *Madrono* 18:250–251.
58. Chapman, V. J., J. M. A. Brown, C. F. Hill, and J. L. Carr. 1974. Biology of excessive weed growth in the hydro-electric lakes of the Waikato River, New Zealand. *Hydrobiologia* 44(4):349–363.
59. Child, L., and M. Wade. 2000. *The Japanese Knotweed Manual*. Chichester, West Sussex, UK: Packard Publishing Limited.
59a. Christophersen, E., and E. L. Caum. 1931. Vascular plants of the Leeward Islands, Hawaii. *Bernice P. Bishop Museum Bulletin* 81:1–41.
60. Clark, D. L., and M. V. Wilson. 2003. Post-dispersal seed fates of four prairie species. *American Journal of Botany* 90:730–735.
61. Clark, D., et al. 2000. Fire, mowing, and hand-removal of woody species in restoring a native wetland prairie in the Willamette Valley of Oregon. *Wetlands* 21(1):135–144.
62. Clark, F. H., C. Mattrick, and S. Shonbrun, eds. 1998. Rogues gallery: New England's notable invasives. *New England Wild Flower*. New England Wildflower Society 2(3):19–26.
63. Clausnitzer, D. W., M. M. Borman, and D. E. Johnson. 1999. Competition between *Elymus elymoides* and *Taeniatherum caput-medusae*. *Weed Science* 47:720–728.
64. Clements, D. R., D. J. Peterson, and R. Prasad. 2001. The biology of Canadian weeds 112: *Ulex europaeus* L. *Canadian Journal of Plant Science* 81:325–337.
65. Coan, E. V., P. V. Scott, and F. R. Bernard. 2000. *Bivalve seashells of Western North America: Marine Bivalve Mollusks from Arctic Alaska to Baja California*. Santa Barbara, CA: Santa Barbara Museum of Natural History.
66. Cohen, A. N., H. Berry, C. E. Mills, D. Milne, K. Britton-Simmons, M. J. Wonham, D. Secord, J. Barkas, J. Cordell, B. Dumbauld, A. Fukuyama, L. Harris, A. Kohn, K. Li,

F. Mumford, R. Vasily, A. Sewell, and K. Welch. 2001. Washington State Exotics Expedition 2000: A rapid survey of exotic species in the shallow waters of Elliot Bay, Totten and Eld Inlets, and Willapa Bay, p. 46. In: Olympia, WA: Washington State Department of Natural Resources, Nearshore Habitat Program.

67. Cohen, A. N., Carlton, J. T., and M. C. Fountain. 1995. Introduction, dispersal and potential impacts of the green crab *Carcinus maenas* in San Francisco Bay, California. *Marine Biology* 122:239–247.
68. Cohen, A. N., C. E. Mills, H. Berry, M. J. Wonham, B. Bingham, B. Bookheim, J. Carlton, J. Chapman, J. Cordell, L. Harris, T. Klinger, A. Kohn, C. Lambert, G. Lambert, K. Li, D. Secord, and J. Toft. 1998. Puget Sound Expedition: A rapid assessment survey of non-indigenous species in the shallow waters of Puget Sound, p. 37. Olympia, WA: Washington State Department of Natural Resources.
69. Connor, E. F., et al. 2003. Insect conservation in an urban biodiversity hotspot: The San Francisco Bay Area. *Journal of Insect Conservation* 6:247–259.
70. Cook, C. D. K. 1990. Seed dispersal of *Nymphoides peltata* (S. G. Gmelin) O. Kuntze (Menyanthaceae). *Aquatic Botany* 37(4):325–340.

70a. Cooper, E. L., ed. 1987. *Carp in North America.* Bethesda, MD: American Fisheries Society.

71. Copley, D., et al. 1999. Purple martins (*Progne subis*): A British Columbian success story. *Canadian Field Naturalist* 113(2):226–229.
72. Counts, C. L. 1981. *Corbicula fluminea* (Bivalvia: Sphaeriacea) in British Columbia. *The Nautilus* 95(1):12–13.
73. Courtenay, W. R., Jr., and J. D. Williams. 2004. Snakeheads (Pisces: Channidae): A Biological Synopsis and Risk Assessment. U.S. Geological Survey Circular 1251. Washington, DC: U.S. Department of the Interior.
74. Cox, G. W. 1999. *Alien Species in North America and Hawaii.* Washington, DC: Island Press.
75. Cranston, R., D. Ralph, and B. Wikeem. 1996. *Field Guide to Noxious and Other Selected Weeds of British Columbia.* 4th ed. Victoria, BC: British Columbia Ministry of Agriculture, Fisheries and Food.
76. Crawford, R. L. 1980. Introduced spiders of Washington. *Proceedings of the Washington State Entomological Society* 42:583–584.
77. Crawford, R. L. 1988. An annotated checklist of the spiders of Washington. Burke Museum Contributions in Anthropology and Natural History, No. 5, 58 pp. Online at www.tardigrade.org/natives/crawford/index.html, May 26, 2005.
78. Creighton, J. H., and D. M. Baumgartner. 2001. *Mammals of the Pacific Northwest.* Pullman, WA: Washington State University Cooperative Extension and USDA.
79. Cronk, Q. C. B., and J. L. Fuller. 1995. *Plant Invaders: The Threat to Natural Ecosystems.* London: Chapman & Hall.
80. Crooks, J. A. 2001. Scale-dependent effects of an introduced, habitat-modifying mussel in an urbanized wetland. In *Proceedings of the 1st National Conference on Marine Bioinvasions*, J. Pederson, ed., pp. 154–156. Cambridge, MA: MIT Sea Grant College Program.
81. Crooks, K. R., and M. E. Soulé. 1999. Mesopredator release and avifaunal extinctions in a fragmented system. *Nature* 400:563–566.
82. Csuti, B., et al. 2001. *Atlas of Oregon Wildlife: Distribution, Habitat, and Natural History.* Corvallis, OR: Oregon State University Press.
83. Daughtrey, M. L., and C. R. Hibben. 1994. Dogwood anthracnose: A new disease threatens two native *Cornus* species. *Annual Review of Phytopathology* 32:61–73.
84. Davidson, J. M., S. Werres, M. Garbelotto, E. M. Hansen, and D. M. Rizzo. 2003. Sudden oak death and associated diseases caused by *Phytophthora ramorum*. Online, May 26, 2005. Plant Health Progress doi:10.1094/PHP-2003-0707-01-DG.
85. De Clerck-Floate, R. 1997. Cattle as dispersers of hound's-tongue on rangelands in southeastern British Columbia. *Journal of Range Management* 50:239–243.
86. De Clerck-Floate, R., and M. Schwarzlaender. 2002. *Cynoglossum officinale* L., Hounds-

tongue (Boraginaceae). In *Biological Control Programmes in Canada* 1981–2000, P. G. Mason and J. T. Huber, eds., pp. 337–342. Oxon, UK:CABI Publishing.

87. Department of the Interior, US Fish and Wildlife Service, and Willapa National Wildlife Refuge. 1997. Environmental Assessment: Control of Smooth Cordgrass (*Spartina alterniflora*) on Willapa National Wildlife Refuge.
88. Devine, R. 1998. That cheatin' heartland. In *Alien Invasion: America's Battle with Non-native Animals and Plants,* pp. 51–71. Washington, DC: National Geographic Society.
89. Deysher L., and R. Norton. 1982. Dispersal and colonization in *Sargassum muticum* (Yendo) Fensholt. *Journal of Experimental Marine Biology and Ecology* 56:179–185.
90. Dill, W. A., and A. J. Cordone. 1997. History and status of introduced fishes in California 1871–1996. California Department of Fish and Game, *Fish Bulletin* 178, pp. 185–189.
91. Dirr, M. 1998. *Manual of Woody Landscape Plants: Their Identification, Ornamental Characteristics, Culture, Propagation and Uses.* 5th ed. Champaign, IL: Stipes Publishing.
92. Docherty, D. E., and M. Friend. 2001. "Newcastle Disease." In *Field Manual of Wildlife Disease: General Field Procedures and Disease of Birds.* United States Geological Survey, National Wildlife Health Center.
93. Dolin, E. J. 2003. *Snakehead: A Fish Out of Water.* Washington, DC: Smithsonian Institution.
94. Donald, W. W. 1994. The biology of Canada Thistle, *Cirsium arvense. Reviews of Weed Science* 6:77–101.
95. Dondale, C. D., and J. H. Redner. 1978. *The Insects and Arachnids of Canada, Part 5: The Crab Spiders of Canada and Alaska.* Agriculture Canada, Biosystematics Research Institute (Ottawa), Publication 1663, 255 pp.
96. Dorgelo, J. 1987. Density fluctuations in populations (1982–1986) and biological observations of *Potamopyrgus jenkinsi* [Syn. of P. *antipodarum*] in two trophically differing lakes. *Hydrobiological Bulletin* 21:95–110.
97. Doyle R. D., M. Grodowitz, R. M. Smart, and C. Owens. 2002. Impact of herbivory by *Hydrellia pakistanae* (Diptera: Ephydridae) on growth and photosynthetic potential of *Hydrilla verticillata. Biological Control* 24:221–229.
98. Dreistadt, S. H., J. K. Clark, and M. L. Flint. 1994. *Pests of Landscape Trees and Shrubs: An Integrated Pest Management Guide.* Berkeley, CA: University of California Publication 3359.
99. Drever, M. C., and A. S. Harestad. 1998. Diets of Norway Rats (*Rattus norvegicus*) on Langara Island, Queen Charlotte Islands, British Columbia: Implications for conservation of breeding seabirds. *Canadian Field Naturalist* 112(4):676–683.
100. Dymond, P., S. Scheu, and D. Parkinson. 1997. Density and distribution of *Dendrobaena octaedra* (Lumbricidae) in aspen and pine forests in the Canadian Rocky Mountains. *Soil Biology and Biochemistry* 29(3–4):265–273.
101. Ebbensmeyer, C., and R. Hinrichsen. 1997. Oceanography of the Pacific Shad invasion. *Shad Journal* 2:4–8.
102. Edwards, C. A., ed. 1998. *Earthworm Ecology.* Boca Raton, FL: St. Lucie Press.
103. Edwards, D. J. 1975. Taking a bite at the waterweed problem. *New Zealand Journal of Agriculture* 130(1):33, 35–36.
104. Elton, C. S. 1958. *The Ecology of Invasions by Animals and Plants.* Chicago: University of Chicago Press.
105. Emmett, R. L., S. A. Hinton, D. J. Logan, and G. T. McCabe, Jr. 2002. Introduction of a Siberian freshwater shrimp to western North American. *Biological Invasions* 4:447–450.
106. Englestoft, C. 2002. Restoration priorities associated with introduced species impacts on Haida Gwaii/Queen Charlotte Islands: Perspectives and strategies. Report prepared for Council of Haida Nation Forest Guardians and Ministry of Water, Land and Air Protection, British Columbia. May.
107. Ernst, C. H., J. E. Lovich, and R. W. Barbour. 1994. *Turtles of the United States and Canada.* Washington, DC, and London: Smithsonian Institution Press.

107a. Everest, J. W., J. H. Miller, D. M. Ball, and M. G. Patterson. 1991. *Kudzu in Alabama*. Alabama Cooperative Extension Circular ANR 65.
108. Feare, C. 1984. *The Starling*. Oxford, UK: Oxford University Press.
109. Forcella, F., and J. M. Randall. 1994. The biology of Bull Thistle, *Cirsium vulgare* (Savi) Tenore. *Reviews of Weed Science* 6:29–50.
109a. Forseth, I. N., and A. H. Teramura. 1987. Field photosynthesis, microclimate and water relations of an exotic temperate liana, *Pueraria lobata*, kudzu. Oecologia 71: 262-267.
109b. Frankel, E. 1989. Distribution of Pueraria lobata in and around New York City. *Bulletin of the Torrey Botanical Society* 116:390-394.
110. Frenkel, R. E. 1987. Introduction and spread of cordgrass *Spartina* into the Pacific Northwest. *Northwest Environmental Journal* 31:152–154.
111. Fuller, P. L., L. G. Nico, and J. D. Williams. 1999. Nonindigenous fishes introduced into inland waters of the United States. Bethesda, MD: American Fisheries Society Special Publication No. 27.
112. See entry 111.
113. Galatowitsch, S. M., N. O. Anderson, and P. D. Ascher. 1999. Invasiveness of wetland plants in temperate North America. *Wetlands* 19(4):733–755.
114. Galle, F. C. 1997. *Hollies: The Genus Ilex*. Portland, OR: Timber Press.
115. Garbelotto, M., D. M. Rizzo, J. M. Davidson, and S. J. Frankel. 2002. How to recognize the symptoms of the diseases caused by *Phytophthora ramorum*, causal agent of Sudden Oak Death. USDA Forest Service, Pacific Southwest Region publication, pp. 1–15.
116. Garlic mustard (*Alliaria petiolata*). 1999. Written Findings of the Washington State Noxious Weed Control Board.
117. Getsinger, K. D., and C. R. Dillon. 1984. Quiescence, growth and senescence of *Egeria densa* in Lake Marion. *Aquatic Botany* 20:329–338.
118. *Giant Hogweed*. 1992. PNW Extension Bulletin No. 429.
119. Gibbons M., H. Gibbons, and M. Systma. 1999. *Invasive, Exotic, Aquatic Plants*, pp. 6–7. Washington State Department of Ecology.
120. Gibbons, H. L. 1984. *Control of Eurasian Water Milfoil*. Pullman, WA: State of Washington Water Research Center.
121. Giudice, J. H., and J. T. Ratti. 2001. Ring-necked Pheasant (*Phasianus colchicus*). In *The Birds of North America*, No. 572, A. Poole and F. Gill, eds. Philadelphia, PA: The Birds of North America, Inc.
122. Giver, K. J. 1999. Effects of the Invasive Seaweed *Sargassum muticum* on Native Marine Communities in Northern Puget Sound, Washington. MS thesis, Western Washington University, Bellingham, WA.
123. Gordon, D. G. 1994. *Western Society of Malacologists Field Guide to the Slug*. Seattle, WA: Sasquatch Books.
124. Grieve, M. 1977. *A Modern Herbal*. London: Penguin Books.
125. Grodowitz, M. J., T. D. Center, A. F. Cofrancesco, and J. E. Freedman. 1997. Release and establishment of *Hydrellia balciunasi* (Diptera: Ephydridae) for the biological control of the submersed aquatic plant *Hydrilla verticillata* (Hydrocharitaceae) in the United States. *Biological Control* 9:15–23.
126. Grosholz, E. D., and G. M. Ruiz. 1995. Spread and potential impact of the recently introduced European green crab, *Carcinus maenas*, in central California. *Marine Biology* 122:239–247.
127. Grzimek, B. 1990. *Grzimek's Encyclopedia of Mammals*. Vol. 3. New York: McGraw-Hill Publishing Company.
128. Guard, J. B. 1995. *Wetland Plants of Oregon and Washington*. Redmond, WA: Lone Pine Press.
129. Gulliun, G. 1965. A critique of foreign bird introductions. *Wilson Bulletin* 77:409–414.

130. Gundale, M. J. 2002. Influence of exotic earthworms on the soil organic horizon and the rare fern *Botrychium mormo. Conservation Biology* 16(6):1555–1561.
131. Haber, E. 1999. Russian-olive – Oleaster. *Elaeagnus angustifolia* L. Oleaster Family – Elaeagnaceae. Invasive Exotic Plants of Canada Fact Sheet No. 14. Ottawa, ON, Canada: National Botanical Services. April.
132. Hacker, S. D., D. Heimer, C. E. Hellquist, T. G. Reeder, B. Reeves, T. J. Riordan, and M. N. Dethier. 2001. A marine plant (*Spartina anglica*) invades widely varying habitats: potential mechanisms of invasion and control. *Biological Invasions* 3:211–217.
133. Hagen, C. A., B. E. Jamison, R. J. Robel, and R. D. Applegate. 2002. Ring-necked pheasant parasitism of lesser prairie-chicken nests in Kansas. *Wilson Bulletin* 114:522–524.
134. Hall, Leslie S. 1977. Feral rabbits on San Juan Island, Washington. *Northwest Science* 51:293–296.
135. Hall, R. O., Jr., J. L. Tank, and M. F. Dybdahl. 2003. Exotic snails dominate nitrogen and carbon cycling in a highly productive stream. *Frontiers in Ecology and the Environment* 1(8):407–411.
136. Hamel, K. S., and J. K. Parsons. 2001. Washington's aquatic plant quarantine. *Journal of Aquatic Plant Management* 39:72–75.
137. Hanna, G. D. 1966. Introduced mollusks of western North America. *Occasional Papers of the California Academy of Sciences* 48:108.
138. Hansen, E. M., and P. B. Hamm. 1996. Survival of *Phytophthora lateralis* in infected roots of Port Orford Cedar. *Plant Disease* 80(9):1075–1078.
139. Hansen, E. M., D. J. Goheen, E. Jules, and B. Ullian. 2000. Managing Port-Orford-Cedar and the introduced pathogen *Phytophthora lateralis. Plant Disease* 84(1):4–14.
140. Harbo, R. M. 1997. *Shells and Shellfish of the Pacific Northwest: A Field Guide.* Madeira Park, BC, Canada: Harbour Publishing.
141. Hartenist, A., N. A. Sloan, and P. M. Bartier. 2002. Living Marine Legacy of Gwaii Haanas III: Marine bird baseline to 2000 and marine bird-related management issues throughout the Haida Gwaii region. Halifax, NS: Parks Canada Technical Reports in Ecosystem Science, No. 036.
141a. Harrington, T. B., L. T. Rader-Dixon, and J. W. Taylor, Jr. 2003. Kudzu (*Pueraria montana*) community responses to herbicides, burning, and high-density loblolly pine. *Weed Science* 51: 965-974.
142. Harris, G. A., and A. M. Wilson. 1970. Competition for moisture among seedlings of annual and perennial grasses as influenced by root elongation at low temperature. *Ecology* 51:530–534.
143. Hartman, L. H., and D. S. Eastman. 1999. Distribution of introduced raccoons *Procyon lotor* on the Queen Charlotte Islands: implications for burrow-nesting seabirds. *Biological Conservation* 88(1):1–13.
144. Haugen, D. A., and E. T. Iede, Wood Borers; Risks of Exotic Forest Pests and their Impact on Trade. Paper from the Exotic Forest Pests Online Symposium, April 16–29, 2001.
145. Hayes, M. P., and M. R. Jennings. 1986. Decline of ranid frog species in western North America: Are bullfrogs (*Rana catesbeiana*) responsible? *Journal of Herpetology* 20:490–509.
146. He, X., and J. F. Kitchell. 1990. Direct and indirect effects of predation on a fish community: a whole lake experiment. *Transactions of the American Fisheries Society* 119:825–835.
146a. Hedgpeth, J. W. 1993. Foreign invaders. *Science* 261:34–35.
147. Hedgepeth, W. 1978. *The Hog Book.* New York: Doubleday & Company.
148. Heidorn, R., and B. Anderson. 1991. Vegetation management guideline: Purple loosestrife (*Lythrum salicaria* L.). *Natural Areas Journal* 11:172–173.
149. Hendrix, P. F., ed. 1995. *Earthworm Ecology and Biogeography in North America.* Boca Raton, FL: Lewis Publishers.
150. Hendrix, P. F., and P. J. Bohlen. 2002. Exotic earthworm invasions in North America: Ecological and policy implications. *BioScience* 52(9):801–811.

151. Hershey, A. E., A. R. Lima, G. J. Niemi, and R. R. Regal. 1998. Effects of *Bacillus thuringiensis israelensis* (Bti) and methoprene on nontarget macroinvertebrates in Minnesota wetlands. *Ecological Applications* 8(1):41–60.
152. Hilken, T. O., and R. F. Miller. 1980. Medusahead (*Taeniatherum asperum* Nevski): A review and annotated bibliography. Corvallis, OR: Oregon State University Agricultural Experiment Station, Station Bulletin 644.
153. Hipps, C. B. 1994. Kudzu: A vegetable menace that started out as a good idea. *Horticulture* 72:36–39.
154. Hitchcock, A. S. 1971. *Manual of Grasses of the United States.* New York: Dover Publications, Inc.
155. Hitchcock, C. L., and A. Cronquist. 1973. *Flora of the Pacific Northwest: An Illustrated Manual.* Seattle, WA: University of Washington Press.
156. See entry 155.
157. Hitchcock, C. L., A. Cronquist, M. Ownbey, and J. W. Thompson. 1961. *Vascular Plants of the Pacific Northwest*, Vol. 3. Seattle, WA: University of Washington Press.
158. Hoi, H., et al. 2003. Post-mating sexual selection in house sparrows: Can females establish good fathers according to their early paternal effort? *Folia Zoologica* 52(3):299–308.
159. Holmes, T. H., and K. J. Rice. 1996. Patterns of growth and soil-water utilization in some exotic annuals and native perennial bunchgrasses of California. *Annals of Botany* 78:233–243.
160. Holthius, L. B. 1980. Shrimps and prawns of the world: An annotated catalogue of species of interest to fisheries. *FAO Fisheries Synopsis* 125(1):1–271.
161. Hopf, A. 1979. *Pigs: Wild and Tame.* New York: Holiday House.
162. Hoshorsky, M. 1989. Element stewardship abstract for *Rubus discolor* (*Rubus procerus*), Himalayan blackberry. Arlington, VA: The Nature Conservancy, Wildland Invasive Species Team.
163. Houston, D. B., et al. 1994. *Mountain Goats in Olympic National Park: Biology and Management of an Introduced Species.* Scientific Monograph NPS/NROLYM/NRSM-94/25, USDI-NPS, 295 pp.
164. Howard-Williams, C., and J. Davies. 1988. The invasion of Lake Taupo by the submerged water weed *Lagarosiphon major* and its impact on the native flora. *New Zealand Journal of Ecology* 11:13–19.
165. Hoy, M. A. 1976. Establishment of gypsy moth parasitoids in North America: An evaluation of possible reasons for establishment or non-establishment. In *Perspectives in Forest Entomology*, J. F. Anderson and H. K. Kaya, eds., pp. 215–232. New York: Academic Press.
166. Hu, S. Y. 1979. Ailanthus. *Arnoldia* 39(2):29–50.
167. Huey, R. B., G. W. Gilchrist, M. L. Carlson, D. Berrigan, and L. Serra. 2000. Rapid evolution of a geographic cline in an introduced species of fly. *Science* 287:308–309.
168. Hunter, J. 2000. *Ailanthus altissima.* In *Invasive Plants of California's Wildlands*, C. C. Bossard, J. M. Randall, and M. C. Hoshovsky, eds. Berkeley, CA: University of California Press.
169. Invasive Plant Atlas of New England (IPANE). 2001. Storrs, CT: University of Connecticut, Department of Ecology and Evolutionary Biology. Online at www.ipane.org, May 26, 2005.
170. Jeffery, W. R. 2001. Determinants of cell and positional fate in ascidian embryos. *International Review of Cytology* 203:3–62.
171. Jenkins, R. E., and N. M. Burkhead. 1993. *Freshwater Fishes of Virginia.* Bethesda, MD: American Fisheries Society.
172. Jensen, G. C., P. S. McDonald, and D. A. Armstrong. 2002. East meets west: Competitive interactions between green crab *Carcinus maenas*, and native and introduced shore crab *Hemigrapsus* spp. *Marine Ecology Progress Series* 225:251–262.
173. Johnson, J. H., A. A. Nigro, and R. Temple. 1992. Evaluating enhancement of Striped Bass

in the context of potential predation on anadromous salmonids in Coos Bay, OR. *North American Journal of Fisheries Management* 12(1):103–108.
174. Johnson, R. E., and K. M. Cassidy. 1997. *Terrestrial Mammals of Washington State: Location Data and Predicted Distributions*, vol. 3. In *Washington State Gap Analysis – Final Report,* K. Cassidy, C. E. Grue, M. R. Smith, and K. M. Dvornich, eds. Seattle, WA: University of Washington, Washington Cooperative Fish and Wildlife Research Unit.
174a. Jules, E. S., M. J. Kauffman, W. D. Ritts, and A. L. Carroll. 2002. Spread of an invasive pathogen over a variable landscape: A nonnative root rot on Port Orford Cedar. *Ecology* 83(11):3167–3181.
175. Karatayev, A. Y., L. E. Burlakova, and D. K. Padilla. 1997. The effects of *Dreissena polymorpha* (Pallas) invasion on aquatic communities in eastern Europe. *Journal of Shellfish Research* 16:187–203.
176. Karatayev, A. Y., L. E. Burlakova, and D. K. Padilla. 1998. Physical factors that limit the distribution and abundance of *Dreissena polymorpha* (Pall.). *Journal of Shellfish Research* 17:1219–1235.
177. Karatayev, A. Y., L. E. Burlakova, and D. K. Padilla. 2002. The impact of zebra mussels on aquatic communities and their role as ecosystem engineers. In: Erkki Leppäkoski (Finland), Sergej Olenin (Lithuania) and Stephan Gollasch (Germany), eds., *Invasive Aquatic Species of Europe: Distributions, Impacts, and Management.* Dordrecht, The Netherlands: Kluwer Scientific Publishers.
178. Karatayev, A. Y., L. E. Burlakova, T. Kesterson, and D. K. Padilla. 2003. Dominance of the Asiatic clam *Corbicula fluminea* (Müller), in the benthic community of a reservoir. *Journal of Shellfish Research* 22(2):487–493.
179. Kendeigh, S. C. 1973. A symposium on the House Sparrow (*Passer domesticus*) and European Tree Sparrow (*P. montanus*) in North America. *Ornithological Monographs* No. 14. American Ornithologists' Union.
180. Kiesecker, J. M., and A. R. Blaustein. 1998. Effects of introduced bullfrogs and smallmouth bass on microhabitat use, growth, and survival of native red-legged frogs (*Rana aurora*). *Conservation Biology* 12:776–787.
181. Kilbride, K. M., and F. L. Paveglio. 1999. Integrated pest management to control reed canarygrass in seasonal wetlands of southwestern Washington. *Wildlife Society Bulletin* 27: 292–297.
182. Klinkhamer, P. G. L., and T. J. De Jong. 1993. Flora of the British Isles: *Cirsium vulgare* (Savi) Ten. *Journal of Ecology* 81:177–191.
183. Knight, A. P., C. V. Kimberling, F. R. Stermitz, and M. R. Roby. 1984. *Cynoglossum officinale* (Hound's-tongue) – a cause of pyrrolizidine alkaloid poisoning in horses. *Journal of American Veterinary Medical Association* 184:647–650.
184. Knopf, F. L., and T. E. Olson. 1984. Naturalization of Russian-olive: Implications for Rocky Mountain wildlife. *Wildlife Society Bulletin* 12:289–298.
185. Kozloff, E. N. 1976. *Plants and Animals of the Pacific Northwest.* Seattle, WA: University of Washington Press.
186. Lajeunesse, S. 1999. Dalmatian and yellow toadflax. In: R. L. Sheley and J. K. Petroff,eds., *Biology and Management of Noxious Rangeland Weeds.* Corvallis, OR: Oregon State University Press.
187. Lapman, B. H. 1946. *The Coming of Pond Fishes.* Portland, OR: Bindfords and Mort.
188. Larrison, E. J. 1976. *Mammals of the Northwest.* Seattle, WA: Seattle Audubon Society.
189. Leaver, C. 1987. *Naturalized Birds of the World.* Essex, UK: Longman Scientific and Technical; New York: John Wiley & Sons, Inc.
190. Lee, D. S., C. R. Gilbert, C. H. Hocutt, R. E. Jenkins, D. E. McAllister, and J. R. Stauffer, Jr. 1980 (et seq.). *Atlas of North American Freshwater Fishes.* Raleigh, NC: North Carolina State Museum of Natural History.

191. Leege, T. A., and G. Godbolt. 1985. Herbaceous response following prescribed burning and seeding of elk range in Idaho. *Northwest Science* 59(2):134–143.
192. Liebhold, A. M., J. Halverson, and G. Elmes. 1992. Quantitative analysis of the invasion of gypsy moth in North America. *Journal of Biogeography* 19:513–520.
193. Lodge, D. M., and J. G. Lorman. 1987. Reductions in submersed macrophyte biomass and species richness by the crayfish, *Orconectes rusticus. Canadian Journal of Fisheries and Aquatic Sciences* 44(3):591–597.
194. Lodge, D. M., C. A. Taylor, D. M. Holdich, and J. Skurdal. 2000. Nonindigenous crayfishes threaten North American freshwater biodiversity. *Fisheries* 25:7–19.
195. Lombardo, C. A., and K. R. Faulkner. 1999. Eradication of feral pigs (*Sus scrofa*) from Santa Rosa Island, Channel Islands National Park, California. In: Minerals Management Service, Proceedings of the Fifth California Islands Symposium, pp. 300–306.
196. Lomer, F., and G. Douglas. 1999. Additions to the vascular flora of the Queen Charlotte Islands. *Canadian Field Naturalist* 113(2):235–240.
197. Long, J. L. 1981. *Introduced Birds of the World.* New York: Universe Books.
198. Louda, S. M., R. W. Pemberton, M. T. Johnson, and P.A. Follett. 2003. Nontarget effects—The Achilles' heel of biological control? Retrospective analyses to reduce risk associated with biocontrol introductions. *Annual Review of Entomology* 48:365–396.
198a. Lowe, S., M. Browne, S. Boudjelas, and M. De Poorter. 2004. 100 of the World's Worst Invasive Alien Species: A selection from the Global Invasive Species Database. In *Aliens 12.* pp. 12. The Invasive Species Specialist Group, Species Survival Commission, World Conservation Union (IUCN). Online at www.issg.org/booklet.pdf, May 26, 2005.
199. Luoma, Jon R., 1997. Catfight. *Audubon* 99(Jul-Aug):84–91.
200. Mack, R. N. 1986. Alien plant invasion into the Intermountain west: a case history. In: H. A. Mooney and J. A. Drake, eds., *Ecology of Biological Invasions of North America and Hawaii*, pp.191–213. New York: Springer-Verlag.
201. Mack, R. N. 2003. Phylogenetic constraint, absent life forms and pre-adapted alien plants: a prescription for biological invasions. *International Journal of Plant Sciences* 164(3 Suppl):S185–S196.
202. Mack, R. N., D. Simberloff, M. W. Lonsdale, H. Evans, M. Clout, and A. F. Bazzaz. 2000. Biotic invasions: causes, epidemiology, global consequences, and control. *Ecological Applications* 10(3):689–710.
203. Madsen, J. D., J. W. Sutherland, J. A. Bloomfield, L. W. Eichler, and C. W. Boylen, 1991. The decline of native vegetation under dense Eurasian watermilfoil canopies. *Journal of Aquatic Plant Management* 29:94–99.
204. Maloy, O. C. 2001. White pine blister rust. Online: Plant Health Progress doi:10.1094/PHP-2001–0924–01–HM.
205. Marten, G., 2001. *Dengue Hemorrhagic Fever, Mosquitoes, and Copepods: Basic Concepts for Sustainable Development*, pp. 184–196. London: Earthscan Publications.
205a. Matlack, G. R. 2002. Exotic plant species in Mississippi, USA: Critical issues in management and research. *Natural Areas Journal* 22: 241-247.
206. Maurer, D. A., R. Lindig-Cisneros, K. J. Werner, S. Kercher, R. Miller, and J. B. Zedler. 2003. The replacement of wetland vegetation by reed canarygrass (*Phalaris arundinacea*). *Ecological Restoration* 21(2):116–119.
207. Mayer, J. J., and I. L. Brisbin, Jr. 1991. *Wild Pigs in the United States: Their History, Comparitive Morphology, and Current Status.* Athens, GA: University of Georgia Press.
208. McDonald, B. 1986. *Practical Woody Plant Propagation for Nursery Growers.* Portland, OR: Timber Press, pp. 558–559.
209. McDonald, G. I., and R. J. Hoff. 2001. Blister rust: an introduced plague. In *Whitebark Pine Communities: Ecology and Restoration*, Tomback, D. F., S. F. Arno, and R. E. Keane, eds. pp. 193–220. Washington, DC: Island Press.
210. McDonald, P. M., and G. O. Fiddler. 2002. Relationship of native and introduced grasses

with and without cattle in a young ponderosa pine plantation. *Western Journal of Applied Forestry* 17(1):31–36.
211. McDonald, P. S., G. C. Jensen, and D. A. Armstrong. 2001. The competitive and predatory impacts of the nonindigenous crab *Carcinus maenas* (L.) on early benthic phase Dungeness crab *Cancer magister* Dana. *Journal of Experimental Marine Biology and Ecology* 258:39–54.
212. McEvoy, P. B., and N. T. Rudd. 1993. Effects of vegetation disturbances on insect biological control of tansy ragwort, *Senecio jacobaea. Ecological Applications* 3:682–698.
213. McEvoy, P. B. 1984. Seedling dispersion and the persistence of ragwort *Senecio jacobaea* (Compositae) in a grassland dominated by perennial species. *Oikos* 442:138–143.
214. McGavin, G. C. 2001. *Essential Entomology: An Order-by-Order Introduction.* Oxford, UK: Oxford University Press, p. 229.
215. McMahon, T. E., and D. H. Bennett. 1996. Walleye and northern pike: boost or bane to Northwest fisheries? *Fisheries* 21(8):6–13.
216. McPhail, J. D., and R. Carveth. 1999. *Field Key to the Freshwater Fishes of British Columbia.* Vancouver, BC: University of British Columbia, Department of Zoology. Online, May 26, 2005.
217. McPhee, J. 2000. *The Founding Fish.* New York: Farrar, Straus and Giroux.
218. Meredith, T. J. 2001. *Bamboo for Gardens.* Portland, OR: Timber Press, pp. 17–75.
219. Merigliano, M. F., and P. Lesica. 1998. The native status of reed canarygrass (*Phalaris arundinacea* L.) in the Inland Northwest, USA. *Natural Areas Journal* 18(3):223–230.
220. Mielke, M. E., and M. L. Daughtrey. 1989. *How to Identify and Control Dogwood Anthracnose.* U.S. Department of Agriculture, Forest Service, Northeastern Area NA–GR-18.
221. Miller, G. K., J. A. Young, and R. A. Evans, 1986. Germination of seeds of perennial pepperweed (*Lepidium latifolium*). *Weed Science* 34:252–255
222. Mitchell, R. G. 1966. Infestation characteristics of the balsam woolly aphid in the Pacific Northwest. USDA Forest Service Research Paper. PNW-35, 18 pp. Portland, OR: PNW Forest and Range Experiment Station.
223. Mitchell, R. G., and P. E. Buffam. 2001. Patterns of long-term balsam woolly adelgid infestations and effects in Oregon and Washington. *Western Journal of Applied Forestry* 16(3):121–126.
223a. Mooney, H. A. and R. J. Hobbs. 2000. *Invasive Species in a Changing World.* Covelo, CA: Island Press.
223b. Mitich, L. W. 2000. Kudzu [*Pueraria lobata* (Willd.) Ohwi]. *Weed Technology* 14: 231-235.
224. Mordukhai-Boltovskoi, F. D., and I. K. Rivier. 1987. *Parthenogenetic and gamogenetic female and male individuals of* Cercopagis pengoi. Helsinki: Finnish Institute of Marine Research.
225. Morse, L., et al. *The Nature Conservancy Element Stewardship Abstract for* Ammophila arenaria. Arlington, VA: The Nature Conservancy.
226. Mosley, J. C., S. C. Bunting, and M. E. Manoukian. 1999. Cheatgrass. In: R. L. Sheley and J. K. Petroff, eds., *Biology and Management of Noxious Rangeland Weeds*, pp. 175–188. Corvallis, OR: Oregon State University Press.
227. Moyle, P. B. 2002. *Inland Fishes of California.* Berkely, CA: University of California Press.
228. Mullen, G., and L. Durden. 2002. *Medical and Veterinary Entomology*, pp. 204–256. New York: Academic Press.
229. Nagorsen, D. W. (in prep.) *The Rodents and Lagomorphs of British Columbia.* Vol. 4. *The Mammals of British Columbia.* Victoria, BC: Royal British Columbia Museum Handbook.
230. Nasci, R. S., et al. 2002. Detection of West Nile Virus-infected mosquitoes and seropositive juvenile birds in the vicinity of virus-positive dead birds. *American Journal of Tropical Medicine and Hygiene* 67(5):492–496.
230a. Naylor, R. L. 2000. *The Economics of Alien Species Invasions. In Invasive Species in a*

Changing World. Mooney, H.A. and R. J. Hobbs, eds. pp. 241–259. Covelo, CA: Island Press.
231. Naylor, R. L., R. J. Goldburg, H. Mooney, M. Beveridge, J. Clay, C. Folke, N. Kautsky, J. Lubchenco, J. Primavera, and M. Williams. 1998. Nature's subsidies to shrimp and salmon farming. *Science* 282:883–884.
232. Naylor, R. L., J. Eagle, and W. L. Smith. 2003. Salmon aquaculture in the Pacific Northwest – a global industry. *Environment* 45:18–21.
232a. Nelson, L. R. 2004. Kudzu eradication guidelines. Clemson Cooperative Extension Publication EC-656.
233. Niering, W. A., and N. C. Olmstead. 1979. *The Audubon Society Field Guide to North American Wildflowers*. New York: Alfred A. Knopf.
234. Norton, T. A. 1977. The growth and development of *Sargassum muticum* (Yendo) Fensholt. *Journal of Experimental Marine Biology and Ecology* 26:41–53.
235. Nowak, R. M. 1991. *Walker's Mammals of the World*, 5th ed., Vol. 2. Baltimore, MD: Johns Hopkins University Press.
236. Nuzzo, V. 2000. *Element Stewardship Abstract for* Alliaria petiolata. Arlington, VA: The Nature Conservancy.
237. Oh, C.-W., H.-L. Suh, K.-Y. Park, C.-W. Ma, and H.-S. Lim. 2002. Growth and reproductive biology of the freshwater shrimp *Exopalaemon modestus* (Decapoda: Palaemonidae) in a lake of Korea. *Journal of Crustacean Biology* 22(2):357–366.
238. Olson, D., E. Dinerstein, E. Wikramanayake, N. Burgess, G. Powell, E. Underwood, J. D'amico, I. Itoua, H. Strand, J. Morrison, C. Loucks, T. Allnutt, T. Ricketts, Y. Kura, J. Lamoreux, W. Wettengel, P. Hedao, and K. Kassem. 2001. Terrestrial ecoregions of the world: a new map of life on Earth. *BioScience* 51:933–938.
239. Page, L. M., and B. M. Burr. *Freshwater Fishes*. Boston, MA: Houghton Mifflin Co.
240. Pannill, P. D. 2000. Tree-of-Heaven Control (Revised). Maryland Department of Natural Resources Forest Service Stewardship Bulletin.
241. Parchoma, G., ed.. 2002. *Guide to Weeds in British Columbia*. 1st ed. Burnaby, BC, Canada: Open Learning Agency.
242. Parish, R., R. Coupe, and D. Lloyd. 1996. *Plants of Southern Interior British Columbia*. Redmond, WA: Lone Pine Press.
243. Pearson, D. E., K. S. McKelvey, and L. F. Ruggiero. 2000. Non-target effects of an introduced biological control agent on deer mouse ecology. *Oecologia* 122:121–128.
244. Petersen, J. H., R. A. Hinrichsen, D. M. Gadomski, D. H. Feil, and D. W. Rondorf. 2003. American shad in the Columbia River. In: K. E. Linburg and J. R. Waldman, eds., *Biodiversity, Status, and Conservation of the World's Shads*. Bethesda, MD: American Fisheries Society.
245. Peterson, D. J., and R. Prasad. 1998. The biology of Canadian weeds 109: *Cytisus scoparius* (L.) Link. *Canadian Journal of Plant Science* 78:497–504
246. Petrak, M. L. 1982. *Diseases of Cage and Aviary Birds*, pp. 519–521. Philadelphia: Lea and Febiger.
246a. Phillips, B. L., and R. Shine. 2004. Adapting to an invasive species: Toxic cane toads induce morphological change in Australian snakes. Proceedings of the National Academy of Sciences 101(49):17150–17155.
247. Pickart, A. 1997. Control of European Beachgrass (*Ammophila arenaria*) on the West Coast of the United States. California Exotic Pest Plant Council, 1997 Symposium Proceedings, Arcata, CA.
248. Pickart, A., and J. Sawyer. 1998. *Ecology and Restoration of Northern California Coastal Dunes*. Sacramento, CA: California Native Plant Society.
249. Pieterse, A. H., and K. J. Murphy, eds. 1990. *Aquatic Weeds: The Ecology and Management of Nuisance Aquatic Vegetation*. London: Oxford University Press.
250. Pilsbry, H. A. 1939–1948. *Land Mollusca of North America (North of Mexico)*. 1939: The

Academy of Natural Sciences of Philadelphia, Monograph 3, 1(1):i–xvii, 1–573, i–ix; 1940: ibid., 1(2):575–994, i–ix; 1946: ibid., 2(1): i–iv, 1–520, i–ix, frontispiece; 1948: ibid., 2(2): i–xlvii, 521–1113.
251. Pimentel, D., L. Mclaughlin, A. Zepp, B. Lakitan, T. Kraus, P. Kleinman, F. Vancini, W. Roach, E. Graap, W. Keeton, and G. Selig. 1991. Environmental and economic effects of reducing pesticide use. *BioScience* 41(6):402–409.
252. Pimentel, D., L. Lach, R. Zuniga, and D. Morrison. 2000. Environmental and economic costs of nonindigenous species in the United States. *BioScience* 50(1):53–64.
253. Pister, E .P. 2000. Wilderness fish stocking: history and perspective. *Ecosystems* 4:279–286.
254. Pojar, J., and A. MacKinnon. 1994. *Plants of the Pacific Northwest Coast.* Redmond, WA: Lone Pine Press.
255. Quayle, D. B. 1964. Distribution of introduced marine Mollusca in British Columbia waters. *Journal of the Fisheries Research Board of Canada* 21(5):1155–1181.
256. Race, M. S. 1982. Competitive displacement and predation between introduced and native mud snails. *Oecologia* 54: 337–347.
257. Rahel, F. J. 2000. Homogenization of fish faunas across the United States. *Science* 288: 854 856.
258. Ramamoorthy, T. P. 1987. The systematics and evolution of Ludwigia sect. *Myrtocarpus sensu lato* (Onagraceae). St. Louis, MO: Missouri Botanical Garden.
259. Ramcharan, C. W., D. K. Padilla, and S. I. Dodson. 1992. Models to predict potential occurrence and density of the zebra mussel, *Dreissena polymorpha. Canadian Journal of Fish and Aquatic Science* 49(12):2611–2620.
260. Reichard, S. 2000. *Hedera helix.* In J. M. Randall and C. Bossard, eds., *Noxious Wildland Weeds of California*, pp. 212–216. Berkeley, CA: University of California Press.
261. Ricciardi, A., and H. J. MacIsaac. 2000. Recent mass invasion of the North American Great Lakes by Ponto-Caspian species. *Trends in Ecology and Evolution* 15:62–65.
262. Rice, K. J., and N. C. Emery. 2003. Managing microevolution: restoration in the face of global change. *Frontiers in Ecology and the Environment* 1:469–478.
263. Richards, D. C., Cazier, L. D., and G. T. Lester. 2001. Spatial distribution of three snail species including the invader *Potamopyrgus antipodarum* in a freshwater spring. *Western North American Naturalist* 61(3):375–380.
264. Rieman, B. E., R. C. Beamesderfer, et al. 1991. Estimated loss of juvenile Salmonids to predation by Northen Squawfish, Walleyes, and Smallmouth Bass in John Day Reservior, Columbia River. *Transactions of the American Fisheries Society* 120:448–458.
265. Rizzo, D. M., and M. Garbelotto. 2003. Sudden oak death: endangering California and Oregon forest ecosystems. *Frontiers in Ecology and the Environment* 1(5):197–204.
266. Roché, C. T., and E. Coombs. 2003. Meadow Knapweed (*Centaurea x pratensis* Thuill.). *Pacific Northwest Extension Publications*, PNW05666.
267. See entry 285–a.
268. Rollo, C. D., and W. G. Wellington. 1975. Terrestrial slugs in the vicinity of Vancouver, British Columbia. *The Nautilus* 89(4):107–115.
269. Rosenzweig, M. 2001. The four questions: What does the introduction of exotic species do to diversity? *Evolutionary Ecology Research* 3:361–367.
270. Roth, B., and Nelson. 1972. Phytophthora Root Rot of Port-Orford-Cedar. USDA, Forest Pest Leaflet.
271. Roth, L. F., R. D. Harvey, Jr., and J. T. Kliejunas. 1987. Port-Orford-Cedar Root Disease, United States Department of Agriculture, Forest Service, R6 FPM-PR-294–87.
272. Rotramel, G. 1972. *Iais californica* and *Sphaeroma quoyanum*, two symbiotic isopods introduced to California (Isopoda: Janiridae and Sphaeromatidae). *Crustaceana* Suppl. 3:193–197.
273. Rotramel, G. 1975. Observations on the commensal relations of *Iais californica* (Richardson, 1904) and *Sphaeroma quoyanum* H. Milne Edwards, 1840 (Isopoda). *Crustaceana* 28:247–256.

274. Rudnick, D. A., K. Hieb, K. F. Grimmer, and V. H. Resh. 2003. Patterns and processes of biological invasion: the Chinese mitten crab in San Francisco Bay. *Basic and Applied Ecology* 4(3):249–262.
275. Ruiz, G. M. and J. A. Crooks. 2001. Marine invaders: patterns, effects, and management of non-indigenous species. In: P. Gallagher and L. Bendell-Young, eds., *Waters in Peril*, Chapter 1. Boston, MA: Kluwer Academic Publishers.
276. Salisbury, E. J., Sir [and J. Fisher, Sir. J. Gilmore, J. Huxley, and L. D. Stamp]. 1961. *Weeds and Aliens.* London: Collins Clear Type Press, pp.29, 216, 237.
277. Salogga, D. S. 1982. Occurrence, Symptoms and Probable Cause of *Discula* Species of Cornus Leaf Anthracnose. MS thesis. University of Washington, Seattle.
278. Saltonstall, K. 2002. Cryptic invasion by a non-native genotype of the common reed, *Phragmites australis*, into North America. *Proceedings of the National Academy of Sciences* 99(4):2445–2449.
279. Saltonstall, K. 2003. Microsatellite variation within and among North American lineages of *Phragmites australis. Molecular Ecology* 12:1689–1702.
280. Sawada, H., H. Yokosawa, and C. C.Lambert. 2001. *The Biology of Ascidians*. Tokyo: Springer Verlag.
281. Scagel, R. F. 1956. Introduction of a Japanese alga, *Sargassum muticum*, into the Northeast Pacific. Washington Department of Fisheries, Fisheries Research Papers 1:1–10
282. Scheffer, V. B. The Olympic goat controversy: a perspective. *Conservation Biology* 7(4):916–919.
283. Scott, W. B., and E. K. Crossman. 1973. Freshwater Fishes of Canada. Fisheries Research Board of Canada, Ottawa, *Bulletin* 184.
284. Shafroth, P. R., G. T. Aubla, and M. L. Scott. 1995. Germination and establishment of the native plains cottonwood (*Populus deltoides* Marshall subsp. *moniifera*) and the exotic Russian-olive (*Elaeagnus angustifolia* L.). *Conservation Biology* 9:1169–1175.
285. Sheley, R. L., L. L. Larson, and D. E. Johnson. 1993. Germination and root dynamics of range weeds and forage species. *Weed Technology* 7: 234–237.
285a. Sheley, R. L., and J. K. Petroff, eds.1999. *Biology and Management of Noxious Rangeland Weeds*. Corvallis, OR: Oregon State University Press.
286. Shurtleff, W., and Aoyagi, A. 1977. *The Book of Kudzu*. Brookline, MA: Autumn Press.
287. Sigler, W. F., and J. W. Sigler. 1987. *Fishes of the Great Basin: A Natural History*. Reno, NV: University of Nevada Press.
288. Silva, P. C. 1955. The dichotomous species of *Codium* in Britain. *Journal of the Marine Biological Association of the United Kingdom* 34:565–577.
289. Skinner, L. C., and J. Taylor. 2001. Impact and management of purple loosestrife (*Lythrum salicaria*) in North America. *Biodiversity and Conservation* 10:1787–1807.
290. Smale, M. C. 1990. Ecological role of Buddleia (*Buddleia davidii*) in stream beds in Te Urewera National Park. *New Zealand Journal of Ecology* 14:1–6.
291. Smith, C., and J. Shireman. 1983. White amur bibliography. Gainesville, FL: University of Florida Center for Aquatic Weeds.
292. Smith, R. H. 1994. *Native Trout of North America*. Portland, OR: Frank Amato Publishing.
293. Smith, W. P. 1985. Plant associations within the interior valley of the Umpqua River basin, Oregon. *Journal of Range Management* 38(6):526–530.
294. *Southern Tablelands Farm Forestry Network Newsletter*, November/December 2001.
295. Spielman, A., Sc.D. and M. D'Antonio. 2001. *Mosquito: A Natural History of Our Most Persistant and Deadly Foe*. New York: Hyperion.
296. See entry 174–a.
297. Starr, F., K. Starr, and L. Loope. 2003. *Buddleia davidii*. Plants of Hawai'i Reports. Maui, HI: United States Geological Survey.
298. Steele, E. N. 1964. *The Immigrant Oyster* (Ostrea gigas). Olympia, WA: Warren's Quick Print.

299. Sutherland, W. J. 1990. Biological flora of the British Isles. *Iris pseudacorus. Journal of Ecology* 78:833–848.
300. Sutton, D. L. 1996. Depletion of turions and tubers of *Hydrilla verticillata* in the North New River Canal, Florida. *Aquatic Botany* 53:121–130.
301. Talley, T. A., Crooks, J. A., and Levin, L. A. 2001. Habitat utilization and alteration by the invasive burrowing isopod *Sphaeroma quoyanum*, in California salt marshes. *Marine Biology* 138(3):561–573.
302. Tan, Q. K. 1984. The ecological study on the anadromous crab *Eriocheir sinensis* going upstream. *Chinese Journal of Zoology* 6:19–22.
303. Taylor, R., and B. MacBryde. 1977. *Vascular Plants of British Columbia: A Descriptive Resource Inventory*. Vancouver, BC: University of British Columbia Press.
304. Taylor, R. J. 1990. *Northwest Weeds: The Ugly and Beautiful Villains of Fields, Gardens, and Roadsides*. Missoula, MT: Mountain Press.
305. The Nature Conservancy. *Purple Loosestrife: Element Stewardship Abstract*. Wildlife Weeds Management and Research Program, Weeds on the Web. http://tncweeds.ucdavis.edu/esadocs/lythsali.html, May 26, 2005.
306. Thieltges, D. W., M. Strasser, and K. Reise. 2003. The American slipper limpet *Crepidula fornicata* (L.) in the northern Wadden Sea 70 years after its introduction. *Helgoland Marine Research* 57:27–33.
307. Thompson, D. Q., R. L. Stuckey, and E. B. Thompson. 1987. Spread, impact and control of purple loosestrife (*Lythrum salicaria*) in North American wetlands. *Fish and Wildlife Research* No. 2. Washington, DC: U.S. Fish and Wildlife Service.
308. Thunhorst, G., and Swearingen, J. 1999. PCA Alien Plant Working Group Fact Sheet: Leafy Spurge. Plant Conservation Alliance. http://www.nps.gov/plants/alien/fact/eues1.htm, May 26, 2005.
309. Tomelleri, J. R., and M. E. Eberle. 1990. *Fishes of the Central United States*. Lawrence, KS: University Press of Kansas.
310. Tomkins, D. M., A. R. White, and M. Boots. 2003. Ecological replacement of native red squirrels by invasive greys driven by disease. *Ecology Letters* 6:189–196.
311. Trowbridge, C. 1999. An assessment of the potential spread and options for control of the introduced green macroalga *Codium fragile* ssp. *Tomentosoides* on Australian shores. CRIMP consultancy report. CSIRO Centre for Research on Introduced Marine Pests. 44 pp.
312. Turner, D. C., and P. Bateson, eds. 2000. *The Domestic Cat: The Biology of Its Behaviour*. 2nd ed. Cambridge, UK: Cambridge University Press.
313. United States Court of Appeals for the Ninth Circuit, No. 00–35667 CV-99–05433. 2002. *Association to Protect Hammersley, Eld, and Totten Inlets v. Taylor Resources, Inc.*
314. Upadhyaya, M. K., and R. S. Cranston. 1991. Distribution, biology, and control of hound's-tongue in British Columbia. *Rangelands* 13:103–106.
315. Upadhyaya, M. K., H. R. Tilsner, and M. D. Pitt. 1988. The biology of Canadian weeds. 87. *Cynoglossum officinale* L. *Canadian Journal of Plant Science* 68:763–774.
316. Upadhyaya, M. K., R. Turkington, and D. McIlvride. 1986. The biology of Canadian weeds. 75. *Bromus tectorum* L. *Canadian Journal of Plant Science* 66:689–709.
317. U.S. Department of Agriculture. 1992. *Potential Impacts of Asian Gypsy Moth in the Western United States*. Washington, DC: Forest Pest Management Department.
318. U.S. Department of Agriculture. 2000. Pest Risk Assessment of Importation of Solid Wood Packing Materials into the United States. http://www.aphis.usda.gov/ppq/pra/swpm/, May 26, 2006.
319. U.S. Department of the Interior, Bureau of Reclamation. 1985. Stop: This Weed Is a Menace.
320. U.S. Geological Service, Biological Resources Division. 1998. South American nutria destroy marsh habitat. http://www.pwrc.usgs.gov/factshts/nutria.pdf, May 26, 2005.

321. U.S. Geological Survey, Center of Aquatic Resource Studies, 2004. Nonindigenous Aquatic Species Web site. Online at nas.er.usgs.gov, May 26, 2005.
321a. Uva, R. H., J. C. Neal, and J. M. DiTomaso. 1997. *Weeds of the Northeast.* Ithaca, NY: Cornell University Press.
322. Vaclav, R., et al. 2002. Badge size, paternity assurance, and paternity losses in male House Sparrows. *Journal of Avian Biology* 33(3):315–318.
323. Vallet C., J. C. Dauvin, D. Hamon, and C. Dupuy. 2001. Effect of the introduced common slipper shell on the suprabenthic biodiversity of the subtidal communities in the Bay of Saint-Brieuc. *Conservation Biology* 15:1686–1690.
324. Van der Velde, G., and Van der Heijden, L. A. 1981. The floral biology and seed production of *Nymphoides peltata* (Gmel.) O. Kuntze (Menyanthaceae). *Aquatic Botany* 10(3):261–293.
325. Van, T. K., G. S. Wheeler, and T. D. Center. 1999. Competition between *Hydrilla verticillata* and *Vallisneria americana* as influenced by soil fertility. *Aquatic Botany* 62:225–233.
326. Vance, D. R., and R. L. Westemeier. 1979. Interactions of pheasants and prairie chickens in Illinois. *Wildlife Society Bulletin* 7:221–225.
327. Van-der-Hock, D. 2004. Nutrient limitation and nutrient-driven shifts in plant species composition in a species rich fen meadow. *Journal of Vegetation Science* 15(3):389–396.
328. Veldhuizen, T. and S. Stanish. 1999. Overview of the Life History, Distribution, Abundance, and Impacts of the Chinese mitten crab, *Eriocheir sinensis.* Publication of California Department of Water Resources, Environmental Services Office, Sacramento, CA.
329. Vigg, S., T. P. Poe, et al. 1991. Rates of consumption of juvenile salmonids and alternative prey fish by Northern Squawfish, Walleyes, Smallmouth Bass, and Channel Catfish in the John Day Reservior, Columbia River. *Transactions of the American Fisheries Society* 120:421–438.
329a. Vitousek, P. M., C. M. D'Antonio, L. L. Loope, and R. Westbrooks. 1996. Biological Invasions as Global Environmental Change. *American Scientist* 84:468–478.
329b. Vitousek, P. M., L. L. Loope, and C. P. Stone. 1987. Introduced species in Hawaii: biological effects and opportunities for ecological research. *Trends in Ecology and Evolution* 2:224–227.
329c. Virginia Native Plant Society. 1999. *Kudzu (Pueraria lobata [Willd.] Ohwi).* Richmond, VA: Virginia Department of Conservation and Recreation, Division of Natural Heritage.
330. Volpe, J. P. 2001. *Super un-natural, Atlantic Salmon in BC Waters.* David Suzuki Foundation publication. 33p. Online at www.davidsuzuki.org, May 26, 2005.
331. Volpe, J. P., E. B. Taylor, E. B. Rimmer, and B. W. Glickman. 2000. Evidence of natural reproduction of aquaculture-escaped Atlantic salmon in a coastal British Columbia river. *Conservation Biology* 14:899–903.
332. Wagner, E. K., and J. H. Martinez. 2004. *Basic Virology*, pp. 241–247. 2nd ed. London: Blackwell Publishing.
333. Wallner, W. E. 1997. *Exotic Pests of Eastern Forests.* Conference Proceedings – April 8–10, 1997, ed. K. O. Britton. Nashville, TN: USDA Forest Service , and Tennessee Exotic Pest Plant Council.
334. Warner, P., C. Bossard, M. Brooks, J. DiTomaso, J. Hall, A. Howald, D. Johnson, J. Randall, C. Roye, M. Ryan, and A. Stanton. 2003. Criteria for Categorizing Invasive Non-Native Plants that Threaten Wildlands. California Exotic Pest Plant Council and Southwest Vegetation Management Association. 24 pp. Online at www.caleppc.org and www.swvma.org, May 26, 2005.
335. Watler, D., and Lake, S. 2001. *Sirex noctilio* F., Sirex Wasp. Canadian Food Inspection Agency Science Branch, Plant Health Risk Assessment Unit. Online at www.inspection.gc.ca/english/sci/surv/data/sirnoce.shtml, May 26, 2005.
336. Weis, J. S., and P. Weis. 2003. Is the invasion of the common reed, *Phragmites australis*,

into tidal marshes of the eastern US an ecological disaster? *Marine Pollution Bulletin* 46:816–820.
337. West Nile Virus: Detection, surveillance, and control. 2001. *Annals of the New York Academy of Sciences* 951:15–35, 84–93.
338. West, R., C. D. Dondale, and R. A. Ring. 1984. A revised checklist of the spiders (Araneae) of British Columbia. *Journal of the Entomological Society of British Columbia* 81:80–98.
339. Whitaker, J. O., Jr. 1988. *The Audubon Society Field Guide to North American Mammals.* New York: Alfred A. Knopf.
340. Whitfield, G. J. 1990. *Life History and Ecology of the Slider Turtle.* Washington, DC: Smithsonian Institution Press.
341. Whitson, T. D., L. C. Burrill, S. A. Dewey, D. W. Cudney, B. E. Nelson, R. D. Lee, and R. Parker. 1999. *Weeds of the West.* Jackson, WY: Grand Teton Lithography.
342. Whitson, T. D.,ed. 1999. *Weeds of the West.* 5th ed. Newark, CA: The Western Society of Weed Science.
343. Wiedemann, A. 1987. The Ecology of European Beachgrass (*Ammophila arenaria* (L.) Link): A Review of the Literature. Oregon Department of Fish and Wildlife, Nongame Wildlife Program Technical Report No. 87–1–01.
344. Wiedemann, A., et al. 1974. *Plants of the Oregon Coastal Dunes.* Corvallis, OR: Oregon State University Press, with OSU Department of Botany, Oregon State University.
345. Williams, T. 1999. The terrible turtle trade. *Audubon* 101(2):44, 46–48, 50–51.
346. Williamson, J., and S. Harrison. 2002. Biotic and abiotic limits to the spread of exotic revegetation species. *Ecological Applications* 12(1):40–51.
347. Winterbourn, M. J. 1970. The New Zealand species of *Potamopyrgus* (Gastropoda: Hydrobiidae). *Malacologia* 10(2):283–321.
348. Woods, C., L. Contreras, G. Wilner-Chapman, and H. P. Whidden. 1992. *Myocastor coypus.* Mammalian Species, No. 398. The American Society of Mammalogists, p. 8.
349. Wootton, A. 1984. *Insects of the World.* pp. 22, 75, 203. London: Blandford.
350. Wydoski, R. S., and R. R. Whitney. 2003. *Inland Fishes of Washington.* 2nd ed. Seattle, WA: University of Washington Press.
351. Yamada, S. B. 2001. *Global Invader: The European Green Crab.* Corvallis, OR: Oregon Sea Grant Publications, Oregon State University.
352. Yeo, P. F. 1985. *Hardy Geraniums.* 2nd ed. Portland, OR:Timber Press.
353. Young, J. A. 1992. Ecology and management of medusahead (*Taeniatherum caput-medusae* ssp. *asperum* [Simk.] Melderis). *Great Basin Naturalist* 52:245–252.
354. Young, J. A., and R. A. Evans. 1969. Control and ecological studies of Scotch thistle. *Weed Science* 17:60–63.
355. Young, J. A., J. D. Trent, R. R. Blank, and D. E. Palmquist. 1998. Nitrogen interactions with medusahead (*Taeniatherum caput-medusae* ssp. *asperum*) seedbanks. *Weed Science* 46:191–195.
356. Young, J. A., C. E. Turner, and L. F. James. 1995. Perennial pepperweed. *Rangelands* 17:121–123.
357. Young, J. A., D. E. Palmquist, and R. R. Blank. 1998. The ecology and control of perennial pepperweed (*Lepidium latifolium* L.). *Weed Technology* 12:402–405.
358. Young, J. A., D. E. Palmquist, and S. O. Wotring. 1997. The invasive nature of *Lepidium latifolium*: A review. In: Brock, J. H., Wade, M., Pysek, P., and Green, D., eds., *Plant Invasions: Studies from North America and Europe.* Leiden, The Netherlands: Backhuys Publishers.
359. Yund, P. O., and Stires, A. 2002. Spatial variation in population dynamics in a colonial ascidian (*Botryllus schlosseri*). *Marine Biology* 141:955–963.
360. Zaranko, D. T., D. G. Farara, and F. G. Thompson. 1997. Another exotic mollusc in the Laurentian Great Lakes: the New Zealand native *Potamopyrgus antipodarum* (Gray 1843) (Gastropoda: Hydrobiidae). *Canadian Journal of Fisheries and Aquatic Sciences* 54:809–814.

361. Zardini, E. M., Gu, H., and P. H. Raven. 1991. On the separation of two species within the *Ludwigia uruguayensis* complex (Onagraceae). *Systematic Botany* 16(2):242–244.
362. Zondag, R., and Nuttall, M. J. 1977. *Sirex noctilio* Fabricius (Hymenoptera: Siricidae), Sirex. New Zealand Forest Service, Forest and Timber Insects in New Zealand No. 20.
363. Zouhar, K. 2001. Centaurea maculosa. In: Fire Effects Information System. U.S. Department of Agriculture, Forest Service, Rocky Mountain Research Station, Fire Sciences Laboratory (Producer). Online at www.fs.fed.us/database/feis/, May 26, 2005.
364. Zueg, S., G. O'Leary, T. Sommer, B. Harrell, and F. Feyer. 2002. Introduced palaemonid shrimp invades the Yolo bypass floodplain. Interagency Ecological Program for the San Francisco Estuary (IEP) *Newsletter* 15(1):13–15.

Additional Sources of Information

1. American Bamboo Society
   www.americanbamboo.org
2. American Bird Conservancy
   Cats Indoors! Campaign
   1834 Jefferson Place NW, Washington, DC 20033
3. American Phytopathological Society
   www.apsnet.org
4. Applied Vegetation Dynamics Laboratory
   www.appliedvegetationdynamics.co.uk
5. Aquatic Invertebrates of Montana Project
   Department of Biology, Montana State University, Bozeman, MT 59717
6. Aquatic Plant Management Society, Inc.
   P.O. Box 821265, Vicksburg, MS 39182
7. *Ascidian News*
   12001 11th Avenue NW, Seattle, WA 98177
8. Atlantic Salmon Watch Program, Fisheries and Oceans Canada
   1–800–811–6010
9. Ayala, F. J., L. Serra, and A. Prevosti. 1989. A grand experiment in evolution: The *Drosophila subobscura* colonization of the Americas. *Genome* 31:246–255.
10. Barkworth, M. 2004. *Cynosurus echinatus*. Draft treatment and distribution map developed for Manual of Grasses for North America project.
11. Barnes, R. D. 1980. *Invertebrate Zoology*. Philadelphia: Saunders College.
12. BC Ministry of Agriculture, Food, and Fisheries
    P.O. Box 9058, STN Provincial Government, Victoria, BC V8W 9E0 Canada
    www.agf.gov.bc.ca
13. Botanical Electronic News
    www.ou.edu/cas/botany-micro/ben
14. Bureau of Land Management, California
    2800 Cottage Way, #W-1834, Sacramento, CA 95825
15. Bureau of Land Management, Oregon and Washington
    P.O. Box 2965, Portland, OR 97208
16. Burke Museum of Natural History and Culture
    University of Washington, Box 353010, Seattle, WA 98195
17. California Department of Fish and Game
    1416 Ninth Street, Sacramento, CA 95814
    www.dfg.ca.gov
18. California Invasive Plant Council
    1442–A Walnut Street, #462, Berkeley, CA 94709
19. California Oak Mortality Task Force
    www.suddenoakdeath.org
20. Calflora
    www.calflora.org
21. Canadian Food Inspection Agency
    BC Office: 4321 Still Creek Drive, #400, Burnaby, BC V5C 6S7
    www.inspection.gc.ca

22. Canadian Wildlife Service
Environment Canada, Ottawa, ON K1A 0H3 Canada
www.cws-scf.ec.gc.ca
23. Cats In Kennels Program
P.O. Box 436, Minden, ON K0M 2K0 Canada
24. Center for Aquatic and Invasive Plants
University of Florida, 7922 NW 71st Street, Gainesville, FL 32653
plants.ifas.ufl.edu
24a. Center for Environmental and Regulatory Systems,
Purdue University
www.ceris.purdue.edu/napis/states/or/imap/orkudzu.html
25. Center for Invasive Plant Management
Montana State University, P.O. Box 173120, Bozeman, MT 59717
www.weedcenter.org
26. Cohen, A., et. al. 2001. *WA State Exotics Expedition 2000: A rapid survey of exotic species in the shallow waters of Elliot Bay, Totten and Eld Inlets, and Willapa Bay*. Nearshore Habitat Program, Washington Department of Natural Resources
26a. Columbia University
www.columbia.edu/itc/cerc/danoff_burg/invasion_bio/inv_spp_summ/Pueraria_montant.html
27. Cornell Laboratory of Ornithology
159 Sapsucker Woods Road, Ithaca, NY 14850
28. Douglas, G. W., D. Meidinger, and J. Pojar, eds., 2001. *Illustrated Flora of British Columbia. Vol. 7: Monocotyledons (Orchidaceae through Zosteraceae)*. Victoria, BC: Ministry of Sustainable Resource Management, Ministry of Forests.
29. Ecology and Management of Invasive Plants Program
www.invasiveplants.net
30. EFISH: The Virtual Aquarium
Virginia Tech, Blacksburg, VA
www.cnr.vt.edu/efish
31. E-Flora – Electronic Atlas of the Plants of British Columbia
www.eflora.bc.ca
32. Environmental Protection Agency, Region 10 (AK, ID, OR, WA)
1200 Sixth Avenue, Seattle, WA 98101
epa.gov
33. ESSA Technologies, Ltd. 2002. Towards a Decision Support Tool to Address Invasive Species in the Garry Oak and Associated Ecosystems of British Columbia. Prepared for the Garry Oak Ecosystems Recovery Team Invasive Species Steering Committee, Victoria, BC.
34. Fish Museum
Department of Zoology, University of British Columbia
6270 University Boulevard, Vancouver, BC V6T 1Z4 Canada
35. FishBase
www.fishbase.org
36. Fisheries and Oceans Canada, Pacific Biological Station
3190 Hammond Bay Road, Nanaimo, BC V9T 6N7 Canada
37. Gaines, X. M., and D. G. Swan. 1972. *Weeds of Eastern Washington and Adjacent Areas*. Davenport, WA: Camp-Na-Bor-Lee Association.
38. Gordon, D. R., J. M. Welker, J. W. Menke, and K. J. Rice. Competition for soil water between annual plants and Blue Oak (*Quercus douglasii*) seedlings. *Oecologia* 79:533–541.
39. Grass Manual on the Web
herbarium.usu.edu/webmanual
40. Gwaii Haanas National Park, Reserve and Haida Heritage Site
P.O. Box 37, Queen Charlotte, BC V0T 1S0 Canada

41. Hannaway, D., et al. 1999. Orchard grass (*Dactylis glomerata* L.). Corvallis, OR: Oregon State University Extension Service Publication No. PNW 502.
42. Hickman, J. C. 1993. *The Jepson Manual: Higher Plants of California*. Berkeley, CA: University of California Press.
43. Idaho Department of Agriculture
Bureau of Vegetation Management
Noxious Weed Program. P.O. Box 7249, Boise, ID 83707
44. Institute for Applied Ecology
227 SW 6th Street, Corvallis, OR 97333–4616
www.appliedeco.org
45. Invasive.org
www.invasive.org
46. Japanese Knotweed Alliance
Bakeham Lane, Egham, Surrey TW20 9TY, UK
47. Jimerson, T. J., and S. K. Carothers. 2002. *Northwest California Oak Woodlands: Environment, Species Composition, and Ecological Status*. Eureka, CA: USDA Forest Service General Technical Report PSW-GTR-184.
48. King County Natural Resources and Parks
Water and Land Resources Division
201 S Jackson Street, #600, Seattle, WA 98104
49. King County Noxious Weed Control Board
201 S Jackson Street, #600, Seattle, WA 98104
dnr.metrokc.gov/wlr/lands/weeds
50. King County Public Health Department
Environmental Health Division
Wells Fargo Center, 999 Third Avenue, #700, Seattle, WA 98104
51. Laskeek Bay Conservation Society
Box 867, Queen Charlotte, BC V0T 1S0 Canada
52. MacDougall, A. S. 2002. Invasive perennial grasses in *Quercus garryana* meadows of southwestern British Columbia: Prospects for restoration. USDA Forest Service General Technical Report PSW-GTR-184.
53. Maryland Department of Natural Resources
580 Taylor Avenue, Tawes State Office Building, Annapolis, MD 21401
www.dnr.state.md.us
54. Ministry of Forests, Forest Practices Branch, Invasive Plant Program
P.O. Box 9513 Stn. Prov. Govt., Victoria, BC V8W 9C2 Canada
55. Montana Weed Control Association
www.mtweed.org
56. Mueller, K., B. Sizemore, and L. Timme. 1997. *Guidelines and requirements for the import and transfer of shellfish, including oysters, clams, and other aquatic invertebrates in Washington State*. Washington Department of Fish and Wildlife, Point Whitney Shellfish Laboratory and Willapa Bay Field Station.
57. National Agricultural Pest Information System (NAPIS)
www.ceris.purdue.edu/napis
58. National Aquatic Nuisance Species Clearinghouse
www.aquaticinvaders.org
59. National Audubon Society, Washington State Office
P.O. Box 462, Olympia, WA 98507
www.audubon.org
60. National Invasive Species Council
Department of the Interior, Office of the Secretary (OS/SIO/NISC)
1849 C Street NW, Washington, DC 20240
www.invasivespecies.gov

61. National Park Service, PNW Regional Office
1111 Jackson Street, #700, Oakland, CA 94607
www.nps.gov
62. National Sea Grant Library
University of Rhode Island
Bay Campus, Pell Library Building, Narragansett, RI 02882
nsgd.gso.uri.edu
63. Natural Lawn and Garden Hotline
Seattle Public Utilities
699 Fifth Avenue, #4900, Seattle, WA 98116 (206) 633–0224
64. Natural Resource Sciences, Cooperative Extension
P.O. Box 646410, Washington State University, Pullman, WA 99162
ext.nrs.wsu.edu
65. NatureWatch
www.naturewatch.ca
66. Nearshore Habitat Program
Washington State Department of Natural Resources
1111 Washington Street SE, P.O. Box 47027, Olympia, WA 98504
www2.wadnr.gov/nearshore/research
67. Nevada Weed Action Committee
agri.state.nv.us/nwac
68. Northwest Horticultural Society
Center for Urban Horticulture
University of Washington, P.O. Box 354115, Seattle, WA 98195
69. Northwest Mosquito and Vector Control Association
521 First Avenue NW, Great Falls, MT 59404
70. Ontario Federation of Anglers & Hunters
4601 Guthrie Drive, P.O. Box 2800, Peterborough, ON K9J 8L5 Canada
71. Oregon Department of Agriculture Noxious Weed Control Program
635 Capitol Street NE, Salem, OR 97301
72. Oregon Department of Agriculture
635 Capitol Street NE, Salem, OR 97301
73. Oregon Department of Fish and Wildlife
3406 Cherry Avenue NE, Salem, OR 97303
74. Oregon Flora Project
www.oregonflora.org
75. Oregon Sea Grant Program, Clackamas County Extension
200 Warner Milne Road, Oregon City, OR 97045
76. Oregon State University Herbarium
Department of Botany and Plant Pathology, 2082 Cordley Hall, Corvallis, OR 97331
oregonstate.edu/dept/botany/herbarium
77. Pheasants Forever
1783 Buerkle Circle, St. Paul, MN 55110
78. Plant Conservation Alliance, Bureau of Land Management
1849 C Street NW, LSB-204, Washington, DC 20240
www.nps.gov/plants/alien
79. Portland State University Center for Lakes and Reservoirs
P.O. Box 751, Portland, OR 97195
80. Progressive Animal Welfare Society (PAWS)
P.O. Box 1037, Lynnwood, WA 98046

81. Puget Sound Action Team
Office of the Governor, P.O. Box 40900, Olympia, WA 98504
www.psat.wa.gov
82. Research Group on Introduced Species
P.O. Box 867, Queen Charlotte, BC V0T 1S0 Canada
83. Royal British Columbia Museum
675 Belleville Street, Victoria, BC V8W 9W2 Canada
84. San Francisco Estuary Institute
7770 Pardee Lane, Oakland, CA 94621
85. Sea Grant Nonindigenous Species
www.sgnis.org
85a. Seattle Weekly
www.seattle.weekly.com/features/0244/news_scigliano.shtml
86. Sugihara, N. G., L. J. Reed, and J. M. Lenihan. 1987. Vegetation of the Bald Hills oak woodlands, Redwood National Park, California. *Madrono* 34(3):193–208.
87. Tamarisk Coalition
P.O. Box 1907, Grand Junction, CO 81502
www.tamariskcoalition.org
88. Tasmania Department of Primary Industries, Water, and Environment. 2003. *Acute Bovine Liver Disease*. New Town, Tasmania: Animal Health and Welfare Branch.
88a. The Ascidian Home Page for the United States
http://depts.washington.edu/ascidian/ (current references)
88b. The Dutch Ascidians Homepage
http://www.ascidians.com (photos of species worldwide)
89. The Invaders Database
invader.dbs.umt.edu
90. The Natural History Museum
Cromwell Road, London, SW7 5BD, UK
www.nhm.ac.uk
91. The Nature Conservancy, Invasive Species Initiative
tncweeds.ucdavis.edu
92. The Nature Conservancy
4245 North Fairfax Drive, #100, Arlington, VA 22203
nature.org
93. The PLANTS Database, Version 3.5.
National Plant Data Center, Baton Rouge, LA 70874
plants.usda.gov
94. Tijuana River National Estuarine Research Reserve
301 Caspian Way, Imperial Beach, CA 91932
95. University of Washington Herbarium
University of Washington, Box 355325, Seattle, WA 98195
herbarium.botany.washington.edu
96. US Department of Agriculture. 1971. *Common Weeds of the United States*. Prepared by the Agricultural Research Service of the USDA. New York: Dover Publications.
97. US Department of Agriculture. 2000. *Draft Supplemental Environmental Impact Statement: Management of Port Orford Cedar in Southwest Oregon.*
98. US Department of Agriculture. 2004. *Final Supplemental Environmental Impact Statement: Management of Port Orford Cedar in Southwest Oregon.*
99. US Department of Agriculture. 2004. *Gypsy Moth Digest*. Morgantown, WV: State and Private Forestry. Online.
100. US Department of Agriculture
APHIS Plant Protection and Quarantine
www.aphis.usda.gov/ppq

101. US Department of Agriculture
APHIS Veterinary Services, Oregon Office
530 Center Street NE, #335, Salem, OR 97300
102. US Department of Agriculture
APHIS Veterinary Services, Washington Office
2604 12th Court SW, #B, Olympia, WA 98501
103. US Department of Agriculture, US Forest Service. 2003. *A Range-Wide Assessment of Port Orford Cedar on Federal Lands.*
104. US Department of Agriculture, US Forest Service
Forest Health Protection
www.fs.fed.us/foresthealth
105. US Department of Agriculture, US Forest Service
Forest Pest Management
Division of State and Private Forestry, P.O. Box 3623, Portland OR 97208
106. US Department of Agriculture, US Forest Service
PNW Research Station, 333 SW First Avenue, Portland, OR 97184
www.fs.fed.us/r6
107. US Department of Agriculture, US Forest Service
St. Paul Field Office, 1992 Folwell Avenue, St. Paul, MN 55106
108. US Fish and Wildlife Service
Aquatic Nuisance Species Task Force
www.anstaskforce.gov
109. US Fish and Wildlife Service
Washington Islands National Wildlife Refuge
C/o Nisqually NWR Complex, 100 Brown Farm Road, Olympia, WA 98506
110. US Fish and Wildlife Service
www.fws.gov
111. US Geological Survey, Nonindigenous Aquatic Species Information
Center for Aquatic Resource Studies, 7920 NW 71st Street, Gainesville, FL 32653
nas.er.usgs.gov
112. Virginia Department of Game and Inland Fisheries
4010 West Broad Street, Richmond, VA 23230
113. Virginia Native Plant Society
400 Blandy Farm Lane, Unit 2, Boyce, VA 22620
114. Washington Department of Fish and Wildlife
Point Whitney Shellfish Laboratory, 1000 Point Whitney Road, Brinnon, WA 98320
115. Washington Department of Fish and Wildlife
Willapa Bay Field Station, 26700 Sandridge Road, Ocean Park, WA 98640
116. Washington Department of Fish and Wildlife
600 Capitol Way North, Olympia, WA 98501
117. Washington Native Plant Society
6310 NE 74th Street, #215E, Seattle, WA 98115
118. Washington Sea Grant Program
3716 Brooklyn Avenue NE, Seattle, WA 98105
119. Washington State Department of Agriculture
Spartina Program, P.O. Box 42560, Olympia, WA 98504
120. Washington State Department of Agriculture
1111 Washington Street SE, P.O. Box 42560, Olympia, WA 98502
121. Washington State Department of Ecology
P.O. Box 47600, Olympia, WA 98504
122. Washington State Department of Health
Office of Environmental Health and Safety, P.O. Box 47825, Olympia, WA 98504

123. Washington State Department of Natural Resources
Commissioner of Public Lands, P.O. Box 47001, Olympia, WA 98504
www.dnr.wa.gov
124. Washington State Department of Natural Resources
Forest Health Section
www.dnr.wa.gov/htdocs/rp/forhealth
125. Washington State Noxious Weed Control Board
P.O. Box 42560, Olympia, WA 98504
www.nwcb.wa.gov
125a. Washington State University Cooperative Extension DNR Sciences, Pullman, WA 99164-6410
126. Weeds BC
Open School BC, 563 Superior Street, Victoria, BC V8V 1T7 Canada
www.weedsbc.ca
127. Western Regional Panel on Aquatic Nuisance Species
answest.fws.gov/home.htm
128. Westland Resource Group. 1999. COSEWIC *Status Report on the Hadley Stickleback Species Pair,* Gasterosteus *sp.* Victoria, BC: Committee on the Status of Endangered Wildlife in Canada. 21 pp.
129. Whatcom County Noxious Weed Control Board
901 W Smith Road, Bellingham, WA 98226
130. Wildlife Habitat Council
1010 Wayne Avenue, #920, Silver Spring, MD 20910

Glossary

abiotic Non-living components of an environment or ecosystem (e.g., chemicals in the air, water, or soil)
adventitious A root growing from a location other than the underground descending portion of the axis of a plant
adventive Growing spontaneously but not native to the locality in which it appears
aeciospores A spore produced in the aecium of a rust fungus
aecium A cuplike structure of some rust fungi that contains chains of aeciospores; aecia usually form on the bottom surface of leaves
allelopathy The inhibition of growth in one plant species by chemicals produced by another species
anadromous A type of fish that spends its adult life at sea but returns to the upper reaches of a river to spawn (e.g., salmon)
auricles Ear-shaped appendages that occur in pairs at the leaf sheath that envelopes the stem
awn The delicate spinous process ("beard") that terminates the grain-sheath of barley, oats, and other grasses; extended in Botany to any similar bristly growth

basal Attached or located at the base of a structure
basidiospores A sexually produced fungal spore borne on a basidium (a typical mushroom produces billions)
benthic Relating to the bottom of a sea or lake or to the organisms that live there
biflagellate Possessing two flagellae (rigid structures that protrude from the cell surface and are used to propel the organism)
bioturbation The disturbance and mixing of sediment by the activity of living organisms; the disturbed state that results

clonal spread Spread through non-sexual reproduction
commensal Animals or plants that live as tenants of others (distinguished from parasitic)
coumarin A fragrant crystalline substance found in the seeds of the Tonka bean, sweet clover, etc.
diatom A distinctive group of microscopic, unicellular plants (algae) that live in fresh and salt water and produce cell walls composed of silica
dioecious Bot.: Having the unisexual male and female flowers on separate plants. Zool.: Having the two sexes in separate individuals; sexually distinct
DO Dissolved oxygen

epiphyte/epiphytic A plant that grows on another plant; usually restricted to those that derive only support, not nutrition, from the plants on which they grow
epizootic Relating to a rapidly spreading disease that affects a large number of animals at the same time within a particular area
extirpation No longer found in an area

forbs Herbaceous plants of a kind other than grass; applied chiefly to any broad-leaved herbs growing naturally on grassland

gametes A reproductive cell containing one copy of each gene (haploid) needed to provide the genetic information required for the development of an offspring. When gametes from the female (egg) are combined with or fertilized by gametes from the male (sperm), a zygote containing two copies of each gene is produced. Zygotes can develop into offspring.
glume One of the two chaffy bracts at the base of a grass spikelet

holdfast An organ for superficial attachment developed by some algae and fungi

inflorescence A group of flowers growing from a common stem, often in a characteristic arrangement; also called a flower cluster
instar Any of the stages between successive skin sheddings in the life of an insect or arthropod, including the stage between hatching from the egg and the first shedding

karyogamy The fusion of cell nuclei

lemma The outer or lower bract enclosing one of the flowers within a grass spikelet
ligules A transparent sheath that projects up from the inside of a leaf blade where it joins the stem

macrophyte A (usually aquatic) plant visible to the naked eye
monotypic A taxonomic category that has only one subordinate taxon within it
motile Moving, or able to move , by itself
mycelium/a The vegetative tissue (thallus) of a fungus—typically consisting of a network of fine filaments (hyphae)
mycorrhiza/al The symbiotic association of the mycelium of a fungus with the roots of plants

obligate Restricted to a particular (specified) mode of life, habitat, or function (e.g., an obligate anaerobe cannot grow, and may not survive, in the presence of oxygen. An obligate parasite cannot survive independently of its host.)

palmate Bot.: Having three or more equal divisions, lobes or veins radiating from a common point like the fingers of an outspread hand, as in a palmately compound, lobed, or veined leaf. Zool.: Having webbed feet
parthenogenesis Reproduction from a gamete without fertilization, occurring most commonly in invertebrates and lower plants
petiole The stalk by which a leaf is attached to the stem of a plant
pinnate Resembling a feather; having lateral parts or branches on each side of a common axis, like the vanes of a feather
piscicide A substance that kills fish
piscivorous Habitually feeding on fish (terns and cormorants are piscivorous birds)
propagule A seed, spore, or other product of a plant which is disseminated to form a new individual

*radula*r A toothed, file-like muscle inside the mouth of a gastropod combining the functions of teeth and tongue in a single organ
riparian Relating to or inhabiting the banks of a natural course of water

scarify To make a number of scratches or slight incisions
sporangium/a A cell or structure in which spores are produced; ferns, fungi, mosses, and algae release spores from sporangia
stolon A horizontal stem or branch that takes root at points along its length, forming new plants (e.g., strawberry); also, an arching stem that forms a new rooted plant at the tip (e.g., blackberry)

substrate, substratum Surface or medium on or in which an organism lives and from which it may derive nourishment
swimmerets In crustaceans, paired abdominal appendages used in part for swimming
sympatric Two species or populations having the same or overlapping geographic areas without interbreeding

talus The rock fragments or debris that accumulate at the base of a hill
talus slope The concave slope formed by an accumulation of rock fragments
teliospore A thick-walled often resistant spore in which karyogamy occurs, produced late in the rust fungus life cycle
trophic Relating to nutrition

ungulates Hoofed mammals, mostly herbivores, such as deer, cows, pigs, horses, etc.

whorls Bot.: A group of 3 or more plant structures arising at the same level on the stem and forming a ring around it; a ring of floral organs around the receptacle of the flower. Zool.: A single turn of a spiral shell of a mollusk

xylem Bot.: The vascular tissue of a plant that has the prime function of water transport

zoospore An independently motile spore, usually asexually reproductive (i.e., not a gamete), found in some algae and fungi

Photography Credits

Introductory essays

Agricultural Inspection : USDA APHIS PPQ Archives, USDA APHIS PPQ
Gypsy Moth (*Lymantria dispar*) : Terry McGovern, USDA APHIS PPQ
Kudzu (*Pueraria montana*) : USDA APHIS PPQ Archives, USDA APHIS PPQ
Longhorned beetles (*Anoplophora glabripennis*) : Gerald J. Lenhard, Louisiana State University
Cane Toad (*Bufo marinus*) : Photo by P. Dee Boersma
Fuller's Teasel (*Dipsacus fullonum*) : © 2003 George W. Hartwell

Freshwater Plants

Fanwort (*Cabomba Caroliniana*) : Photo by Karen Hahnel, Maine Dept. of Environmental Protection
Brazilian Elodea (*Egeria densa*) : Photo courtesy of California Department of Food and Agriculture
Hydrilla (*Hydrilla verticillata*) : Raghavan Charudattan, University of Florida
Yellow Flag Iris (*Iris pseudacorus*) : Photo courtesy of Paul Busselen, Katholieke Universiteit Leuven, campus Kortrijk, Belgium
Water Primrose (*Ludwigia hexapetala*) : © 2000 Joe DiTomaso
Eurasian Watermilfoil (*Myriophyllum spicatum*) : Alison Fox, University of Florida
Yellow Floatingheart (*Nymphoides peltata*) : Photo courtesy of California Department of Food and Agriculture
Common Reed (*Phragmites australis*) – seed head : Joseph McCauley, U.S. fish and Wildlife Service
Saltcedar (*Tamarix ramosissima*) – in flower : Steve Dewey, Utah State University
Swollen Bladderwort (*Utricularia inflata*) : Photo courtesy of Barry Rice/sarracenia.com

Marine Plants

Spartina spp. – *Spartina alterniflora, S. anglica, S. densiflora,* and *S. patens*) – Willapa Bay, Washington, USA : Photo courtesy of Dr. Marjorie Wonham, University of Alberta, Canada
Sargassum (*Sargassum muticum*) : Photo courtesy of Kevin Britton-Simmons
Japanese Eelgrass (*Zostera japonica*) : Photo by Jamie Clark/Planet Magazine

Terrestrial Plants

Evergreen Blackberry (*Rubus laciniatus*) : © Br. Alfred Brousseau, Saint Mary's College
Scotch Broom (*Cytisus scoparius*) : Photo by P. Dee Boersma
Cirsium Thistles
  Canada Thistle (*Cirsium arvense*) – flowers : Chris Evans, The University of Georgia; http://www.forestryimages.org
  Bull Thistle (*Cirsium vulgare*) – flowers : Kenneth M. Gale; http://www.forestryimages.org
English Ivy (*Hedera helix*) : Photo by P. Dee Boersma
Diffuse Knapweed (*Centaurea diffusa*) – flowers : USDA APHIS – Oxford, North Carolina Archives, www.forestryimages.org
Japanese Knotweed (*Polygonum cuspidatum*) – flowering plants : Jil M. Swearingen, USDI National Park Service; http://www.forestryimages.org
Musk Thistle (*Carduus nutans*) : Wendy VanDyk Evans; http://www.forestryimages.org
Tree-of-heaven (*Ailanthus altissima*) : Photo by Tihomir Kostadinov

Garlic Mustard (*Alliaria petiolata*) – flowers : Victoria Nuzzo, Natural Area Consultants; http://www.forestryimages.org
European Beachgrass (*Ammophila arenaria* – Poaceae) : © Br. Alfred Brousseau, Saint Mary's College
Slender False-brome (*Brachypodium sylvaticum*) : Photo by Thomas N. Kaye
Cheatgrass (*Bromus tectorum*) : Photo by Richard N. Mack
Butterfly Bush (*Buddleia davidii*) : Photo courtesy of Forest and Kim Starr, USGS
Spotted Knapweed (*Centaurea biebersteinii*) – flowers : Photo courtesy of Cindy Roche; http://www.forestryimages.org
Traveler's Joy (*Clematis vitalba*) : Photo courtesy of Paul Busselen, Katholieke Universiteit Leuven, Kortrijk, Belgium
Houndstongue (*Cynoglossum officinale*) : Photo courtesy of Mark Schwarzlaender, University of Idaho
Hedgehog Dogtail (*Cynosurus echinatus*) : Photo by Keir Morse
Orchard Grass (*Dactylis glomerata*) : Photo by G. F. Hrusa
Spurge laurel (*Daphne laureola*) : Photo by Amadej Trnkoczy
Fuller's Teasel (*Dipsacus fullonum*) : © 2003 George W. Hartwell
Russian Olive (*Elaeagnus angustifolia*) – fruits : Paul Wray, Iowa State University; http://www.forestryimages.org
Leafy Spurge (*Euphorbia esula*) : John M. Randall, The Nature Conservancy; http://www.forestryimages.org
Fennel (*Foeniculum vulgare*) : Photo by P. Dee Boersma
Storksbill (*Erodium – Geranium cicutarium* [var. genus]) : Photo courtesy of Alvin J.Bussan, Montana State University
Herb Robert (*Geranium robertianum*) : © 2003 Monty Rickard
Giant Hogweed (*Heracleum mantegazzianum*) : Donna R. Ellis, University of Connecticut; http://www.forestryimages.org
Common Velvet-grass (*Holcus lanatus*) : Photo by Steve Matson
St. Johnswort (*Hypericum perforatum*) : Norman E. Rees, USDA ARS ; http://www.forestryimages.org
English Holly (*Ilex aquifolium*) : © 1986 Joe DiTomaso
Kudzu (*Pueraria montana*) : David J. Moorhead, University of Georgia; http://www.forestryimages.org
Whitetop (*Lepidium draba* ssp. *draba* [*Cardaria draba*]) : © Br. Alfred Brousseau, Saint Mary's College
Perennial Pepperweed (*Lepidium latifolium*) : © 2002 Jennifer Forman
Dalmation Toadflax (*Linaria dalmatica* ssp. *Dalmatica*) : © Br. Alfred Brousseau, Saint Mary's College
Purple Loosestrife (*Lythrum salicaria*) – flowers : Linda Wilson, University of Idaho; http://www.forestryimages.org
Tansy Ragwort (*Senecio jacobaea*) : © Br. Alfred Brousseau, Saint Mary's College
Medusahead (*Taeniatherum caput-medusae* ssp. *asperum*) : Photo courtesy of California Department of Food and Agriculture
Reed Canarygrass (*Phalaris arundinacea*) : Chris Evans, University of Georgia; http://www.forestryimages.org

Freshwater Invertebrates

Red Swamp Crayfish (*Procambarus clarkii*) : George W. Robinson, © California Academy of Sciences
Asian Freshwater Clam (*Corbicula fluminea*) – from Lake Washington, Seattle, WA : Photo by Alan C. Trimble

Siberian Prawn (*Exopalaemon modestus*): Steven Slater, California Department of Fish and Game
New Zealand Mudsnail (*Potomopyrgus antipodarum*) – from the Missouri River near Holter Lake : Photo courtesy of Trevor R. Anderson

Marine Invertebrates

Star Ascidian (*Botryllus schlosseri*) : Judith Pederson, MIT Sea Grant College Program
Marine Clams
Eastern Softshell Clam (*Mya arenaria*) : Photo by Alan C. Trimble
Purple Varnish Clam (*Nuttallia obscurata*) : Photo by Alan C. Trimble
Manila Clam (*Venerupis philippinarum*) : Photo by Alan C. Trimble
Asian Mudsnail (*Batillaria attramentaria*) : Photo Courtesy of Marjorie Wonham, University of Alberta, Canada
Oyster Drills
Asian Oyster Drill (*Ocinebrellus inornatu*) : Photo by Jennifer Ruesink
Eastern Oyster Drill : Photo by Jennifer Ruesink
European Green Crab (*Carcinus maenas*) : Dr. Sylvia Behrens Yamada holding crab; photo by Jim Carlton
Pacific Oyster (*Crassostrea gigas*) : Photo by Dr Peter Dyrynda; javascript:ol ('http://www.solaster-pd.net')
Atlantic Slipper Limpet (*Crepidula fornicata*) : David Byres, Florida Community College, Jacksonville
Japanese Mussel (*Musculista senhousia*) : SFBay:2K, California Academy of Sciences
Mediterranean Mussel (*Mytilus galloprovincialis*) : Photo courtesy of Dr. Marjorie Wonham, University of Alberta, Canada
New Zealand Isopod (*Sphaeroma quoyanum*) : SFBay:2K, California Academy of Sciences

Terrestrial Invertebrates

Common Night Crawler (*Lumbricus terrestris*) : Photo by P. Dee Boersma
Multi-colored Asian Ladybeetle (*Harmonia axyridis*) – adult with pupa : Photo courtesy of Joyce Gross
Terrestrial Mollusks
Leopard Slug (*Limax maximus*) : © 2000 William Leonard
Balsam Wooly Adelgid (*Adelges picea*) – eggs : Scott Tunnock, USDA Forest Service; http://www.forestryimages.org
Asian Longhorned Beetle (*Anoplophora glabripennis*) – adult : Michael Bohne, USDA Forest Service, www.forestryimages.org
European Fruit Fly (*Drosophila subobscura*) : Photo by G. W. Gilchrist
Asian/European Gypsy Moth (*Lymantria dispar*) : John H. Ghent, USDA Forest Service; http://www.invasive.org
European Yellowjacket (*Paravespula germanica*) : Photo by Hans Arentsen, The Garden Safari; http://www.gardensafari.net
Spider (*Philodromus dispar*) – Philodromus dispar crab spiders, male (top) and female : Photos © Ed Nieuwenhuys
Asian Bush Mosquito (*Ochlerotatus japonicus*) : Photo by Mike Sardalis, USAMRIID

Freshwater Vertebrates

Rock Bass (*Ambloplites rupestris*) : Photo courtesy of The Nature Conservancy, Ohio chapter
Common Carp (*Cyprinus carpio*) : Photo courtesy of Shedd Aquarium; http://www.fishphotos.org
Black Bullhead (*Ameiurus melas*) : Photo courtesy of Al Staffen, Ohio Division of Wildlife

Pickerel (*Esox americanus*) : Photo by Mark Binkley; http://www.jonahsaquarium.com
Green Sunfish (*Lepomis cyanellus*) : Mark Binkley; http://www.jonahsaquarium.com
Pumpkinseed (*Lepomis gibbosus*) : Photo courtesy of Steffen Zienert
Striped Bass (*Morone saxatilis*) : Photo courtesy of EPA; http://www.epa.gov/gmpo/education/photo/birds-animals.html
Cutthroat Trout (*Oncorhynchus clarki*) : Chris Schnepf, University of Idaho; http://www.forestryimages.org
Red-eared Slider (*Trachemys scripta*) : France (09-08-04) © Jelger Herder; http://www.DigitalNature.com
American Shad (*Alosa sapidissima*) : © 2004 by The Marine Biological LaboratoryTM
Nutria (*Myocastor coypus*) : Photo by Lee Foote, USGS National Wetlands Research Center
Yellow Perch (*Perca flavescens*) : Mark Binkley; http://www.jonahsaquarium.com
American Bullfrog (*Rana catesbeiana*) : Photo courtesy of Thomas Tyning, Berkshire Community College and MA Audubon

Marine Vertebrate

Atlantic Salmon (*Salmo salar*) : Photo by J. Moreau

Terrestrial Vertebrates

Mountain Goat (*Oreamnos americanus*) : Photo by Kimber Owen; courtesy of Sea Wolf Adventures
Ring-necked Pheasant (*Phasianus colchicus*) : het Twiske, Oostzaan (11-05-04); The Netherlands; © Maaike Pouwels; http://www.digitalnature.org
European Starling (*Sturnus vulgaris*) : Rotterdam (15-10-03), The Netherlands; © Jelger Herder; http://www.digitalnature.org
Black Rat (*Rattus rattus*) : © 2004 Larry Jon Friesen
Fox Squirrel (*Sciurus niger*) : Photo courtesy of Phil Meyers at www.animaldiversity.ummz.umich.edu
Domestic Cat (*Felis catus*) : © 2004 Larry Jon Friesen (purchased from istock photos)
European Rabbit (*Oryctolagus cuniculus*) : Photo by Andrei Mikhailov
European House Sparrow (*Passer domesticus*) : Abisko (15-09-04), The Netherlands; © Jelger Herder; http://www.digitalnature.org
Domestic Pig (*Sus scrofa*) : Photo by P. Dee Boersma

Diseases

White Pine Blister Rust (*Cronartium ribicola*) – symptoms : Susan K. Hagle, USDA Forest Service
Dogwood Anthracnose (*Discula destructiva*) – damage : Robert L. Anderson, USDA Forest Service
Newcastle Disease (genus *Rubulavirus*, family Paramyoviridae) – cormorants infected : USGS photo by Linda Glaser, USGS National Wildlife Health Center
Port Orford Cedar Root Rot (*Phytophthora lateralis*) – symptoms : Donald Owen, California Department of Forestry and Fire Protection
Sudden Oak Death (*Phytophthora ramorum*) – sign : Joseph O'Brien, USDA Forest Service
West Nile Virus – transmission cycle : Photo courtesy of the Greater Vancouver Regional District

Threats

Fish Hook Water Flea (*Cercopagis pengoi*) : [?]
Zebra Mussel (*Dreissena polymorpha*) : Randy Westbrooks, U.S. Geological Survey; http://www.forestryimages.org

African Waterweed (*Lagarosiphon major*) : Rohan Wells, National Institute of Water and Atmospheric Research; http://www.forestryimages.org
Northern Snakehead (*Channa argus*) : USGS Archives, U.S. Geological Survey; http://www.forestryimages.org
Wood Wasp (*Sirex noctilio*) : Photo courtesy of Forest Research, New Zealand
Bamboo (*Bamboo*) : Photo by P. Dee Boersma
Dead Man's Fingers (*Codium fragile* ssp. *tomentosoides*) – with Bryozoans attached : Photo by Luis A. Solorzano; http://www.californiabiota.com
Chinese Mitten Crab (*Eriocheir sinensis*) : Photo by Wolfgang Rabitsch, Vienna

Contributing Authors

Carmen M. Albert, Department of Biology, University of Washington, Seattle, WA
Trevor R. Anderson, Department of Biology, University of Washington, Seattle, WA
Jennifer E. Andreas, Plant, Soil, and Entomological Sciences Department, University of Idaho, Moscow, ID
Joseph Arnett, Tetra Tech EC, Inc., Bothell, WA
Katie Barnas, NOAA Fisheries, Northwest Fisheries Science Center, Seattle, WA
J. Katie Barndt, Center for Urban Horticulture, University of Washington, Seattle, WA
Kathryn Beck, Beck Botanical Services, 1708 McKenzie Avenue, Bellingham, WA 98225
Matt Bennett, Center for Urban Horticulture, University of Washington, Seattle, WA
P. Dee Boersma, Department of Biology, University of Washington, Seattle, WA
Kevin Britton-Simmons, Department of Ecology and Evolution, University of Chicago, Chicago, IL
Francis S. Brown, Department of Biology, University of Washington, Seattle, WA
Eric R. Buhle, Department of Biology, University of Washington, Seattle, WA
D. Shallin Busch, Department of Biology, University of Washington, Seattle, WA
Joan Cabreza, Region 10 EPA Invasive Species Coordinator, Seattle, WA
Sergio Camacho, College of Forest Resources, University of Washington, Seattle, WA
Jackie L. Carter, Department of Biology and School of Aquatic and Fishery Sciences, University of Washington, Seattle, WA
J. Alan Clark, Department of Biology, University of Washington, Seattle, WA
Rod Crawford, Burke Museum, University of Washington, Seattle, WA
Jeffrey Crooks, Tijuana River National Estuarine Research Reserve, Imperial Beach, CA
Lisa Crozier, Department of Ecology and Evolution, University of Chicago, Chicago, IL, and NOAA Fisheries, Northwest Fisheries Science Center, Seattle, WA
Demetrius Fletcher, College of Forest Resources, University of Washington, Seattle, WA
Tara S. Fletcher, Department of Biology, University of Washington, Seattle, WA
Tessa B. Francis, Department of Biology, University of Washington, Seattle, WA
Tracy Fuentes, USDA Forest Service, North Bend, WA
Rebecca Gamboa, Department of Biology, University of Washington, Seattle, WA
Diane P. Genereux, Departments of Biology, University of Washington, Seattle, WA, and Emory University, Atlanta, GA
Wendy Gibble, College of Forest Resources, University of Washington, Seattle, WA
David Giblin, University of Washington Herbarium, University of Washington, Seattle, WA
George W. Gilchrist, Department of Biology, College of William & Mary, Williamsburg, VA
Martha J. Groom, Interdisciplinary Arts & Sciences and Department of Biology, University of Washington, Seattle, WA

Kathy Hamel, Department of Ecology, University of Washington, Seattle, WA
Dale R. Herter, Raedeke Associates, Inc., Seattle, WA
Melisa L. Holman, College of Forest Resources, University of Washington, Seattle, WA
Gordon W. Holtgrieve, Department of Biology, University of Washington, Seattle, WA
Allisyn Hudson, Department of Biology, University of Washington, Seattle, WA
Raymond B. Huey, Department of Biology, University of Washington, Seattle, WA
Chad C. Jones, Department of Biology, University of Washington, Seattle, WA
T. N. Kaye and M. Blakeley-Smith, Institute for Applied Ecology, Corvallis, OR
Delia R. Kelly, Department of Biology, University of Washington, Seattle, WA
Ericka Kendall, Department of Biology, University of Washington, Seattle, WA
Nancy LaFleur, Department of Ecology and Evolutionary Biology, University of Connecticut, Storrs, CT
Darren A. Linker, Department of Environmental and Occupational Health Sciences, University of Washington, Seattle, WA
Devin R. Malkin, College of Forest Resources, University of Washington, Seattle, WA
Samantha Martin, College of Forest Resources, University of Washington, Seattle, WA
P. Sean McDonald, School of Aquatic and Fishery Sciences, University of Washington, Seattle, WA
Joshua C. Misenar, Department of Biology, University of Washington, Seattle, WA
Jonathan W. Moore, Department of Biology, University of Washington, Seattle, WA
Peter Morrison, Pacific Biodiversity Institute, PO Box 298, Winthrop, WA 98862
Jennifer M. Moslemi, Ecology and Evolutionary Biology, Cornell University, Ithaca, NY
Margo Murphy, Center for Urban Horticulture, University of Washington, Seattle, WA
Colin A. Olivers, Department of Biology, University of Washington, Seattle, WA
Dawn Olmsted, Department of Biology, University of Washington, Seattle, WA
Fernanda X. Oyarzun, Department of Biology, University of Washington, Seattle, WA
Dianna K. Padilla, Stony Brook University, Stony Brook, NY
Wendy J. Palen, Department of Biology, University of Washington, Seattle, WA
Rachel Price-Rayner, Center for Urban Horticulture, University of Washington, Seattle, WA
Casey L. Ralston, Department of Biology, University of Washington, Seattle, WA
Ginger A. Rebstock, Department of Biology, University of Washington, Seattle, WA
Sarah H. Reichard, Center for Urban Horticulture, University of Washington, Seattle, WA
Jennifer L. Ruesink, Department of Biology, University of Washington, Seattle, WA
Daniel E. Schindler, School of Aquatic and Fishery Sciences and Department of Biology, University of Washington, Seattle, WA
Bridget Simon, Department of Biology, University of Washington, Seattle, WA
Elizabeth A. Skewes, Department of Biology, University of Washington, Seattle, WA
Maureen P. Small, WDFW, Conservation Biology Unit, Olympia, WA
Joanna L. Smith, Department of Biology, University of Washington, Seattle, WA
Stacey Solie, The Nature Conservancy and Columbia University, New York, NY
Carson Sprenger, Entomology and Tropical Forestry Lab, University of Washington, Seattle, WA

Amanda G. Stanley, Department of Biology, University of Washington, Seattle, WA
David Stokes, Department of Environmental Studies and Planning, Sonoma State University, Rohnert Park, CA
Heather M. Tallis, Department of Biology, University of Washington, Seattle, WA
Christopher N. Templeton, Department of Biology, University of Washington, Seattle, WA
Diana Thayer, Department of Biology, University of Washington, Seattle, WA
Patricia A. Townsend, Department of Biology, University of Washington, Seattle, WA
Alan C. Trimble, Department of Biology, University of Washington, Seattle, WA
Mandy Tu, The Nature Conservancy, Arlington, VA
Lauren S. Urgenson, College of Forest Resources, University of Washington, Seattle, WA
Allison A. Van, Department of Biology, University of Washington, Seattle, WA
Amy N. Van Buren, Department of Biology, University of Washington, Seattle, WA
Sacha N. Vignieri, Department of Biology, University of Washington, Seattle, WA
Carly H. Vynne, Department of Biology, University of Washington, Seattle, WA
Eric Wagner, Department of Biology, University of Washington, Seattle, WA
Bertie J. Weddell, Distance Degree Program, Washington State University, and Draba Consulting, 1415 NW State Street, Pullman, WA
David L. Wilderman, Natural Areas Program, Washington Department of Natural Resources, Ellensburg, WA
Monika Winder, School of Aquatic and Fisheries Science, University of Washington, Seattle, WA
Marjorie J. Wonham, Department of Biological Sciences, University of Alberta, Edmonton, AB, Canada

About the Editors

P. D. BOERSMA holds the Wadsworth Endowed Chair in Conservation Science in the Department of Biology at the University of Washington. She is a past president of the Society for Conservation Biology and the executive editor of *Conservation in Practice*, a publication of the Society for Conservation Biology. She thrives on studying Magellanic penguins in the South Atlantic. Professor Boersma brings a passion for natural history and native communities to her mission of slowing the spread of invasive species.

SARAH REICHARD is an associate professor affiliated with the Center for Urban Horticulture at the University of Washington. For more than a decade she has used science to stem the invasive species problems on the ground in the United States and has helped develop policy at both the state and federal levels. She served three terms on the federal Invasive Species Advisory Committee and is a past secretary of the Society for Conservation Biology.

AMY VAN BUREN is a PhD candidate in the Department of Biology at the University of Washington. She has spent time fishing, farming, and studying the landscapes of the Pacific Northwest, where she has watched invasive perch, bass, and bullfrogs change the community composition of lakes by eliminating native salamanders. In the Falkland Islands, she has seen how introduced species affect the seabird colonies she studies.